21st Latin American Symposium on Solid State Physics

(SLAFES XXI)

Journal of Physics: Conference Series Volume 480

Villa de Leyva, Columbia
30 September – 4 October 2013

Editors:

J. Albino Aguiar **Carlos Arturo Parra Vargas**

ISBN: 978-1-63266-702-1
ISSN: 1742-6588

Printed from e-media with permission by:

Curran Associates, Inc.
57 Morehouse Lane
Red Hook, NY 12571

CURRAN ASSOCIATES INC. proceedings.com

Some format issues inherent in the e-media version may also appear in this print version.

Printed by Curran Associates, Inc. (2014)

For permission requests, please contact the Institute of Physics at the address below.

Institute of Physics
Dirac House, Temple Back
Bristol BS1 6BE UK

Phone: 44 1 17 929 7481
Fax: 44 1 17 920 0979

techtracking@iop.org

Additional copies of this publication are available from:

Curran Associates, Inc.
57 Morehouse Lane
Red Hook, NY 12571 USA
Phone: 845-758-0400
Fax: 845-758-2634
Email: curran@proceedings.com
Web: www.proceedings.com

TABLE OF CONTENTS

Journal of Physics: Conference Series **480** (2014) 011001 doi:10.1088/1742-6596/480/1/011001

Preface

The Latin American Symposium on Solid State Physics (SLAFES) started in Caracas-Venezuela, and over time the symposia have taken place in 9 different Latin American countries. The last five events took place in Mérida-Venezuela (2002), Havana-Cuba (2004), Puebla-Mexico (2006), Puerto Iguazú-Argentina (2008) and Maragogi-Brazil (2011).

During the last years, in the different SLAFES editions, the aim has been to bring together researches from Latina America and invited renowned scientists from around the world to a unique forum to discuss the latest developments regarding Solid state Physics.

The 21st Latin American Symposium on Solid State Physics (SLAFES XXI) was held in Villa de Leyva-Colombia, from September 30 to October 04, 2013. The 21st SLAFES version featured the participation of experts in various areas of Solid State Physics from countries such as Belgium, Germany, United States, Spain, Ireland, Chile, Argentina and Brazil, had 270 submitted works and was attended by 140 researchers.

The development of this event was made possible by financial support from of the Universidad Pedagógica y Tecnológica de Colombia, Universidad Nacional de Colombia, Universidad del Norte-CO, Universidad de Magdalena-CO, Universidade Federal de Pernambuco-BR and the Academia Colombiana de Ciencias Exatas, Naturales y Física.

Editors

Professor J Albino Aguiar
Departamento de Física Universidade Federal de Pernambuco
50670-901 Recife PE
Brazil
e-mail: albino@df.ufpe.br Professor Jairo Roa-Rojas

Grupo de Física de Nuevos Materiales
Departamento de Física
Universidad Nacional de Colombia A.A.
5997 Bogotá DC, Colombia
e-mail: jroar@unal.edu.co

Professor Carlos Arturo Parra Vargas
Grupo Física de Materiales
Escuela de Física
Universidad Padagógica y Tecnológica de Colombia
Tunja
Colombia
e-mail: carlos.parra@uptc.edu.co

Professor David A Landí nez Téllez
Grupo de Física de Nuevos Materiales
Departamento de Física
Universidad Nacional de Colombia
A.A. 5997 Bogotá DC
Colombia
e-mail: dalandinezt@unal.edu.co

Professor Laura T Corredor Bohórquez
Departamento de Física
Universidade Federal de Pernambuco
50670-901 Recife PE
Brazil
e-mail: ltcorredorb@df.ufpe.br

Professor Arkady Shanenko
Departamento de Física
Universidade Federal de Pernambuco
50670-901 Recife PE
Brazil
e-mail: arkadyshanenko@df.ufpe.br

Professor Renato F Jardim
Instituto de Física
Universidade de São Paulo
CP 66318
São Paulo SP
Brazil
e-mail: rjardim@if.usp.br

Professor Francois Peeters
Department Fysica
Universiteit Antwerpen
Groneneborgerlann 171
B-2020, Antwerpen
Belgium
e-mail: francois.peeters@uantwerpen.be

Organizing committee

Chairman	Carlos Arturo Parra Vargas
Proceedings Editor	José Albino Aguiar
Program Chair	Jairo Roa-Rojas
Secretary	Aura Janeth Barón González
Treasurer	Armando Sarmiento Santos
Speaker Chair	Rafael González Hernández
	Fernando Naranjo Mayorga
	David A Landínez Téllez
	Jesús Oswaldo Morán
	José Sierra Ortega

Journal of Physics: Conference Series **480** (2014) 011002 doi:10.1088/1742-6596/480/1/011002

Peer review statement

All papers published in this volume of *Journal of Physics: Conference Series* have been peer reviewed through processes administered by the proceedings Editors. Reviews were conducted by expert referees to the professional and scientific standards expected of a proceedings journal published by IOP Publishing.

21st Latin American Symposium on Solid State Physics (SLAFES XXI) IOP Publishing
Journal of Physics: Conference Series **480** (2014) 012001 doi:10.1088/1742-6596/480/1/012001

1s-like state binding energy of a donor impurity bound to Xz valleys of GaAs/AlAs type II quantum dots with an uniform applied magnetic field

Pablo Villamil[1*] and Carlos L. Beltrán Ríos[2]

[1] Departamento de Matemáticas y Física: Grupo de Teoría de la Materia Condensada, Universidad de Sucre, Sincelejo A.A. 406, Colombia.
[2]Escuela de Física, Universidad Industrial de Santander, A.A. 678, Bucaramanga, Colombia

*Author E-mail: pabloe1954@yahoo.es

Abstract. In this paper we calculate the binding energy of a donor impurity in the 1s-like state bound to X valleys of GaAs/AlAs type II quantum dots, using an anisotropic variational method, which allows us to take into account the anisotropy of the effective mass and quantum confinement. The binding energy is plotted against the radius of the cylinder and the uniform magnetic field applied along the cylinder axis. The binding energies have a monotone behavior with the cylinder radius for different magnetic fields and almost constant behavior with increasing magnetic field.

1. Introduction

The The study of low dimensional heterostructures type II has increased in recent years especially for potential applications such as infrared detectors[1], devices for use in spintronics[2], and including the observation of the called Aharonov-Bohm effect[3] and it is also use for creating a laser device in the infrared[4] and green[5]. Also it has been observed phononic effects in type II superlattices that would allow the development of sound amplification devices by stimulated emission[6]. The study of type II quantum dots has been expanded in recent years. Najjar et al.[7], studied the exciton recombination in quantum dots and wells of ZnTe/ZnSe. Kai Cui et al.,[8] observed emission spectrum of multilayer self-assembled quantum dots GaSb/GaAs at room temperature. Dag Wang et al.,[9] studied the binding energy of the ground state of shallow donor impurities in type II materials depending on the width of the well, considering mixing the Γ and X valleys in a structure of AlAs/GaAs, which leads to a differences in the behavior of the energy compared with type I. Carneiro and Weber[10] calculated the energy of the ground state and some excited states for donor impurities in the X valley of a type II quantum wells, they obtained a strong dependence of the binding energy with the effective mass of the carriers and the symmetry of the valley. Gulyaev et al.,[11] studied the recombination of carriers in a type II superlattice doped with acceptor and donor impurities through photoluminescence, and they found that the increase in the binding energy of the impurities results in a decrease in the rate of recombination.

In this paper we report the calculation of the binding energy for a donor impurity centered in a type II quantum dot of AlAs/GaAs, with cylindrical geometry and an applied magnetic field. We study the energy behavior in function of geometrical parameters of the dot and the magnetic field variation.

2. Theoretical Model

We consider a heterostructure based on two concentric cylinders, the inner radius R and height L correspond to the AlAs cylindrical structure, the outer radius R + a and height $L + 2a$ correspond to the GaAs cylindrical structure and they are located within a matrix of $Ga_{1-x}Al_xAs$, the magnetic field (B) is applied in the direction of the cylinders axis, as shown in Figure 1

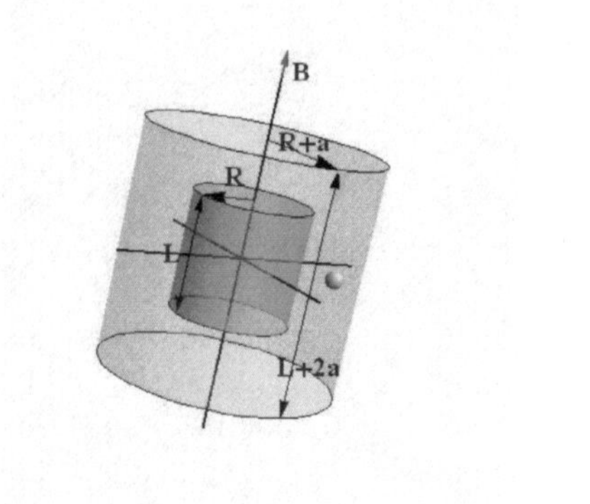

Figure 1. Heterostructure scheme of a type II quantum dot with a donor impurity in (sphere) and a uniform magnetic field applied, the inner cylinder is considered AlAs, the outward is GaAs, and the entire system is within a matrix of $Ga_{1-x}Al_xAs$.

We used a Hamiltonian in cylindrical coordinates (ρ,φ,z) given by

$$H = \frac{1}{2}(P_i - eA_i)m_{ij}^{-1}(P_j - eA_j) + V_{eff}(r) \quad (1)$$

Where P_i is the momentum operator, A_i is the vector potential, associated with the magnetic field (B), m_{ij} is the effective mass tensor, i,j run over ρ,φ,z coordinates, and V_{eff} is the effective potential due to coulomb and geometrical confinement potential. The last one is given by

$$V(\rho) = \begin{cases} V_0\left(\frac{\rho}{R}\right)^2, & \rho < R \\ V_0, & R \le \rho < R + a \\ V_1, & \rho \ge R + a \end{cases}, \quad V(z) = \begin{cases} 0, & 0 \le z \le L \\ V_0, & 0 > z > -a, \wedge, L < z \le L + a \\ V_1, & z < -a \wedge z > L + a \end{cases} \quad (2)$$

The coulomb interaction, in arbitrary units, is $V = -\frac{2\tau}{|\vec{r}-\vec{r}_{impl}|}$.When the coulomb interaction is not considered (τ=0) in V_{eff} the Schrödinger equation, for an electron in the structure, have analytic solution given by ψ_0 and its energy is E_0. When the coulomb interaction is taking in account (τ=1) in V_{eff} the Schrödinger equation ($H\psi=E\psi$) is solved by a variational method. We considered a Bastard[12] type function given by

$$\Psi(\rho, z, \phi) = \Psi_0 Exp\left(-\sqrt{\frac{(\vec{\rho}-\vec{\rho}_i)^2}{\lambda_1^2} + \frac{(z-z_i)^2}{\lambda_2^2}}\right) \quad (3)$$

where λ_1 y λ_2 are variational parameters, $\vec{\rho}_i$ and z_i, are the impurity position along the perpendicular and parallel direction to the magnetic field. We calculated the binding energy, $E_b=E_0-E$, as a function of the geometrical parameters and the magnetic field, $\vec{B} = B\hat{k}$, symmetric gauge for the potential vector is considered.

3. Results and Analysis

In the numerical calculations it has been used the following parameters for AlAs material, dielectric constant ε = 10.06, well barrier height 316 meV., alignment of the conduction bands and valence in 35% and 65%, respectively. From the bottom of the Γ conduction band, in the GaAs, to the bottom of the X valley, in the AlAs, there is a difference of 145 meV. The values of the effective mass in the equation 1 depend on the X valley under consideration. The two sets of effective masses in volume of AlAs ($m^*_\perp = m^*_{x,y,}$, m^*_z) and their corresponding values used for each valley are the same used in ref 10.

In Figure 2 is plotted the binding energy of a donor impurity in function of the length of the cylinder with an applied magnetic field of 20T and 50 T, with different values of the radii. It is observed that for a given radius, for example $1.4a_0^*$, the binding energy decreases when the cylinder height increases. Also it is observed that the binding energy of the impurity increases when the cylindrical radius is diminishing, for all values of the cylindrical height. For r_c = R, r_c is the cyclotron radius of a carrier in a magnetic field and R the radius of the cylinder, there is a limit for the transition from geometric confinement regime to the magnetic confinement regime of the carrier. When B = 20T, the cyclotron radius of an electron equals to a $r_c = \sqrt{\frac{1}{\gamma}} = 4.742\ a_0^*$, in that way for the cylinder radii, R<r_c, the geometric confinement is greater than the magnetic confinement. The smaller radius of AlAs considered in this problem is $1.4a_0^*$, because for smallest values, the X_z valley is aligned with the first sublevel e_1 of the well GaAs and appears a mixture of bands, which is not considered in this paper.

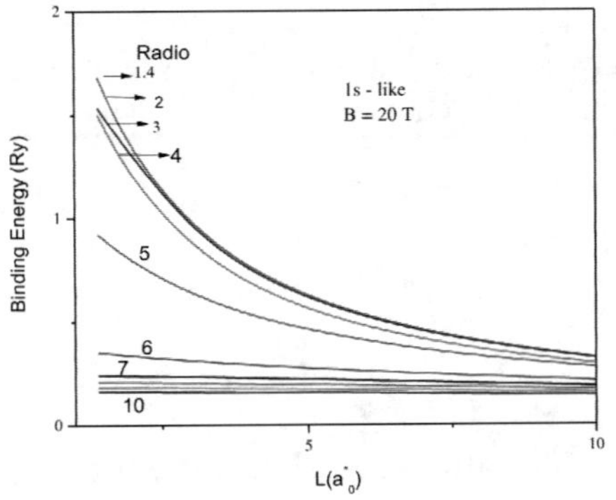

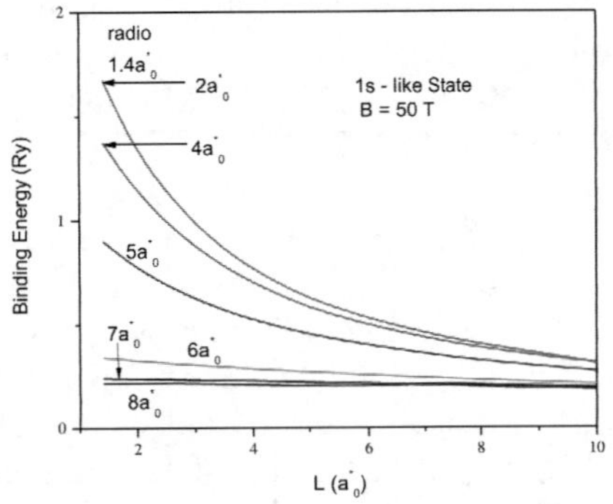

Figure 2. Graphic of the binding energy E_b of a donor impurity vs length of a quantum well for various radii and with an applied magnetic field of 20 T (left) and 50 T (right).

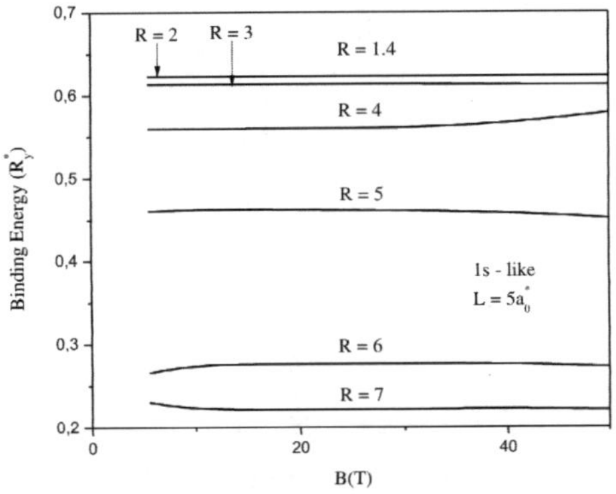

Figure 3. Ground state Binding energy for an impurity centered in the structure as a function of the applied magnetic field and different values for the wire radii, with cylindrical height L=$5a_0^*$.

Figure 3 shows the binding energy as a function of the magnetic field for a cylinder height of $5a_0^*$ and different values of the cylindrical radii. For a given magnetic field, the energy increases with decreasing radius of the cylinder and geometric confinement is more effective for smaller radii. With small cylinder radii the binding energy remains approximately constant.

4. Conclusion

Through the proposed model were calculated the energies E_z and E_ρ for the first level of a carrier in a quantum well in the z and directions and the total energy in the quantum well when applying different magnetic fields. We calculated binding energies of a donor impurity in a type II quantum dot in function of the well height for different radii of the cylinder and applied magnetic fields of 20T and 50T. The binding energy of the impurity, for a given radius of the cylinder decreases as the height of the cylinder increases. The binding energy of the impurity, for a given cylinder height increases as the radius of the cylinder decreases.

References

[1] Z. M. Zhu, P. Bhattacharya, E. Plis, X. H. Su and S. Krishna, 2006 J. Phys. D: Appl. Phys. **39** 4997–5001.

[2] Ya. V. Terent'ev, M. S. Mukhin, A. A. Toropov, M. O. Nestoklon, B. Ya. Meltser,A. N. Semenov,V. A. Solov'ev, and S. V. Ivanov, 2013 Phys. Rev. B **87** 045315
[3] Ian R. Sellers, Vincent R. Whiteside, Igor L. Kuskovsky, Alexander O. Govorov, Bruce D. McCombe, 2008 Physica E **40** 1819–1823
[4] R.Q. Yang, 1999 Microelectronics Journal **30** 1043–1056
[5] G. P. Yablonskii, E. V. Lutsenko, A. G. Vainilovich, V. N. Pavlovskii, S. V. Ivanov, I. V. Sedova, S. V. Sorokin, P. S. Kop'ev, 2010 Journal of Non-Crystalline Solids **356** 1928–1934
[6] A. J. Kent, R. N. Kini, N. M. Stanton, M. Henini, B. A. Glavin, V. A. Kochelap, and T. L. Linnik, 2006 PRL **96** 215504
[7] Rita Najjar , Régis André, Lucien Besombes, Catherine Bougerol, Serge Tatarenko, Henri Mariette, 2009 Superlattices and Microstructures **46** 253-257
[8] Kai Cui , Wenquan Ma, Jianliang Huang , Yang Wei , Yanhua Zhang , Yulian Cao , Yongxian Gu , Tao Yang, 2012 Physica E **45** 173–176
[9] Dag Wang, E. A. de Andrada e Silva and I. C. da Cunha Lima, 1992 Phys. Rev. B **46** 7304
[10] Gleise N. Carneiro and Gerald Weber, 1998 Phys. Rev. B **58** 7829
[11] D.V. Gulyaev, A.M. Gilinsky, A.I. Toropov, A.K. Bakarov, K.S. Zhuravlev, (001 Physica B **308–310** 784–787
[12] G. Bastard 1981 Phys. Rev. B **24** 4714

21st Latin American Symposium on Solid State Physics (SLAFES XXI) IOP Publishing
Journal of Physics: Conference Series **480** (2014) 012002 doi:10.1088/1742-6596/480/1/012002

Microstructural and Morphological Properties of Nanocrystalline $Cu_2ZnSnSe_4$ Thin Films: Identification New Phase on Structure

Heiddy P. Quiroz[1], N. J. Seña[1], A. Dussan[1]

Nano-structured Materials Group, Department of Physics, Universidad Nacional de Colombia, Colombia

hpquirozg@unal.edu.co

Abstract. This paper presents a study of the structural and morphological properties of thin films of compound $Cu_2ZnSnSe_4$. Mass (M_X) and temperature of the substrate ($T_{S(Cu)}$) of compound copper (Cu), were varied. All samples were deposited by co-evaporation method in three stages. From measurements of X-ray diffraction it was possible to establish with TS increasing the presence of associated binary phases quaternary compound during the growth process of the material. It was found that the main peak around, $2\theta = 27.1°$, predominate binary phases $Cu_{1.8}Se$ and ZnSe. Measurements of X-ray diffraction were performed to pure binary compounds, showing a peak corresponding to the main peak found around the compound. Raman shifts showed associated binary compounds with the observed by XRD. In this work, we report for the first time the binary phase identification $Cu_{1.8}Se$ and ZnSe as part of the structure of the stannite CZTSe. Since the Scherrer equation was found that the crystallite sizes ranged between 30 and 40 nm. A correlation between structure and topography superficial is presented.

1. Introduction

Over the last decades, new photovoltaic materials have been studied on search for new renewable clean energy sources. $Cu(In,Ga)Se_2$ (CIGS) is one of the most promising thin film solar cell materials, with an efficiency about 20% [1]; however their components such as indium (In) and gallium (Ga), have high toxicity levels and limited availability in the earth. Cu_2ZnSnS_4 (CZTS) and $Cu_2ZnSnSe_4$ (CZTSe) compounds are very promising candidates to replace CIGS in thin film solar cell applications thanks to their optical properties (band gap range from1to1.5eV) and absorption coefficient higher than 10^4 cm^{-1}. [2,3]

Quaternary CZTS(Se) thin films have been produced by pulsed laser ablation [4], one step RF magnetron sputtering [3], co-evaporation of elements [5], sequential evaporation [6] as well as synthesis from melts [7]. However, the understanding of the properties of the different phases is still unclear; there are studies that have addressed their structural, electric, morphological and defect properties of CZTSe thin films grown by others methods. [8,9,10]

In this work we presented a study of the microstructural and morphological properties of thin films of compound $Cu_2ZnSnSe_4$ deposited by co-evaporation method in three stages. Parameters deposition such as mass (M_X) and temperature of the substrate, $T_{S(Cu)}$, were varied on large range. Influence of synthesis parameters on properties were investigated from X-ray diffraction (XRD), Raman and Scanning Electron Microscope (SEM) measurements. In this work, we report for the first time the binary phase identification $Cu_{1.8}Se$ and ZnSe as part of the structure of the stannite CZTSe. A correlation between structure and topography superficial is presented.

2. Experimental procedure

Thin films $Cu_2ZnSnSe_4$ were prepared on Soda Lime-type glass substrates through physical vapor deposition (PVD) in the presence of Se. The substrates were subjected to previous cleaning following a specific process, reported previously [11]. The masses of the precursors used were 0.118 g for Cu, 0.097 g for Sn, and 0.153 g for ZnSe; all had a high degree of purity (99.95 ± 0.05%). The deposition process was conducted in three stages. During the first stage, Cu was evaporated in a Se atmosphere at a substrate temperature of $T_{S(Cu)}$ = 673 K. This stage formed the Cu_2Se composite, as a result of the reaction of the two precursors. During the second stage, Sn was evaporated, again, in a Se atmosphere, but maintaining $T_{S(Sn)}$ = 523 K; this new layer was deposited over the existing Cu_2Se layer. This stage formed the $SnSe_2$ composite, which in turn reacted with Cu_2Se to give way to the formation of the Cu_2SnSe_3 ternary composite. During the third stage ZnSe was evaporated, maintaining $T_{S(ZnSe)}$ = 673 K, and this was deposited on the ternary layer. The final reaction gave way to the formation of the $Cu_2ZnSnSe_4$ quaternary composite. The reactions occurring have been considered by other researchers in the fabrication of the composite. [12]

The samples were characterized through XRD measurements by using a Philips X'Pert Pro diffractometer by PANalytical, equipped with a Cu-Kα source: 1.540598Å, 40 kV, 40 mA and an X´Celerator detector. The software used to compare the samples was the X'Pert High-Score Plus through Rietveld refinement in the range of $10° 2\theta \leq 80°$ with angular steps of 0.026°. All the samples, including the binary samples for their comparison with the analyses presented, were scanned at the same spectral range. The Raman measurements were carried out in the high-resolution Horiba Jobin Yvon Confocal Raman spectrometer, Labram HR high-resolution model equipped with a CCD detector (1024x256-pixel resolution), a 633-nm He/Ne laser, and a 785-nm diode laser. The Labspec Software was used to operate the equipment, as well acquire and treat the data, varying the Raman displacement from 100 to 500 cm^{-1}.

The samples were about 0.6 µm thick. Sample thickness was obtained from Uv-Vis-IR spectrophotometry measurements and corroborated with profilometry techniques. SEM images were obtained in a scanning electronic microscope FEI QUANTA 200.

3. Results and discussion

Fig. 1-a shows the X-ray diffraction spectrum obtained for the sample with optimal reproducible conditions of CZTSe quaternary compounds prepared by co-evaporation method in three stages: $T_{S(Cu)}$ = 673 K, M_{Cu} = 0,118 g, $T_{S(ZnSe)}$ = 673 K, M_{ZnSe} = 0,153g. It is observed, that characteristic phase have a preferential orientation on the plane (1 1 2). Crystal structure simulated is presented in the inset of figure 1-a.

Diffractogram obtained (see figure 1-a) presents a Stoichiometry coinciding with that of $Cu_2ZnSnSe_4$. Position of the peaks and diffraction planes were identified using Rietvelt method and X'Pert HighScore Plus program. A single phase is observed with Stannite-type structure [13,14] and spatial group of conventional symmetry $I-\bar{4}2m$ [15,16].

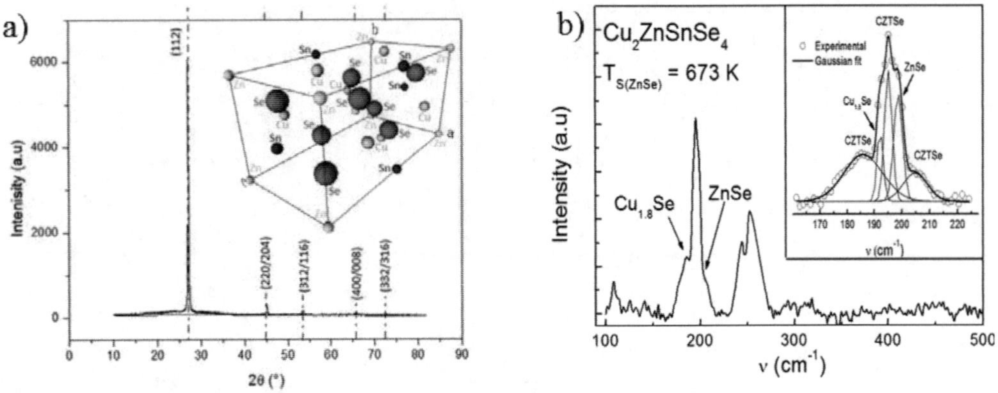

Figure 1. a) Diffraction pattern of $Cu_2ZnSnSe_4$ quaternary sample. Crystallographic structure in reference to the atomic location of the composite's atoms can be observed in the inset. b) Raman spectrum for the sample in a). Inset of the figure detail of the peak form ν = 195 cm^{-1} for the CZTSe sample with Gaussian fit.

Figure 2 shows the diffractograms of the CZTSe films when was varied the substrate temperature during the Cu evaporation stage. It was observed, that new phases are evident in the spectra on structure of the quaternary when $T_{S(Cu)}$ change from 573 to 773 K. Main peak 2θ = 27.1° is characterized by a splitting with the change of $T_{S(Cu)}$ from optimal value. Crystal size was calculated using the Scherrer equation and reported in table 1.

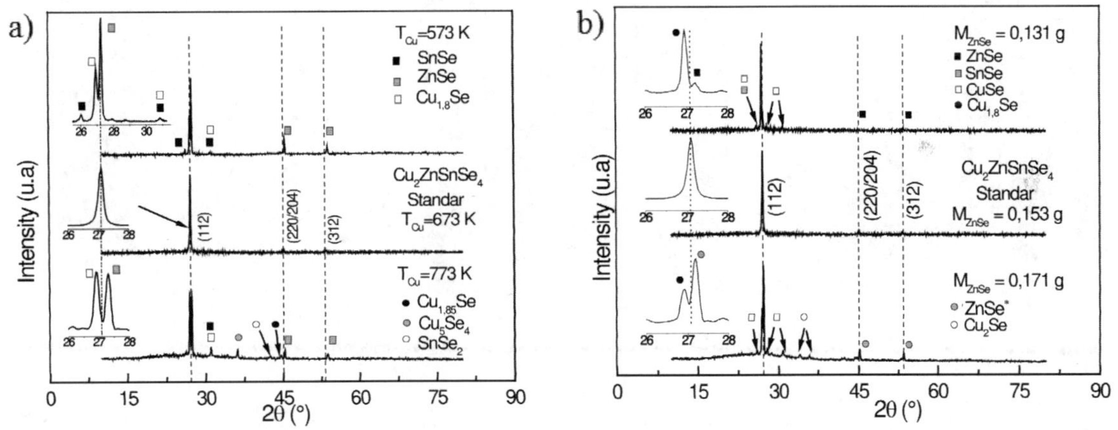

Figure 2. XRD spectra of the $Cu_2ZnSnSe_4$ composite varying: a) Substrate temperature during Cu evaporation and b) Mass of ZnSe. Figure inset shows detail of double peak for 2θ = 27.1°.

From analysis of binary compounds used as precursors in the synthesis of $Cu_2ZnSnSe_4$ allowed determining that the phases accompanying quaternary phase, are associated to secondary phases of ZnSe and Cu1.8Se (see Figure 3). It is was observed, that, changes in temperature influence the formation double peak around 2θ = 27.1° evidencing a contribution of ZnSe and $Cu_{1.8}Se$ at very close angles.

Raman measurements permitted evidencing the presence of these binary phases previously observed in the figure 3, identified for shifts Raman around peak 195 cm^{-1} (see figure 1-b). In this figure, an overlapping is clearly observed of peaks associated to the binary composites, $Cu_{1.8}Se$ ~192,2 cm^{-1} and ZnSe ~198,7 cm^{-1}.

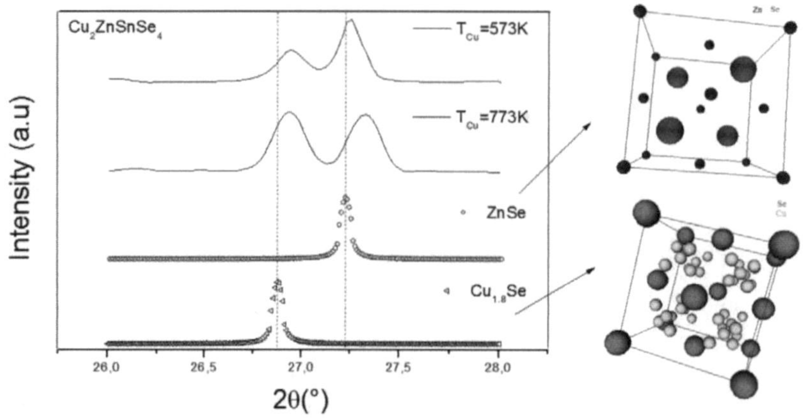

Figure 3. Broadening of peak $2\theta = 27.1°$ for the samples shown in Figure 2-a. The XRD spectra of the ZnSe and $Cu_{1.8}Se$ binary composites are presented comparatively and the crystal structures of the binary phases. The dash line red is indicating the main peak for binary compounds ZnSe and $Cu_{1.8}Se$.

Figure 4 shows micrographic SEM of CZSTe thin films varying the $T_{S(Cu)}$ between 573K and 773K.These images evidenced the grains formation on the surface that change of size with increasing of $T_{S(Cu)}$. Table 1 shows the crystal size and mean size grain obtained from XRD spectra and SEM measurements, respectively.

a) b) c)

Figure 4. SEM images of thin films CTZSe varying $T_{S(Cu)}$. **a)** 573K, **b)** 673K and **c)** 773K.

Table 1. Crystal and grain size obtained varying the substrate temperature during the Cu evaporation stage.

$T_{S(Cu)}$	Crystal size ± 5% nm	Grain size ± 5% nm
573	31,45	680
673	45,38	610
773	37,11	1187

When we analyze the effect of the mass of ZnSe in the formation of compound (see figure 2) shows that changes in the concentration of the ZnSe about to M_{ZnSe}= 0,153 g (constant for the change of the other parameters) significantly affect the formation of the compound to the point of losing that characterizes phase. This can be attributed to processes of re-evaporation of the material, contributing to the formation of the phase for CZTSe compound.

4. Conclusions

In this work CZTSe nanostructured thin films were fabricated, varying both T_S and M_x in the Cu and ZnSe composites. It was observed, that variation of these parameters sensibly influences the formation of the CZTSe characteristic phase with presences of secondary phases on structure. From a detailed study of the binary composites by XRD and Raman measurements, was possibly to establish the presence of two new phases around the characteristic peak given for $2\theta = 27{,}1°$. Additionally, Raman shifts were identified for $Cu_{1,8}Se$ and ZnSe at 195.1 cm^{-1} and 204.8 cm^{-1}, respectively. Grain size in CZTSe thin films, was increased with change of temperature around of optimal value for which was obtained the CZTSe characteristic phase.

5. Acknowledgments

This work was supported by Universidad Nacional de Colombia - DIB. We special thank Semiconductor Materials and Solar Energy Group for Laboratory support in sample preparation.

References

[1] Arnulf Jager Waldau, Sol. Energy Mat. Sol. Cells, 95 (2011), 1509-1517.
[2] F. Luckert, D. I. Hamilton, M. V. Yakushev, N. S. Beattie, G. Zoppi, M. Moynihan, I. Forbes, A. V. Karotki, A. V. Mudryi, M. Grossberg, J. Krustok, R. W. Martin, Appl. Phys. Lett., 99 (2011), 062104.
[3] Kentaro Ito, Tatsuo Nakazawa, Jpn. J. Appl. Phys., 27 (1988), 2094-2097.
[4] K.Moriya, K.Tanaka, H.Uchiki, Jpn. J. Appl. Phys., 47 (2008), 602-604.
[5] T.Tanaka,D.Kawasaki, M.Nishio, Q.Guo, H.Ogawa, Phys. Stat. Sol. 3 (2006), 2844–2847.
[6] J.Raudoja, E.Mellikov, M.Grossberg, S.Bereznev, R.Traksmaa, O. Volobujeva, J. Phys. and Chem. Solids, 70 (2009), 567–570.
[7] Louis Grenet, SergioBernardi, DavidKohen, Christophe Lepoittevin, Sebastien Noe, Nicolas Karst, Arnaud Brioude, Simon Perraud, Henri Mariett, Sol. Energy Mate. Sol Cells, 101(2012), 11-14.
[8] Narjes Beigom Mortazavi Amiri, Andrei Postnikov, J. Appl. Phys. 112 (2012), 033719.
[9] O. Volobujeva, S. Bereznev, J. Raudoja, K. Otto, M. Pilvet, E. Mellikov, Thin Solid Films 535 (2013) 48–51.
[10] P.M.P. Salomé, P.A. Fernandes, A.F. da Cunha, Thin Solid Films 517 (2009) 2531–2534.
[11] F. Mesa, Ph.D. Thesis, Departamento de Física, Universidad Nacional de Colombia, Colombia, 2010.
[12] Alex Redinger, Marina Mousel, Rabie Djemour, Levent Gütay, Nathalie Valle, Susanne Siebentritt, Prog. Photovolt: Res. Appl., (2013) 1-7, Doi: 10.1002/pip.2324
[13] G Suresh Babu, Y B Kishore Kumar, P Uday Bhaskar and V Sundara Raja, J. Phys. D: Appl. Phys. 41 (2008), 205305, Doi: 10.1088/0022-3727/41/20/205305.
[14] Rachmat Adhi Wibowo, Woo Seok Kim, Eun Soo Lee, Badrul Munir, Kyoo Ho Kim, J. Phys. and Chem. Solids, 68 (2007), 1908–1913, Doi: 10.1016/j.jpcs.2007.05.022.
[15]. F. Hergert, R. Hock, Thin Solid Films 515 (2007), 5953–5956.
[16]. I.D. Olekseyuk, L.D. Gulay, I.V. Dydchak, L.V. Piskach, O.V. Parasyuk, O.V. Marchuk, J. Alloys Compd. 340 (2002) 141.

Journal of Physics: Conference Series **480** (2014) 012003 doi:10.1088/1742-6596/480/1/012003

Bose-Hubbard model with local two- and three-body interaction under off-diagonal confinement

G J Cruz, R Franco, J Silva-Valencia

Departamento de Física, Universidad Nacional de Colombia, Bogotá, Colombia

E-mail: jsilvav@unal.edu.co

Abstract. We calculate the density profiles of a system of bosons with off-diagonal confinement and local interaction between two or three particles, using the density matrix renormalizaction group method. We found different ground-states: Mott insulator, superfluid and diverse mixed states. We observe that the Mott insulator lobes are surrounded by Mott insulator-superfluid coexistence state regardless the kind of the local interaction. The area of the Mott insulator lobes decreases for the two-body interactions and remain the same for three-body interactions.

1. Introduction

The experiments realized with boson system about phase transitions have opened novel theoretical studies for proposing deeper insights in the Bose-Hubbard model. In 2002, Greiner *et al.*[1] observed a phase transition from superfluid to Mott Insulator. Initially each atom is spread out over the entire lattice, however when the hopping parameter or the density changes, the atoms can be localized at individual lattice sites leading to the Mott insulator phase. The above experimental observations were interpreted using the Bose-Hubbard model which describes a system of bosons in a lattice, where the bosons interact under a two-body term. The phase diagram and the main features of this model have been studied widely [2, 3, 4].

Recently Will*et al.* [5] highlighted the importance of considering more than two-body interaction by an experimental observation of the local wave-function. With an interferometric technique, they measured the absolute energies of atoms at a lattice site. The results demonstrated the presence of ultracold bosonic-system multi-body interaction in a three-dimensional optical lattice.

Recently, a modified Bose-Hubbard model (with local interaction terms between two and three bodies) was studied by B. L. Chen *et al.* [6] and K. Zhou *et al.* [7] with the mean-field aproximation. Silva-Valencia and Souza [8, 9] used the density matrix renormalization group method to determine the phase diagram of the Bose-Hubbard model with two- and three-body interactions, finding that the Mott insulator lobes increases with the density and the three-body interaction.

When atoms are confined in optical lattices they undergo a harmonic potential, which leads to a Mott insulator state without gap i.e. this state is compresible. With the aim of finding true Mott insulator states in experiments, Rousseau *et al.* [10] proposed the off-diagonal confinement (ODC), which leads to an inhomogeneous hopping parameter. Using quantum Monte Carlo,

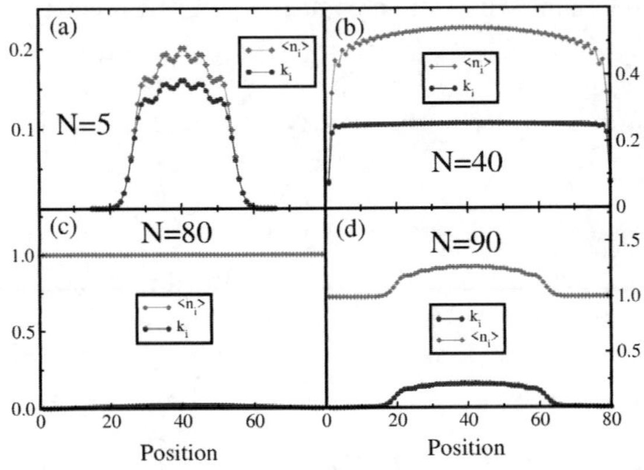

Figure 1. The density profile and local compressibility of bosons under ODC potential and two-body interaction. Here $t = 0.05$

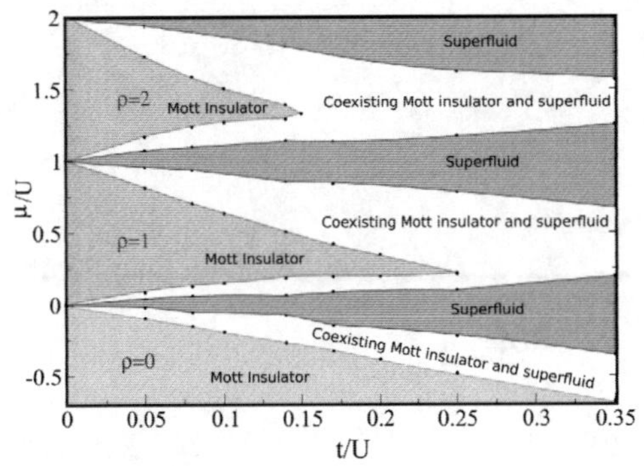

Figure 2. Phase diagram of bosons under ODC potential and two-body interaction. The lattice size of the system is $L = 80$.

this new confinement method allows to obtain incompressible Mott insulator states and a novel Mott-superfluid coexistence phase.

This paper is focused in the study of a one-dimensional system with off-diagonal confinement considering two- or three-body interaction, using the canonical ensemble and the density matrix renormalization group method to find the ground-state and his properties.

The present work is organized as follows: in Section 2, we explain the Hamiltonian models. In section 3 we show the results such as one-dimensional density profiles with two- or three-body interaction and the ODC phase diagrams. Finally, in Section 4, we summarize our results.

2. Hamiltonian models

The Hamiltonian of the Bose-Hubbard model that describes a system of bosons under ODC potential and two-body interaction is given by

$$H_1 = -\sum_{\langle ij \rangle} t_{ij}(b_i^\dagger b_j + b_j^\dagger b_i) + \frac{U}{2}\sum_i n_i(n_i - 1), \qquad (1)$$

where $b_i^\dagger(b_i)$ creates (annihilates) a boson on site i and $n_i = b_i^\dagger b_i$ is the local density operator. The sum in the above Hamiltonian runs in the range $[0; L-1]$, with L representing the lattice size, and the symbol $\langle ij \rangle$ restricts the interactions to the nearest neighbors. The two-body interaction between the bosons is given by the second term of the above Hamiltonian, and his strength is $U = 1$.

The first term in the Hamiltonian (Eq. 1) represents the kinetic energy, where the hopping parameter t_{ij} is given by $t_{ij} = t(i + j + 1)(2L - i - j - 1)/L^2$, which changes like as an inverted parabola, i. e. there is a maximum value t in the lattice center and a null value in both ends of the lattice. This non-constant hopping parameter is due to the ODC confinement.

When we consider the three-body interaction between the bosons, the Hamiltonian of the systems is

$$H_2 = -\sum_{\langle ij \rangle} t_{ij}(b_i^\dagger b_j + b_j^\dagger b_i) + \frac{W}{6}\sum_i n_i(n_i - 1)(n_i - 2), \qquad (2)$$

where the strength of the three-body interaction is $W = 1$.

Journal of Physics: Conference Series **480** (2014) 012003 doi:10.1088/1742-6596/480/1/012003

The ground-state above Hamiltonians is studied using the density matrix renormalization group (DMRG) [11], we consider 200 states and obtain an error of 10^{-8}.

The chemical potential to add a particle to the system of size L is given by $\mu(L) = E(N+1, L) - E(N, L)$, where $E(N, L)$ is the ground-state energy of a system of lattice size L, with N particles.

3. Results

The density profiles of a system of bosons that interact under a two-body term are shown in Fig. 1 for a maximum hopping parameter $t = 0.05$. Considering that the real systems of bosons confined in optical lattices are finite, we consider lattice of $L = 80$ sites [1]. In Fig. 1(a), we show a typical profile for low density, for example $N = 5$. ODC potential generates a charge redistribution where empty sites in both ends of the lattice and a superfluid phase in the center are obtained. So, we can stablish a coexistence phase between superfluid and band insulator. The insulator regions can be confirm with the local compressibility, which is given by $\kappa_i =< (n_i)^2 > -(< n_i >)^2$, and we note that the local compressibility in the band insulator sites is zero because the particle number fluctuations are insignificant.

In Fig. 1(b), a superfluid state appears with $N = 40$ particles, where the local density and the local compressibility vary in the lattice. A Mott insulator state characterized by a null local compressibility appears in Fig. 1(c), with $N = 80$ particles, therefore, for integer density and two-body interaction a truly Mott insulator is obtained. In Fig. 1(d), with $N = 90$ particles the coexistence phase between the superfluid phase and the Mott insulator phase is presented, in this figure the local compressibility in the lattice ends is zero and the local density in the extremes of the lattice shows one particle for site.

The phase diagram of a system of bosons under the ODC potential and two-body interaction term is shown in Fig. 2. We observe the already-known Mott insulator lobes and the superfluid regions, however a new Mott insulator-superfluid coexistence regions surrounding the Mott insulator lobes appear. Also, we found that the area of the Mott insulator and superfluid phases are smaller than in the harmonic potential case (diagonal confinement).

In Fig. 3, we show the density profiles for the Bose-Hubbard model with ODC potential as well as the interaction between three bodies (Eq. 2). Again, the coexistence between superfluid and band insulator regions appear for low densities when we consider $N = 20$ bosons in a lattice of $L = 80$ sites, and a maximum hopping parameter $t = 0.05$ (see Fig. 3(a)).

A superfluid profile is presented in Fig. 3(b), the local compressibility is nonzero for each lattice site except at both ends. Also there is not a Mott insulator for $N = 80$ contrary to the two-body interaction profiles, because the system with one particle for site and three-body interaction do not generate a insulator region.

With $N = 100$, there is another superfluid profile, where the lattice ends are not empty and the compressibility is nonzero, this is presented in Fig. 3(c). In Fig. 3(d), for $N = 160$ a truly Mott insulator phase appears characterized for a local compressibility equal to zero.

In Fig. 3(e), with $N = 170$ we show the coexistence between Mott insulator and superfluid. Equally, in Fig. 3(f), for $N = 220$ there is a coexistence phase but this profile has a minimum value in the superfluid cloud.

The phase diagram of the system of bosons interacting with a three-body term under a ODC potential is shown in Fig. 4. We note that the Mott insulator lobes are surrounded by coexistence states alike than the two-body case, however the area of the Mott lobes remain the same, this is due to strong localization generates by the three-body interaction term. Also, we observe that the Mott lobes area increases with the density, and the superfluid regions of the phase diagram are smaller than in the homogeneous lattice.

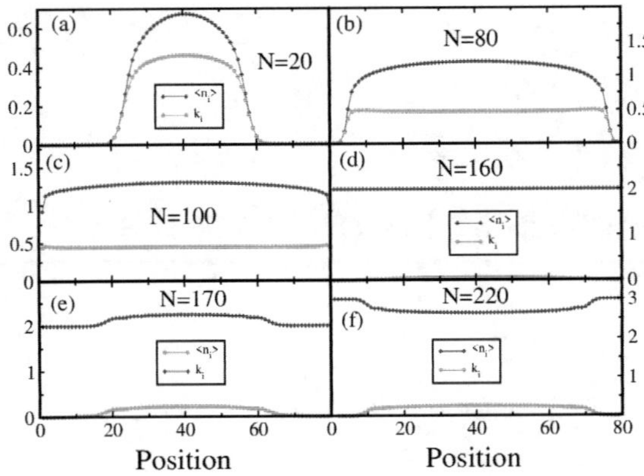

Figure 3. The density profile and local compressibility of bosons under ODC potential and three-body interaction. Here $t = 0.05$.

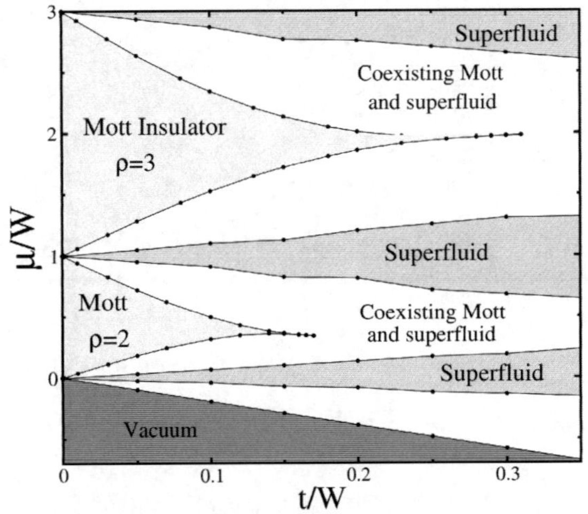

Figure 4. Phase diagram of bosons under ODC potential and three-body interaction. The lattice size of the system is $L = 80$.

4. Conclusions

In this paper we have studied the novel scheme for confining atoms in lattices based on off-diagonal confinement, for which the hopping parameter t is maximum in the lattice center and vanish at the lattice extremes. This confinement permits us to obtain truly Mott phases, which are observed when we calculate the local density and local compressibility for system with interaction between two or three bodies by DMRG.

The ODC potential generates different ground-state configurations, such as: truly Mott insulator, superfluid and states where coexist superfluid regions with Mott insulator ones.

We obtained that the area of the Mott insulator lobes decrases due to the ODC potential for the two-body interaction, whereas remain the same for the three-body interaction.

References

[1] Greiner M, Mandel O, Esslinger T, Hnsch T W and Bloch I 2002 *Nature* **415** 39
[2] Kühner T D and Monien H 1998 *Phys. Rev. B* **58** R14741
[3] Kühner T D, White S R and Monien H 2000 *Phys. Rev. B* **61** 12474
[4] Ejima S, Fehske H and Gebhard F 2011 *Europhys. Lett.* **93** 30002
[5] Will S, Best T, Schneider U, Hackermller L, Lhmann D S and Bloch I 2010 *Nature* **465** 09036
[6] Chen B L, Huang X B, Kou S P and Zhang Y 2008 *Phys. Rev. A* **78** 043603
[7] Zhou K, Liang Z and Zhang Z 2010 *Phys. Rev. A* **82** 013634
[8] Silva-Valencia J and Souza A M C 2011 *Phys. Rev. A* **84** 065601
[9] Silva-Valencia J and Souza A M C 2012 *Eur. Phys. J. B* **85:** 161
[10] Rousseau V G, Batrouni G G, Sheehy D E, Moreno J and Jarrell M 2010 *Phys. Rev. Lett.* **104** 167210
[11] White S R 1992 *Phys. Rev. Lett.* **69** 2863
[12] Belitz D and Kirkpatrick T R 1998 *arXiv:cond-mat/9811058v1*
[13] Wada N, Taniguchi J, Ikegami H, Inagaki S and Fukushina Y 2001 *Phys. Rev. Lett.* **86** 4322
[14] Fazio R and van der Zant H 2001 *Phys. Rep.* **355** 235

Journal of Physics: Conference Series **480** (2014) 012004 doi:10.1088/1742-6596/480/1/012004

Morphological and electrical properties of SnS:Bi thin films

C Calderón, E Banguero and G Gordillo

Departamento de Física, Universidad Nacional de Colombia, Bogotá, Colombia

E-mail: clcalderont@unal.edu.co

Abstract. This work presents results regarding the influence of the Bi content on the morphology and electrical transport properties of SnS:Bi thin films grown by sulfurization of the metallic precursors on a soda-lime glass substrate. The SnS:Bi films were characterized through atomic force microscopy and electrical conductivity measurements. The results showed that the Bi content affects the grain size and roughness of the SnS:Bi films. The grain size affects the carrier mobility and therefore the electrical conductivity at room temperature. Measurements of conductivity as a function of temperature, carried out in the range between 90K and 630K, revealed the electrical transport in SnS:Bi thin films is affected by two different mechanisms: at temperatures greater than 350K, the conductivity is predominantly affected by free carrier transport in extended states of the conduction/valence band, whereas at temperatures below 350K the conductivity is mainly determined by the variable range hopping transport mechanism.

1. Introduction

The chalcogenide materials such as SnS, SnS_2, $Sn2S_3$ and Bi_2S_3 are currently of great interest due to its applications in the fabrication of optoelectronic, thermoelectric and photovoltaic devices, and as a holographic recording medium. Much of this interest is due to S and Sn being inexpensive, abundant in nature and less toxic than most of the materials used in the industry for manufacturing these types of devices. Particularly, the SnS presents suitable properties for an absorbent layer in solar cells, with conversion efficiencies of up to 25% theoretically [1], thanks to its properties: energy band gap of 1.3 eV, fundamental absorption coefficient greater than 10^4 cm^{-1} and p-type electrical conduction. The Bi_2S_3 is widely studied because of their physical and chemical properties, suitable for the development of solar cells, luminescent devices, pigments, IR detectors, and thermoelectric devices [2, 3].

SnS compound has been synthesized by different techniques such as chemical bath deposition [4], two step processes [5], electrochemical deposition [6] and thermal evaporation [7]. In this work the SnS:Bi thin films were grown by sulfurization of the metallic precursors, which is a novel method for synthesizing thin films based on SnS. The influence of the concentration of Bi on the electrical transport properties and on the morphology of the SnS:Bi films was investigated by electrical conductivity and Atomic Force Microscope (AFM) measurements. The substitution of Sn atoms by Bi atoms generates free electrons in the SnS lattice, and hence, a change of conductivity type from p to n in the SnS can be induced. This is important because it opens the possibility of in situ fabricating p-SnS/n-SnS:Bi solar cells, which would reduce the photocurrent losses due to recombination at the interface, as well as the manufacturing cost of the device.

2. Experimental

The deposition of the SnS:Bi films was accomplished using a system constituted by an evaporation chamber connected to a vacuum system working at pressures of about 10^{-5} Torr, two tungsten boats (used to evaporate Sn and Bi respectively), a tantalum effusion cell to evaporate S and a thickness monitor (Maxtec TM-400) with a quartz crystal as sensor, used to measure the evaporated metallic elements flow. The substrate temperature and the evaporation temperature of S were controlled with a programmable PID controller (Eurotherm 900C). The SnS:Bi thin films were grown by sulfurization of the metallic precursors, which are sequentially evaporated on a glass substrate keeping the substrate temperature at 350 ºC; in a second stage the layer containing the metallic precursors is annealed in a S environment at the same substrate temperature, the evaporation temperature of S is kept at 140 ºC. Details of the growth routine of the films are presented in Ref. [8]. The Bi concentration (x) in the SnS:Bi films was varied between 0 and 1 according to the relation:

$$x = \frac{mols\ of\ Bi}{mols\ of\ Sn + mols\ of\ Bi}$$

In particular, when x=0 SnS films were obtained, and Bi_2S_3 films when x=1.

The SnS:Bi thin films were electrical and morphologically characterized to study the effect of the content of Bi on the electric conductivity, and on the grain size and average roughness. Morphological characterization of the deposited films was performed using an Atomic Force Microscope; the AFM measurements were done with a 0100 model PSI Auto-probe CP microscope. The conductivity type was determined from thermoelectric power measurements. The film thicknesses were measured with a Veeco Dektak 150 surface profiler and their values fell in the range between 800 nm and 1 μm.

3. Results and discussion

3.1. Morphological characterization

Figure 1 shows typical AFM images of SnS:Bi thin films deposited varying the Bi concentration (x = 0, 0.3, 0.5, 0.7 and 1). From the results of figure 1 it is clear that the Bi content significantly affects the films' morphology. In general, the increase of the Bi concentration (x) leads to a decrease of both the grain size and roughness. The curves depicted in figure 2 show the variation of the average roughness and of the average grain size as functions of x.

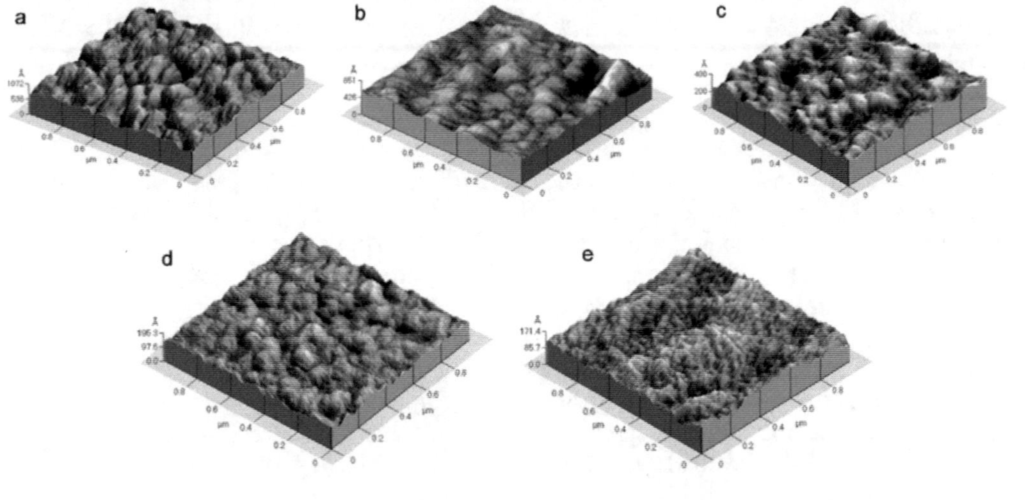

Figure 1. Typical AFM images of SnS:Bi films deposited with different Bi concentrations: a) x=0, b) x=0.3, c) x=0.5, d) x=0.7 and e) x=1.

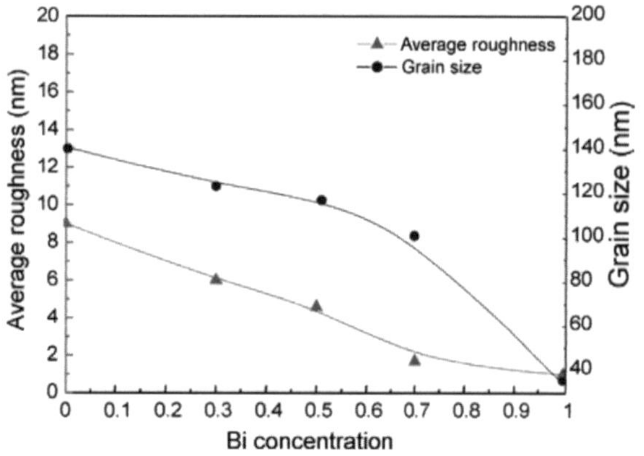

Figure 2. Variations of the average values of grain size and roughness as functions of the Bi concentration in Sn:Bi thin films.

Decreasing the roughness causes a decrease in light absorption, which in turn leads to an increased transmittance [8], while decreasing grain size causes an increase in the density of grain boundaries which results in a decreased mobility of the carriers and thus the electrical conductivity (σ). The SnS films (x=0) present higher grain size and electrical resistivity of the order of 1.1x10^2 Ωcm, while films of SnS:Bi with high concentrations of Bi (x=1 corresponding to Bi_2S_3) present lower size grain and resistivity of the order of 1.2x10^4 Ωcm.

It must be realized that although the grain size influences the mobility (μ) of the charge carriers and this one in turn on the resistivity (ρ), the charge carrier concentration (n) also affects the behavior of the electrical resistivity of the films; particularly for the films deposited, values of μ between 26.14 and 4.11 cm^2/Vs and values of n between 2.04x10^{15} and 1.26x10^{14} cm^{-3} (at room temperature) were obtained for SnS and Bi_2S_3 films, respectively. Both the mobility and carrier concentration at a determined temperature depend on the compound; SnS:Bi thin films deposited in this work have different phases, in accordance to previous studies [8]: SnS films (x=0) grow in the orthorhombic phase, while the films synthesized by sulfurization of a Sn and Bi (x=0.3, x=0.5 and x=0.7) grow with a mixture of several phases including the orthorhombic Sn_2S_3, SnS and Bi_2S_3 and the hexagonal SnS_2; when x=1 the Bi_2S_3 compound is formed in the orthorhombic phase, therefore the increase in electrical resistivity observed when the Bi concentration raises is caused by the increase of Bi_2S_3 in the SnS:Bi samples, since, in general, the Bi_2S_3 phase has greater resistivity than the other phases present in the compound.

It was also found from thermoelectric power measurements that the SnS:Bi thin films with Bi content less than or equal to 0.5 have p type conductivity and films with Bi content greater than 0.5 have type n.

3.2. Electrical properties

Electric conductivity measurements as a function of temperature in a range between 90K and 630K were used to identify the scattering processes affecting electrical transport. The conductivity was measured using electrical contacts of Ag deposited by sputtering DC. Figure 3 shows typical curves of electric conductivity as a function of temperature, for SnS:Bifilms deposited wit h different content of Bi (x=0, 0.5, 1). The increase in conductivity with increasing temperature indicates the semiconducting nature of the films.

The σ vs. T curves show two different slopes, indicating that the conductivity is affected by two different transport mechanisms predominating at different temperature ranges. The results of figure 3 were analyzed assuming predominance of one mechanism at a given temperature range in an attempt to identify the mechanisms affecting the electrical transport of the SnS:Bi thin films. We found that at temperatures above 350K, the conductivity can be expressed by

$$\sigma = \sigma_0 \exp[-(E_C - E_F)/(kT)] \tag{1}$$

indicating in this range of temperatures the conductivity is predominantly affected by free carrier transport in extended states of the conduction/valence band. Since the factor σ_0 (given by $qNC\ \mu(T)$) varies slowly with temperature and the term $exp[-(E_C-E_F)/kT]$ depends strongly on T, σ_0 can be assumed as approximately constant. Under this approximation, the slope of the graph $ln\sigma$ vs. $1/T$ gives the activation energy $E_\sigma = E_C - E_F$; the calculated values of the corresponding activations energies are 0.22 eV for x=0, 0.24 eV for SnS:Bi x=0.5 and 0.43 eV for x=1.

In figure 4 are depicted curves of $ln\sigma$ vs. $1000/T$ using the experimental data of figure 3, measured between 350K and 630K. The linear behavior of the $ln\sigma$ vs. $1000/T$ curves demonstrates that, in the high temperature range, the SnS:Bi film conductivity is dominated by free carrier transport in extended states of the conduction/valence band (for n/p type conductivity respectively).

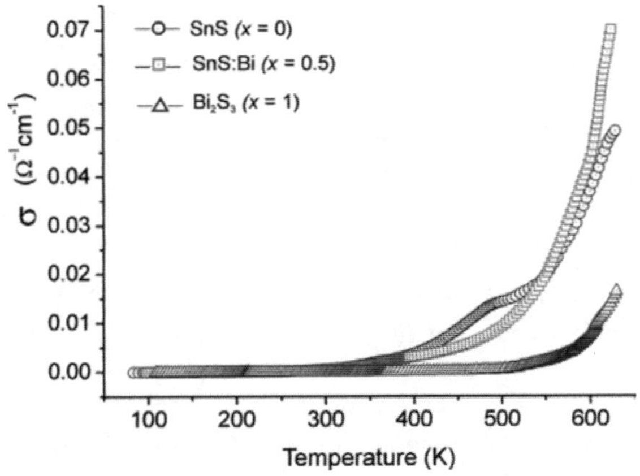

Figure 3. Curves of σ vs. T corresponding to SnS:Bi films deposited with different content of Bi.

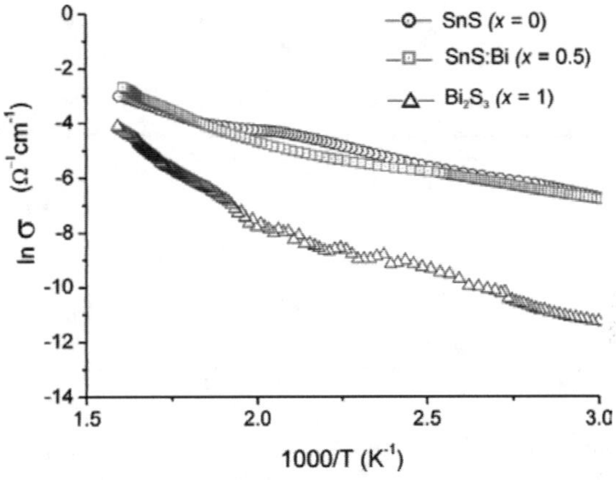

Figure 4. Curves of $ln\sigma$ vs. $1000/T$ corresponding to SnS:Bi films deposited with different content of Bi.

In the range of low temperatures (T< 350 K), the curve of σ vs T can be expressed by a relation of the form

$$\sigma = \sigma_0(T)\exp(-\frac{A}{T^{1/4}}) \tag{2}$$

This behavior of σ vs. T indicates in this range of temperature the conductivity is predominantly affected by transport in a band of localized states as predicted by the Davis-Mott model, where the carriers can move between states via a phonon-assisted tunneling process [9]. This transport mechanism is known as Variable Range Hopping (VRH).
In Mott's derivation:

$$\sigma_0 = \frac{e^2}{2(8\pi)^{1/2}}\nu_{ph}[\frac{N(E_F)}{\beta kT}]^{1/2} \tag{3}$$

and $$A = [\frac{\beta^3}{kN(E_F)}]^{1/4} \tag{4}$$

where $N(E_F)$ is the density of localized states, and β a quantity representing the fall-off of the wave function on the site. In principle the two parameters β and $N(E_F)$ can be evaluated from the slope of a plot of $\ln[\sigma(T)T^{1/2}] vs T^{-1/4}$ and from the intercept at $T^{1/4} = 0$, respectively.

Figure 5 shows plots of $\ln[\sigma(T)T^{1/2}] vs T^{-1/4}$ performed with the experimental data of figure 3, measured between 90K and 350K. The linearity of the $\ln[\sigma(T)T^{1/2}] vs T^{-1/4}$ curves confirms that the conductivity of the SnS:Bi films in the low temperature range is dominated by the VRH transport mechanism.

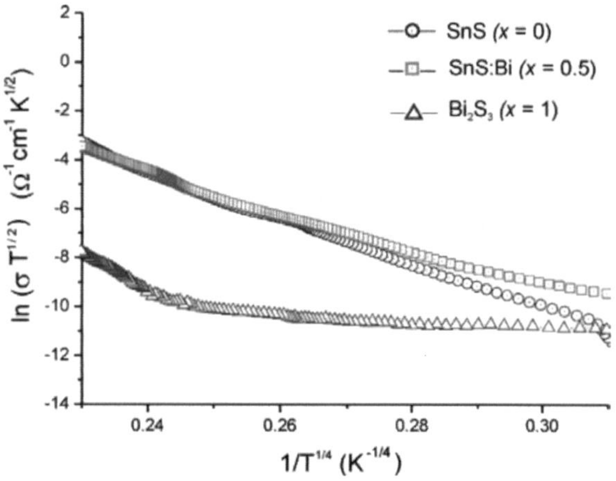

Figure 5. Curves of $\ln[\sigma(T)T^{1/2}] vs T^{-1/4}$ corresponding to SnS:Bi films deposited with different content of Bi.

4. Conclusions

SnS:Bi thin films were deposited using a novel procedure based on sulfurization of the metallic precursors. The Bi concentration in the films was varied between x=0 and x=1.

AFM measurements indicated that the morphology of the SnS:Bi films is affected by Bi concentration: the increase of the Bi concentration leads to a decrease of both the grain size and roughness. The decrease of grain size leads to a decrease of the mobility of free carriers when the Bi concentration in the SnS:Bi films increases, probably due to grain boundary effect.

Measurements of conductivity as a function of temperature revealed that the electrical transport in SnS:Bi films is affected by two different mechanisms: in the range of temperatures above 350K, the conductivity is predominantly affected by carrier transport in extended states of the conduction/valence band, whereas in the range of temperatures below 350K, the conductivity is dominated by the Variable Range Hopping transport mechanism.

Thermoelectric power measurements revealed that it is possible to grow SnS:Bi films with *p* or *n* conductivity controlling adequately the Bi content in thefilms. This result indicates it is possible to fabricate *p*-SnS/*n*-SnS:Bi solar cells in-situ, which facilitates industrial production.

Acknowledgments

We acknowledge the financial support of this work by the DIB-Universidad Nacional de Colombia.

References

[1] J. J. Lofersky 1956 *J. Appl. Phys.* **27** 77.
[2] Zhen-Hua Ge, Bo-Ping Zhang, Peng-Peng Hang, Yi-Qiang Yu, Chen Chen and Jing-Feng Li 2011 *J. Electron. Mater.* **40** (5) 1087.
[3] Y. Xiao, H. Cao, K. Liu, S. Zhang and V. Chernow 2010 *Nanotechnology* **21** (14) 5601.
[4] Y. Jayasree, U. Chalapathi and V. Sundara Raja 2013 *Thin Solid Films* **537** 149.
[5] M. Leach, K.T. Ramakrishna Reddy, M.V. Reddy, J.K. Tan, D.Y. Jang and R.W. Miles 2012 *Energy Procedia*, **15** 371.
[6] R. Mariappan, T. Mahalingam and V. Ponnuswamy 2011 *Optik* **122** 2216.
[7] A. Schneikart, H-J Schimper, A. Klein and W. Jaegermann 2013 *J. Phys. D Appl. Phys.* **46** (30) 5109.
[8] C. Calderón, E. Banguero, P. Bartolo Pérez and G. Gordillo 2011 *Braz. J. Phys.* **41** 15.
[9] N.F. Mott and R.A. Davis, 1979 *Electronic Processes in Non-Crystalline Materials*, 2nd. ed. Oxford University Press, Oxford.

Journal of Physics: Conference Series **480** (2014) 012005 doi:10.1088/1742-6596/480/1/012005

Upconversion emission in (Ln,Yb):KLu$(WO_4)_2$ nanocrystals for white light generation

E W Barrera[1], M C Pujol[1], J J Carvajal[1], X Mateos[1], R Solé[1], J Massons[1], A Speghini[2], M Bettinelli[2], C Cascales[3], M Aguiló[1] and F Díaz[1]

[1] Física i Cristal·lografia de Materials i Nanomaterials (FiCMA-FiCNA-EMaS), Universitat Rovira i Virgili (URV), Campus Sescelades, c/ Marcel.lí Domingo, s/n, E-43007 Tarragona, Spain

[2] Dipartimento di Biotecnologie, Università di Verona and INSTM, UdR Verona, Ca' Vignal, Strada Le Grazie 15, 37134 Verona, Italy

[3] Instituto de Ciencia de Materiales de Madrid, CSIC, calle Sor Juana Inés de la Cruz, Cantoblanco, E-28049 Madrid, Spain

Email: mariacinta.pujol@urv.cat

Abstract. Optical active lanthanides ions in double tungstates exhibit high emission cross section with low concentration quenching. Production of double tungstates nanocrystals doped with lanthanide ions is suitable for applications in white light emitters if high quantum yield is reached. In this work lanthanide doped (Ln:Yb):KLu$(WO_4)_2$ nanocrystals (Ln = Er, Tm, Ho) upconversion emission properties have been study after 980 nm diode laser excitation at 14-334 W/cm². From CIE chromaticity theory a possible mixture weight ratio to obtain a white light emitter powder was predicted. Luminescence under 980 nm of the (Ln:Yb):KLu$(WO_4)_2$ nanocrystals was study and decay times under 460 nm was measured. Reduction of the upconversion quantum yield was observed in the case of physical mixture of nanocrystals.

1. Introduction

In the last years, there has been a great interest in the development of lanthanide (Ln) doped white light emitters based in the upconversion (UC) process in which near infrared photons are converted in visible photons by a multiphoton absorption process [1]. The suitability of optical active trivalent ions Ln^{3+} cations for the UC processes is due to their abundant energy levels and narrow emission spectral lines [1]. Yb^{3+} is used to act as sensitizer due to its high absorption cross section, allowing the efficient excitation around 980 nm with cheap diode lasers and followed by energy transfer to other trivalent Ln ions, such as Er^{3+}, Tm^{3+} and Ho^{3+}. A white light source can be tailored by mixing red-green-blue UC phosphors in proportions determined by the UC emission spectra of each phosphor at a specific chemical composition and pumping power. It is important to consider the excitation conditions of the phosphors because local high pumping induces temperature increment modifying UC emission spectra and consequently the color of the emitter [2]. Monoclinic KLu$(WO_4)_2$ (KLuW) host is a well-known material for laser applications. Its main advantages are the admittance of a high doping level of active

Ln^{3+} ions in this host and the possibility to be highly doped with lanthanides ions, with low concentration quenching due to the relatively large Ln-Ln distances in the structure [3], and the high absorption and emission cross sections of these Ln ions in this host. In this work codoped lanthanide doped KLuW nanocrystals (NCs) were synthesized by the sol-gel modified Pechini method to obtain a red, green and blue phosphors. Emission spectra under 980 nm near infrared (NIR) radiation as a function of excitation pump power were characterized to study the UC luminescence dynamics and for calculation of CIE color coordinates.

2. Experimental

(Ln:Yb):KLuW (Ln = Ho, Er, Tm) NCs were synthesized by the modified Pechini Method [4]. Er^{3+}, Ho^{3+}, Tm^{3+} atomic content was fixed to 1 at. % and Yb^{3+} to 10 at. %. Analytic grade purity reagents of Er_2O_3 (99.99%), Ho_2O_3 (99.9999%), Tm_2O_3 (99.9999%), Yb_2O_3 (99.9%), Lu_2O_3 (99.9999%), were dissolved in hot nitric acid to form the nitrate precursor. We used citric acid (CA) as the chelation agent and ethylenglycol (EG) as the esterification agent. The precursors were dissolved in distilled water solution with citric acid. Ammonium tungstate $(NH_4)_2WO_4$ (99.99%) and potassium carbonate K_2CO_3 (99.99%) were added to the aqueous solution and it was fixed at 353 K under magnetic stirring during 24 h until complete dissolution. Further, EG was added to the mixture and the solution was heated at 373 K in order to remove water and generate the polymeric gel. The polymer gel was calcinated at 573 K for 3 h to eliminate the organic compounds and finally annealed at 1023 K to obtain a nanocrystalline white powder that was putted in clear glass vials, especially for UC and PL measurements.

3. Results and Discussion

Figure 1 shows the X-ray powder diffraction patterns of doped KLuW samples obtained by annealing at 1023 K for 2h. All the peaks were indexed to the monoclinic phase of $KLu(WO_4)_2$ with the space group *C2/c* by comparison with the JCPDS 54-1204 card. After annealing, single phase material was obtained indicating that doping cations were incorporated into the host lattice, as expected. The morphology was characterized for faceted and irregular shapes as showed in Figure 2. The grain size values were estimated fitting the particle size distributions obtained by contrast method. Mean particle size of 130±70, 100±30 and 90±30 nm were calculated for Er-doped, Ho-doped and Tm-doped sample respectively.

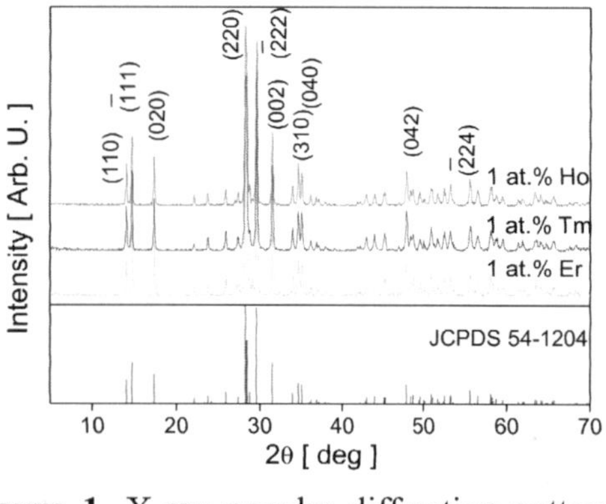

Figure 1. X-ray powder diffraction pattern of the synthesized doped KLuW NCs. JCPDS 54-1204 of $KLu(WO_4)_2$ is presented for comparison

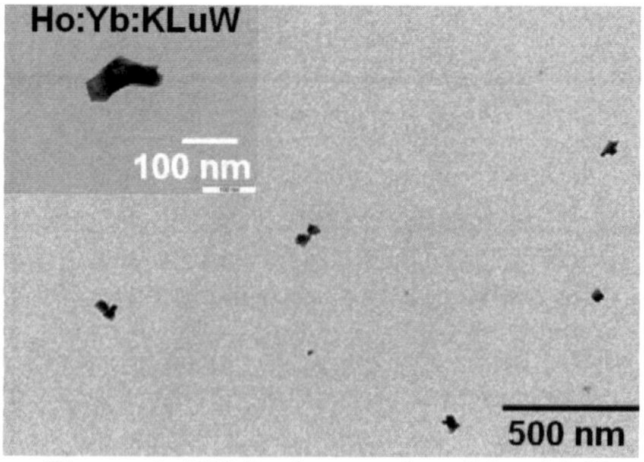

Figure 2. TEM micrographs of 1 at.% Ho 10 at.% Yb KLuW nanocrystals.

Figure 3 shows the 980 nm excited UC emission spectra in the range of 450-725 nm, of Yb, Ln-doped nanocrystals and the observed emission bands labelled with the corresponding electronic transition in Er^{3+}, Ho^{3+} and Tm^{3+} ions. UC emission intensities (I_{UC}) versus power excitation (I_0) follow the proportionality $I_{UC}\ \alpha\ (I_0)^n$ [5] as it shows in the log-log graphs in Figure 4. The calculated n slope values for each emission band corresponds to the number of photons involved in the upconversion process. In that way upconversion dynamics corresponds to previous descriptions reported for Er: KYbW [6] and Ho:Tm:Yb:KLuW nanocrystals [7]. The experimental emission decay times under 460 nm excitation were evaluated by proposed model by Inokuti and Hirayama [5]. Table 1 summarizes the measured decay times from photoluminescence decay time curves of Ln-doped NCs

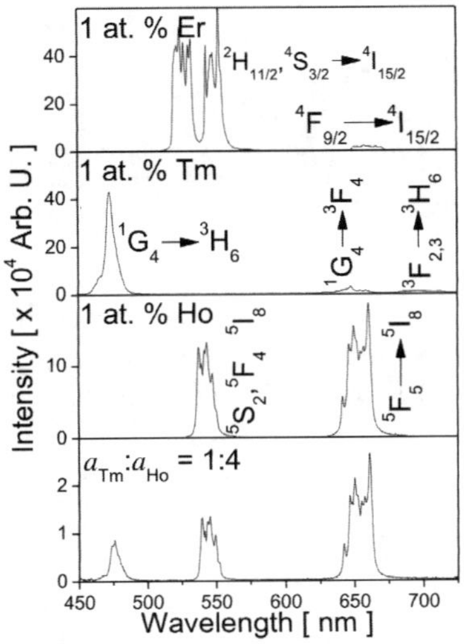

Figure 3. NIR to visible UC luminescence spectra of codoped KLuW NCs at 74 W/cm² excitation power.

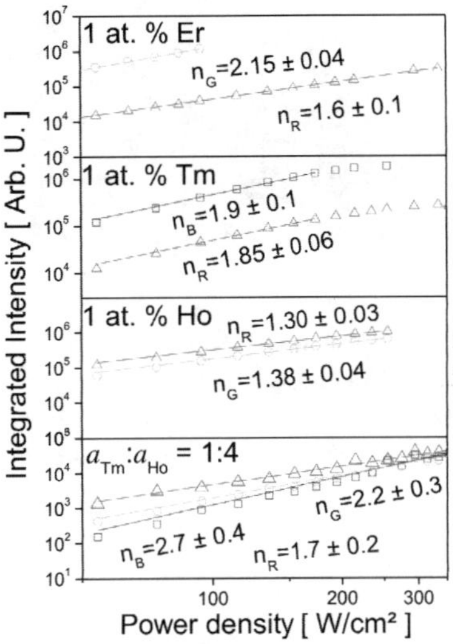

Figure 4. Integrated intensities versus excitation power density. The emissions correspond to 475 nm (squares), 540 nm (circles) and 650 nm (up triangles).

UC emitted colors have been quantified in terms of the CIE 1931 color diagram [8,9]. The dependence of CIE coordinates for different power density values for 1 at. % Ln, 10 at. % Yb:KLuW nanocrystals (Ln = Er, Ho, Tm) is shown in Figure 5. UC emitted colors are visible by the naked eye as shown in the inset pictures in Figure 5. Chromaticity coordinates are near the boundary, which means high color purity. The dominant wavelength and color purity were calculated for different values of excitation power and they are listed in Table 2. Restricted excitation power density of Yb, Er-doped nanocrystals in Table 2 was a consequence of the observed saturation of the CCD detector under the fixed setup, for the Er^{3+} $^2H_{11/2}$, $^4S_{3/2}$→$^4I_{15/2}$ green emission

Table 1. Summary of measured decay times for 1 % Ln^{3+} doped 1% Ln 10 % Yb:KLuW nanocrystals after 460 nm excitation.

	Emitting State	λ_{EMI} [nm]	τ [±1 μs]
Er^{3+}	$^2H_{11/2}$, $^4S_{3/2}$	550	9
	$^4F_{9/2}$	660	5
	$^4I_{9/2}$	800	11
Ho^{3+}	$^5S_2$	540	5
	$^5F_5$	650	7
Tm^{3+}	$^1G_4$	475	76
	$^3H_4$	800	137

Table 2. Chromaticity properties of 1 at. % Ln, 10 at. % Yb:KLuW (Ln = Er, Ho, Tm) nanocrystals

Active Ion	Excitation power density	Dominant wavelength	Color purity
Ln^{3+}	[W/cm²]	[nm]	[%]
Er^{3+}	15-100	547 - 540	96 – 97
Ho^{3+}	50 – 250	574 - 563	96 – 97
Tm^{3+}	50 – 250	473 - 472	84 – 87

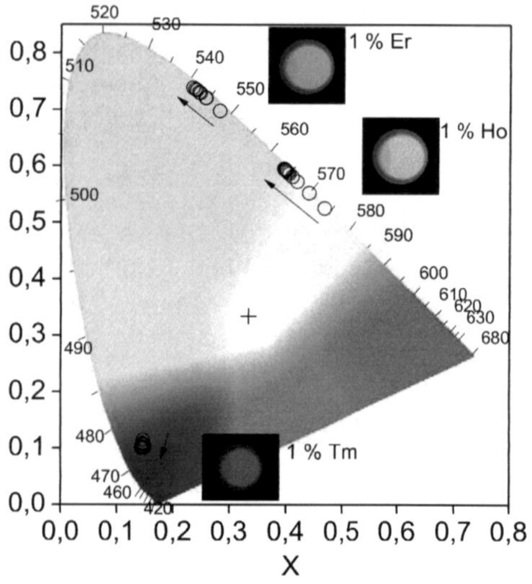

Figure 5. Chromaticity diagram for 1 at. % Ln, 10 at. % Yb:KLuW (Ln= Er, Ho, Tm) nanocrystals for excitation power densities in the range 10-250 W/cm², in the indicated arrow direction.

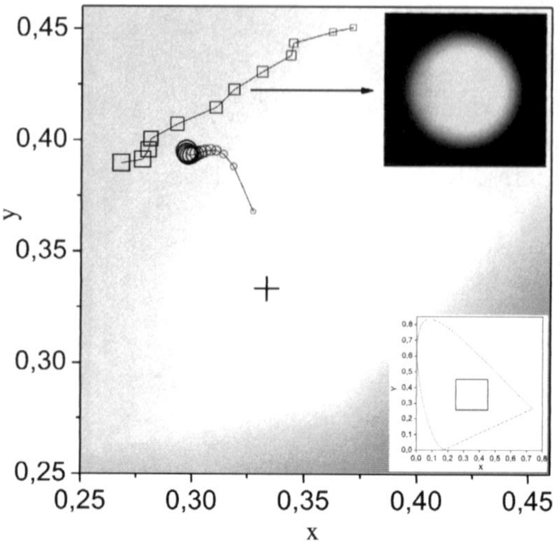

Figure 6. Chromaticity diagram for the mixture of 1 % Ho, 10 at. % Yb: KLuW and 1 at. % Tm, 10 at. % Yb: KLuW nanocrystals, with a_{Tm}:a_{Ho}=1:4. Symbol size is proportional to excitation power density, from 94 to 354 W/cm². Simulated data (circles), experimental data (squares).

For the calculation of a white light emitter, we have defined a sub-space base in the color *XYZ* space which is formed by linear combination of the tristimulus functions obtained for each sample expressed as vectors (X_i, Y_i, Z_i) where *i* indicates the dopant cation Er^{3+}, Tm^{3+} or Ho^{3+}. In this sub-space any vector is expressed as,

$$\begin{bmatrix} X \\ Y \\ Z \end{bmatrix} = \begin{bmatrix} X_{Er} & X_{Tm} & X_{Ho} \\ Y_{Er} & Y_{Tm} & Y_{Ho} \\ Z_{Er} & Z_{Tm} & Z_{Ho} \end{bmatrix} \begin{bmatrix} a_{Er} \\ a_{Tm} \\ a_{Ho} \end{bmatrix} \tag{1}$$

where the a_i coefficients determine the relative intensity of each (X_i, Y_i, Z_i) vector in the linear combination. Here we suppose that each a_i coefficient corresponds experimentally to relative doping concentration of sample doped with cation *i* in the mixture. Chromaticity coordinates were calculated by application of the normalization condition $X+Y+Z=1$.

In the mixture of Yb,Ln-doped nanocrystals, a reduction of the total Ln^{3+} to Yb^{3+} ratio is realized, and therefore a proportional quenching of the UC emission could be expected. The excitation is produced through Yb^{3+} ions. Individual samples of nanocrystals have the same 1 at. % Yb^{3+} content, in consequence, the mixture has 1 at. % Yb^{3+} content. Before mixture the total Tm-to-Yb (Ho-to-Yb) ratio is 1/10 (1/10). After a mixture with the relative molar a_{Tm}:a_{Ho}=1:4, it changes to 0.2/10 (0.8/10), so we expect an intensity of 20 % (80 %) in comparison with the corresponding emissions in the Ln,Yb:KLuW samples. Figure 3 shows the UC emission spectra for mixture at low excitation power. As the intensities are reduced after mixture, and the UC quantum yield is proportional to the emission UC intensity we can indicate that quantum yield is reduced after physical mixture of nanocrystals.

For the larger quenching of the Tm^{3+} $^1G_4 \rightarrow {}^3H_6$ blue emission, we propose a self-absorption process of this radiation through the Ho^{3+} $^5F_{2,3}$ multiplet. The temperature can play an important role by modifying the UC efficiency of each emission. However interactions between Yb,Ho-doped and Yb,Tm-doped nanocrystals cannot be discarded. The UC process is nonlinear in nature and tailoring of a UC white light emitter by mixture of powders does not follow a simple addition of the spectra of each powder.

The chromaticity coordinates for the mixture of Yb, Ho-doped and Yb, Tm-doped nanocrystals start moving across the upper part of the white region in the chromaticity diagram in Figure 6, from the yellow to the blue region, as the power density increases in the 94-354 W/cm² range. Simulated data only coincide in following the direction path traced by experimental chromaticity coordinates. In the 94-354 W/cm² power density range, deviations of the simulated chromaticity coordinates correspond to a large contribution of the Tm^{3+} $^1G_4 \rightarrow ^3H_6$ blue emission at low power excitation and the saturation of this emission at high power excitation. These phenomena are not observed in the UC emission spectra of the nanocrystals mixture, indicating that there is an energy interaction between Yb, Ho- and Yb, Tm-doped nanocrystals for the UC process and the proposed model cannot explain these results.

4. Conclusions

1 at. % Ln:10 at. %Yb:$KLu(WO_4)_2$ (Ln=Er, Tm, Ho) nanocrystals with ~100 nm particle size have been prepared by the modified Pechini method, with monoclinic *C2/c* crystalline phase. Chromaticity coordinates and tristimulus functions were calculated from excitation power dependence of the upconversion spectra. A white light emitter was predicted for a physical mixture of 1 at. % Tm^{3+} ,10 at. % Yb^{3+} $KLu(WO_4)_2$ and 1 at. % Ho^{3+} ,10 at. % Yb^{3+} $KLu(WO_4)_2$ in the 1:4 ratio. Comparison with a physical mixture of nanocrystals shows partial approximation for excitation power densities below 74 W/cm². The upconversion quantum yield was reduced after physical mixture of nanocrystals showing that resulting spectra of a mixture of upconversion emitting powders does not match simple addition of the spectra of each emitting powder, .

Acknowledgements

This work is supported by the Spanish Government under projects, MAT2011-29255-C01-02, TEC2010-21574-C02-02 and the project PI09/90527; and the Catalan Authority under project 2009SGR235 and FP7-SPA-2010-263044. E. W. Barrera is supported by Catalan Government for the fund provided through the fellowship 2012FI_B200136. The support of the COST Action CM1006 by the STSM-CM1006-050512-017263 grant is gratefully acknowledged.

References

[1] Mahalingam V, Mangiarini F, Vetrone F, Venkatramu V, Bettinelli M, Speghini A and Capobianco J A 2008 *J. Phys. Chem. C* **112** 17745.
[2] Milliez J, Rapaport A, Bass M, Cassanho A, Jenssen H P 2006 *J. Display Technol* **2** 307.
[3] Petrov V, Pujol M C, Mateos X, Silvestre O, Rivier S, Aguiló M, Solé R M, Liu J, Griebner U and Díaz F 2007 *Laser & Photon. Rev.* **1** 179.
[4] Galceran M, Pujol M C, Aguiló M and Díaz F 2007 *J. Sol-Gel Sci Techn* **42** 79.
[5] Gamelin D R and Güdel H U 2001 *Top. Curr. Chem.* **214** 1.
[6] Galceran M, Pujol M C, Zaldo C, Díaz F and Aguiló M 2009 *J. Phys. Chem C* **113** 15497.
[7] Barrera E W, Cascales C, Pujol M C, Carvajal J J, Mateos X, Aguiló M and Díaz F 2011 *Phys Status Solidi C* **8** 2676.
[8] Comission Internationale d' Eclairage CIE Technical report 2004 No. 15.3 *Colorimetry*.
[9] Schanda J 2007 *Colorimetry: understanding the CIE system*, (New Jersey, John Wiley & Sons).

21st Latin American Symposium on Solid State Physics (SLAFES XXI) IOP Publishing
Journal of Physics: Conference Series **480** (2014) 012006 doi:10.1088/1742-6596/480/1/012006

Theoretical calculations of valence states in Fe-Mo compounds

F. Estrada[1,2], H. Noverola[3], J.R. Suárez[4], O. Navarro[1,*] and M. Avignon[5]

[1] Instituto de Investigaciones en Materiales, Universidad Nacional Autónoma de México, Apartado Postal 70-360, 04510 México, D. F., México.
[2] Centro de Investigación en Materiales Avanzados, S. C., Miguel de Cervantes 120, Complejo Industrial Chihuahua, 31109 Chihuahua, Chih., México.
[3] División Académica de Ingeniería y Arquitectura, Universidad Juárez Autónoma de Tabasco, Cunduacán-Jalpa de Méndez, Col. La Esmeralda, Cunduacán, 86690 Tabasco, México.
[4] Unidad Académica de Física, Universidad Autónoma de Zacatecas, Calzada Solidaridad esq. Paseo, La Bufa s/n, 98060 Zacatecas, México.
[5] Institut Néel, CNRS and Université Joseph Fourier-BP 166, 38042 Grenoble Cedex 9, France.

E-mail: navarro@unam.mx

Abstract. The half-metallic ferromagnetic double perovskite compound Sr_2FeMoO_6 is considered as an important material for spintronic applications. It appears to be fundamental to understand the role of electronic parameters controlling the half-metallic ground state. Fe-Mo double perovskites usually present some degree of Fe/Mo disorder which generally increases with doping. In this work, we study the valence states of Fe-Mo cations in the off-stoichiometric system $Sr_2Fe_{1+x}Mo_{1-x}O_6$ $(-1 \leq x \leq 1/3)$ with disorder. Our results for Fe and Mo valence states are obtained using the Green functions and the renormalization perturbation expansion method. The model is based on a correlated electron picture with localized Fe-spins and conduction Mo-electrons interacting with the local spins via a double-exchange-type mechanism.

1. Introduction

Half-metallic materials, in which only one-spin direction is present at the Fermi level, have been investigated in view of their attractive properties for potential spintronic applications [1] and as potential candidates for memory devices by virtue of their large magnetoresistance (MR). The double perovskite Sr_2FeMoO_6 (SFMO) combines half-metallic ferromagnetic character with a high Curie temperature $T_c = 400K$ and substantial low-field MR [2]. The enhanced T_c, complete spin polarization and substantial low-field magnetoresistance (LFMR) at room temperature are compared to those of manganites and generate a great interest for magnetoelectronic applications [3].

In the fully ordered double perovskite structure, Fe and Mo occupy two interpenetrating sublattices, respectively α and β, in a rock-salt structure. Oxygen atoms bridge the Fe and Mo ions to form alternating FeO_6 and MoO_6 octahedra. The ferromagnetism and half-metallicity in ordered SFMO have been understood [4] on the basis of a strongly correlated picture, with Fe^{3+} $(3d^5)$ localized high-spin $S = 5/2$ and Mo^{6+} cores together with one conduction electron (per formula unit) which can hop between Mo and Fe sites (S) only with an

Journal of Physics: Conference Series **480** (2014) 012006 doi:10.1088/1742-6596/480/1/012006

antiparallel orientation to the localized spins (Hund's coupling). The Fe-Mo hopping stabilizes a ferromagnetic arrangement of the local spins and also leads to the opposite spin polarization of the conduction electrons, in agreement with *ab* initio calculations [2, 5, 6]. The energy Δ needed to transfer an electron from Fe to Mo (charge transfer energy in unit of the bandwidth w) is the key parameter determining the Curie temperature T_c. Therefore ordered SFMO is expected to have a full polarization of the conduction electrons ($P = 1$) (half-metallic) and a saturation magnetization of $4\mu_B$, however, experimental measurements give $P \approx 0.85$ in a SFMO-based tunnel junction device [7] and a saturation magnetization usually lower than the expected value in the ordered system [2, 8, 9]. These deviations can be attributed to antisite disorder in which some Fe and Mo atoms interchange their positions and is consistent with a lowering of T_c. To understand how the presence of AS affects the magnetic properties, the off-stoichiometric system $Sr_2Fe_{1+x}Mo_{1-x}O_6$ ($-1 \leq x \leq 0.25$) has been investigated [10] based on the simple idea that off-stoichiometry generates Fe(S)-O-Fe(AS) (Fe-rich regime $x > 0$) or Mo(S)-O-Mo(AS) (Mo-rich regime $x < 0$) nearest neighbors (n.n.) pairs even in the absence of disorder. Disorder produces both Fe(S)-O-Fe(AS) and Mo(S)-O-Mo(AS) pairs, the Mo(S)-O-Mo(AS) pairs play a fundamental role as they open new hopping channels available to both spin directions, thereby depolarizing the conduction band [11, 12]. The saturation magnetization shows the remarkable behavior of first increasing with x for $x < 0$ up to $x = 0$ and then decreasing for $x > 0$ [10], which can be qualitatively understood in terms of Fe local spins. The decrease for $x > 0$ has a very strong implication as it can be explained unambiguously considering that n.n. Fe(S) and Fe(AS) local spins are coupled antiferromagnetically. The electronic structure of the system $Sr_2Fe_{1+x}Mo_{1-x}O_6$ has been studied [13] using first-principles density functional theory DFT+U, with an effective on-site correlation energy only on Fe sites, though it has been shown [12] that electronic interaction in Mo sites is also important when disorder is included. In this work, we focus on the calculation of Fe and Mo valence states resulting from disorder in the above off-stoichiometric system $Sr_2Fe_{1+x}Mo_{1-x}O_6$ ($-1 \leq x \leq 1/3$) following the correlated picture of Carvajal et al.[4] for SFMO. We calculate the number of conduction electrons on Fe and Mo sites, n_α^{Fe}, n_β^{Mo}, the valence being defined as the core charge minus the local electronic charge.

2. Model details

We assume the simple uncorrelated distribution of AS defects (no short range order) and define the order parameter a as the probability for Fe and Mo to occupy their correct positions, respectively sites α and β in the ordered structure. For $x \geq 0$, we have the occupation probabilities, $p_{Mo}^\beta = a(1-x)$, $p_{Fe}^\beta = 1 - a(1-x)$, $p_{Fe}^\alpha = x + a(1-x)$ and $p_{Mo}^\alpha = (1-a)(1-x)$. Similarly for $x \leq 0$, we have $p_{Fe}^\alpha = a(1+x)$, $p_{Mo}^\alpha = 1 - a(1+x)$, $p_{Mo}^\beta = a(1+x) - x$ and $p_{Fe}^\beta = (1-a)(1+x)$. Of course a is irrelevant for $x = -1$. As in SFMO we consider that Fe remains in the Fe^{3+} ($3d^5$) $S = 5/2$ localized spins state and Mo as Mo^{6+} ($4d^0$) cores. In addition, $n = 1-3x$ electrons ensure charge neutrality, these electrons are itinerant providing the metallic behavior of the system. This explains the limit $x = 1/3$ above which Fe can no longer retain its 3+ valency as Mo cannot take a valency larger than 6+. We incorporate intra-atomic correlations both for Fe and Mo itinerant electronic states[12, 14]. For Mo both intra and inter-orbitals correlations $U^{Mo} + 2J^{Mo}$, U^{Mo}, $U_{eff}^{Mo} = U^{Mo} - J^{Mo}$ are present while for Fe intervene only interactions between the same spin direction on different orbitals U_{eff}^{Fe} since only one spin direction is possible on Fe sites. In mean-field the correlations renormalize self-consistently the Fe and Mo site energies E_{Fe}, E_{Mo} as $\widetilde{E}_{Fe}$, $\widetilde{E}_{Mo}$and consequently the charge transfer energy $\Delta \longrightarrow \widetilde{\Delta}$. All the values of parameters are taken as a previous work[14]. In particular, the charge transfer energy Δ is taken in order to reproduce the mixed-valence character 2.5+ for Fe and 5.5+ for Mo in the ordered SFMO ($x = 0$, $a = 1$) as suggested experimentally[15, 16], therefore we choose $\widetilde{\Delta} = 0$.

Journal of Physics: Conference Series **480** (2014) 012006 doi:10.1088/1742-6596/480/1/012006

In order to obtain the density of states (DOS) per orbital for itinerant electrons on Fe and Mo sites, we calculate the local Green's functions within the renormalized perturbation expansion method [17]. In an alternating Bethe lattice in the limit $z \longrightarrow \infty$ and zt^2 scaled as $w^2/4$, the local average Green's functions take the dynamical mean-field (DMF) form $G_{ii,\sigma}^{-1} = \omega - \varepsilon_i - \sum_{l \neq i} t^2 G_{ll,\sigma}$, where ε_i is the corresponding on-site energy and the summation is over all nearest neighbors sites. In the presence of disorder this requires to take an average over the different n.n. atomic configurations, thus simply considering the probabilities p_{Fe}^{α}, p_{Mo}^{α}, etc. defined above. This averaging procedure is an extension similar to the usual coherent potential approximation in the case of random alloys [18]. It includes both the disorder of the site energies and the difference in hopping parameters.

3. Results and discussion

In Figure 1 we show the density of states over the whole range of disorder for $x = 0.1$ and $x = -0.1$. In Figure 1(a), for the Fe-rich regime ($x = 0.1$) and in the order case, $a = 1$, we observed that spin up electrons cannot move because all α-sites are occupied by Fe ions (localized spin up), so the ground state remains half-metallic. However, when disorder increases, $1 < a \leq 0.5$, states appear with spin up resulting from Mo(AS)-Fe(AS) hoppings and the system losing its half-metallic character. In Fe-deficient regime ($x = -0.1$), Figure 1(b) shows a quite different behavior, states with spin up appear as a result of Mo-Mo hoppings, the system losing its half-metallic character.

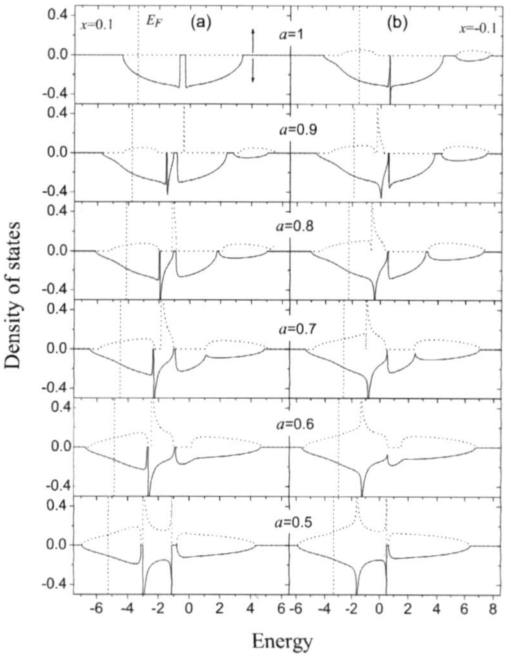

Figure 1. Evolution of the density of states as a function of disorder, a: (a) in the Fe-rich regime, $x = 0.1$; (b) in the Fe-deficient regime, $x = -0.1$. Dashed lines indicate the Fermi energy, E_F. Parameters of electronic correlation are defined by $U_{eff}^{Mo} = w$, $U_{eff}^{Fe} = 3w$ and $U^{Mo} = U_{eff}^{Mo} + \frac{w}{6}$.

We calculate the numbers of conduction electrons, n_{α}^{Fe}, n_{β}^{Mo}, through expressions n_{α}^{Fe}, $n_{\beta}^{Mo} = 3 \int_{-\infty}^{E_F} \rho(\omega) d\omega$, where a factor of 3 takes into account the orbital degeneracy of t_{2g} states, $\rho(\omega)$ is the local density of state (Fe or Mo) and E_F is the Fermi energy of the system.

Figure 2 shows the number of conduction electrons in Fe^{α} sites as a function of disorder. In Figure 2(a), Fe-rich regime ($x \geq 0$), the number of Fe ions increases in the system while the number of Mo ions decreases, we observe that the number of electrons decreases when disorder

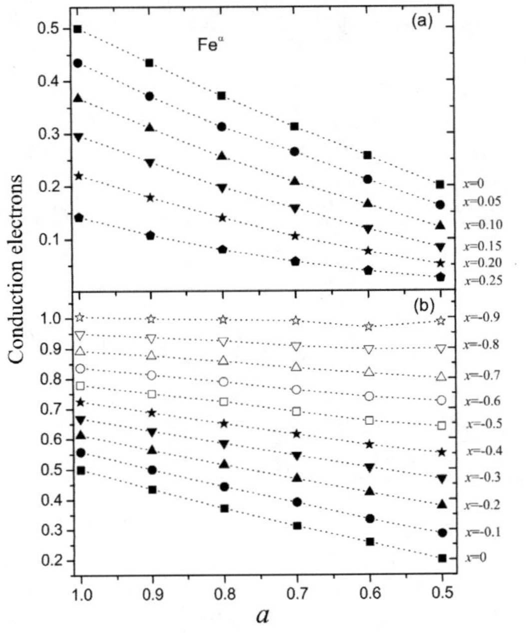

Figure 2. Conduction electrons in Fe sites (Fe^{α}) as a function of disorder: (a) in the Fe-rich regime ($x \geq 0$); (b) in the Fe-deficient regime ($x \leq 0$).

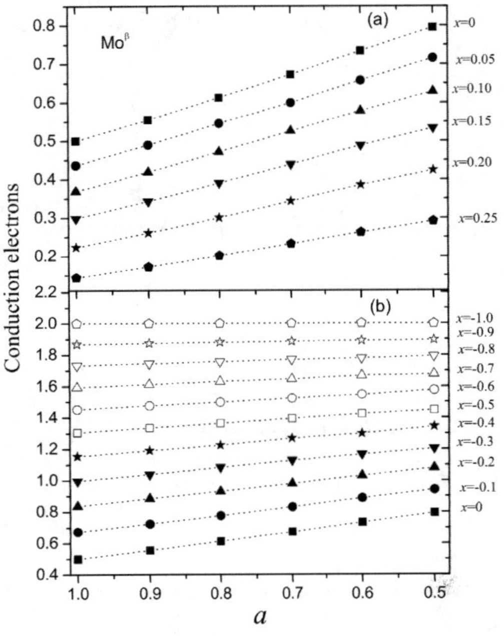

Figure 3. Conduction electrons in Mo sites (Mo^{β}) as a function of disorder: (a) in the Fe-rich regime ($x \geq 0$); (b) in the Fe-deficient regime ($x \leq 0$).

and x increase. As x approaches 1/3 and disorder increases, the number of conduction electrons becomes 0, therefore the valence state in Fe sites goes from 2.5+ to 3+. In case of Fe-deficient regime ($x \leq 0$), see Figure 2(b), we observe for $a = 1$ that the number of conduction electrons increases as x goes to -1 because the number of Mo ions increases in the system, therefore, the valence state in Fe sites goes from 2.5+ ($x = 0$) to 2+ ($x = -1$). Increasing of disorder generates both types of bonding Fe(S)-O-Fe(AS) and Mo(S)-O-Mo(AS) which allow spin up electrons itinerant, so there is an increases in the valence state in Fe sites from 2.8+ ($a = 0.5$) to 2+ ($a = 1$). Figure 3 shows the number of conduction electrons in Mo^{β} sites as a function of disorder. Figure 3(a) shows that the number of conduction electron increases with disorder, it also should be noted that n decreases in the Fe-rich regime. In this regime the number of itinerant electrons decreases since the Mo sites are occupied by Fe ions. For $x = 0$ and $a = 1$, the valence state of Mo sites is 5.5+, which tends to 6+ as x goes to 1/3. When disorder increases and $x \to 0$ the Mo valence state goes to 5+. The Fe-deficient regime is shows in Figure 3(b), where it is observed that in the ordered system the valence state in Mo changes from 5.5+ ($x = 0$) to 4+ ($x = -1$) and with disorder goes from 5.2+ ($a = 0.5$) to 4+ ($a = 1$). In the limit case $x = -1$ we obtain the $SrMoO_3$ structure, where the valence state for Mo is 4+ as it should.

In summary, for Fe-rich regime the number of conduction electrons decreases so valence states in Fe sites go up from 2.5+ to 3+, whereas for Fe-deficient regime the number of conduction electrons increases so that valence states in Fe sites tend to 2+. Regarding Mo sites, valence states go from 5.5+ to 6+ when the system is Fe-rich and from 5.2+ to 4+ in the Fe-deficient regime.

4. Acknowledgments

This work was partially supported by CONACyT Grant 131589 and PAPIIT-IN100313 from UNAM.

References

[1] T K Mandal and M Greenblatt 2010 *in Functional Oxides* (edited by D.W. Bruce, D. O'Hare, R.I. Walton: Wiley) p 257-293
[2] K I Kobayashi, T Kimura, H Sawada, K Terakura and Y Tokura 1988 *Nature* **395** 677
[3] D Page 1999 *Electronics Now* December 01
[4] E Carvajal, O Navarro, R Allub, M Avignon and B Alascio 2005 *Eur. Phys. J.* B **48** 179
[5] D D Sarma, P Mahadevan, T Saha-Dasgupta, S Ray and A Kumar 2000 *Phys. Rev. Lett.* **85** 2549
[6] H Wu 2001 *Phys. Rev.* B **64** 125126
[7] M Bibes, K Bouzehouane, A Barthélémy, M Besse, S Fusil, M Bowen, P Seneor, J Carrey, V Cros, A Vaurès and A Fert 2003 *Appl. Phys. Lett.* **83** 2629
[8] Y Tomioka, T Okuda, Y Okimoto, R Kumai, K I Kobayashi and Y Tokura 2000 *Phys. Rev.* B **61** 422
[9] S Ray, A Kumar, S Majumdar, E V Sampathkumaran and D D Sarma 2001 *J. Phys.: Condens. Matter* **13** 607
[10] D Topwal, D D Sarma, H Kato, Y Tokura and M Avignon 2006 *Phys. Rev.* B **73** 94419
[11] T Saha-Dasgupta and D D Sarma 2001 *Phys. Rev.* B **64** 064408
[12] B Aguilar, O Navarro and M Avignon 2009 *Europhys. Lett.* **88** 67003
[13] X F Zhu, Q F Li and L F Chen 2008 *J. Phys.: Condens. Matter* B **20** 075218
[14] J R Suárez, F Estrada, O Navarro and M Avignon 2011 *Eur. Phys. J.* B **84** 53-58
[15] J Lindén, T Yamamoto, M Karppinen, H Yamauchi and T Pietari 2000 *Appl. Phys. Lett.* **76** 2925
[16] O Chmaissem, R Kruk, B Dabrowski, D E Brown, X Xiong, S Kolesnik, J D Jorgensen and C W Kimball 2000 *Phys. Rev.* B **62** 14197
[17] E N Economou 2006 *in Greenś Functions in Quantum Physics* (Berlin: Springer)
[18] H Shiba 1971 *Prog. Theor. Phys.* **46** 77

Journal of Physics: Conference Series **480** (2014) 012007 doi:10.1088/1742-6596/480/1/012007

Preparation and characterization of $Cu_2ZnSnSe_4$ thin films grown from ZnSe and Cu_2SnS_3 precursors in a two stage process

R Moreno[1], A Leguizamon[2], M Hurtado[1], F Guzmán[1], G Gordillo[1]

[1]Departamento de Física, Universidad Nacional de Colombia, Bogotá

[2]Universidad Libre - Bogotá

E-mail: rmorenom@unal.edu.co

Abstract. Compound of the kesterite familie has been considered as an alternative absorber layer in the manufacture of thin film solar cells due to its earth abundant and environmental friendly constituents and high absorption coefficient. In this work we propose a new route to grow single phase Cu2ZnSnSe4 (CZTSe) thin films with tetragonal-kesterite type structure; this consist in sequential evaporation of thin films of CuSe, SnSe and ZnSe in a two stage process. Measurements of X-ray diffraction (XRD) revealed the formation of the Cu2ZnSnSe4 compound, grown with tetragonal Kësterite type structure. Optical characterization performed through spectral transmittance measurements established that this compound has high absorption (absorption coefficient> 104 cm-1) and a forbidden energy gap of 1.46 eV; these results indicate that the CZTSe thin films we have prepared has properties suitable for later use as absorber layer in solar cells. Results regarding electrical transport properties determined from temperature dependent conductivity measurements are also reported

1. Introduction

Recent growth of the thin-film PV industry and champion cell efficiencies of 20% for Cu(InGa)Se2 [1], demonstrate that thin film photovoltaics (PV) is becoming a viable option for large-scale power generation. However, due to the limited availability of In, some concern exists that high material expenses restrict the capacity to lower production costs, especially in view of the desired industrial mass production. Kesterites used as thin film solar cell absorbers have attracted much attention in the past years [2-4] due to the fact that this material has several promising attributes for efficient, inexpensive solar cells made from abundant and nontoxic elements. Efficiencies of up to 11.1 % [5] have been reported for Cu2ZnSn(S,Se)4-based photovoltaic devices. Clearly, CZTSe is a new material with very high potential for future PV industrial mass production, but advances in both material quality and device structures are needed.

Several groups have reported the fabrication of CZTSe thin films using a variety of methods such as co-evaporation [6,7] and selenization of elemental and binary chalchogenides precursors, deposited with different techniques such as, evaporation, sputtering and non-vacuum processes [8-12]. In this work, we report on studies performed to get conditions to grow Cu2ZnSnSe4 (CZTSe) thin-films by means of a solid state chemical reaction between the binary chalcogenide precursors sequentially

Journal of Physics: Conference Series **480** (2014) 012007 doi:10.1088/1742-6596/480/1/012007

deposited, followed by annealing in selenium atmosphere at temperatures around 500°C.The optical, structural and electrical properties of CZTSe thin films has also been investigated.

2. Experimental

Thin films of Cu2ZnSnSe4 were grown by means of a solid state chemical reaction between the binary chalcogenide precursors sequentially deposited on a soda lime glass substrate in a two stage process. In the first stage the binary CuSe, SnSe and ZnSe precursors are sequentially deposited in high vacuum environment; The CuSe and SnSe layers are deposited by co-evaporation of its elemental precursors and the ZnSe layer is deposited by evaporation of the compound. In the second stage the CZTSe compound is formed by annealing the stacked CuSe/SnS/ZnSe system at a temperature of 500 °C, under Se atmosphere.

In Fig.1 is shown the system used to prepare the CZTSe films This is equipped with two tungsten boats to evaporate Cu and Sn, and two effusion cells used to evaporate ZnSe and Se respectively. The background pressure of the vacuum chamber was around 1x10-5 mbar. The temperatures of each effusion cells are controlled using PID temperature controllers. The deposition rates were monitored with a thickness monitor which uses a quartz sensor as transducer.

To find conditions to grow single phase CZTSe films with kesterite structure, a broad number of samples were deposited on glass substrates under different sequences (CuSe/SnSe/ZnSe, SnSe/CuSe/ZnSe) and varying the main deposition parameters (post deposition annealing temperature, deposition rate, mass ratio of the elemental precursors) in a wide range. XRD measurements carried out to each one of the prepared samples allowed us to find the sequence and deposition parameters which lead to the growth of Cu2ZnSnSe4 thin films with kesterite structure. The study revealed that single phase CZTSe films can be grown using the CuSe/SnSe/ZnSe sequence and the deposition routine displayed in fig. 1b. Samples prepared using any other sequence and/or deposition routines, grow in general with a mixture of the CZTSe phase and secondary phases

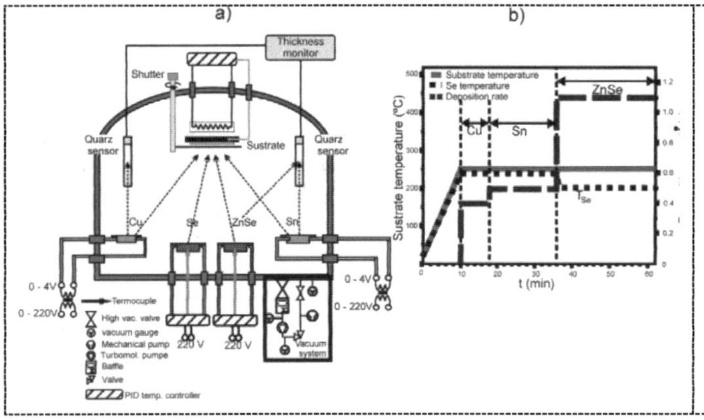

Figure 1. a) System used to prepare $Cu_2ZnSnSe_4$ thin films by sequential deposition of CuSe/SnSe/ZnSe thin films followed by annealing in Se atmosphere and b) Routine used to prepare single phase $Cu_2ZnSnSe_4$ thin films.

The electrical properties of the CZTSe films were studied through temperature dependent conductivity measurements and the optical properties through transmittance measurements carried out on a Oriel VIS-NIR spectrofotometer. Further characterization involved X-ray diffraction on a Shimadzu-6000 diffractometer.The film thickness was determined using a Veeco Dektak 150 surface profiler.

3. Results and Discussion

3.1. Structural characterization

Fig. 2, shows the XRD pattern of a CZTSe thin film prepared by sequential deposition of thin films of CuSe, SnSe and ZnSe, using a preparation routine like that plotted in Fig.1b and evaporated mass of Cu, Sn and ZnSe of 0.01 gr., 0.07 gr. and 0.15 gr. respectively. In order to identify with a greater degree of accuracy the reflexions corresponding to secondary phases that could be present in the

Journal of Physics: Conference Series **480** (2014) 012007 doi:10.1088/1742-6596/480/1/012007

diffractogram of the CZTS films, this is compared with those of thin films of CuSe, SnS, ZnSe and ternary chalcogenides of Cu and Sn (CTSe). The CuSe and SnSe layers were deposited by co-evaporation of its elemental precursors at 250°C followed by annealing at 500° in Se atmosphere and the ZnSe layer was deposited by evaporation of the compound at 250°C followed by annealing at 500°C. The ternary CTSe layer was prepared by sequential deposition of CuSe/SnSe at 250°C followed by annealing at 500°C in Se atmosphere.

It is observed that the diffractogram of the CZTSe film shows only reflections corresponding to the Cu2ZnSnSe4 phase grown with tetragonal-kesterite type structure (PDF cart #00-052-0868), oriented preferentially along the (112) plane. The diffractograms taken on thin films of binary chacogenides show that these samples grow in the Cu2Se (PDF card #03-065-7737), SnSe (PDF card # 00-032-1382), SnSe2 (PDF card # 00-023-0602), and ZnSe PDF card #01-080-0021) phases respectivelly. The XRD pattern of the ternary compound shows that this compound grow with a mixture of the Cu2SnSe3 (PDF card #03-065-7524) and Cu2Se phases being Cu2SnSe3 the majority phase.

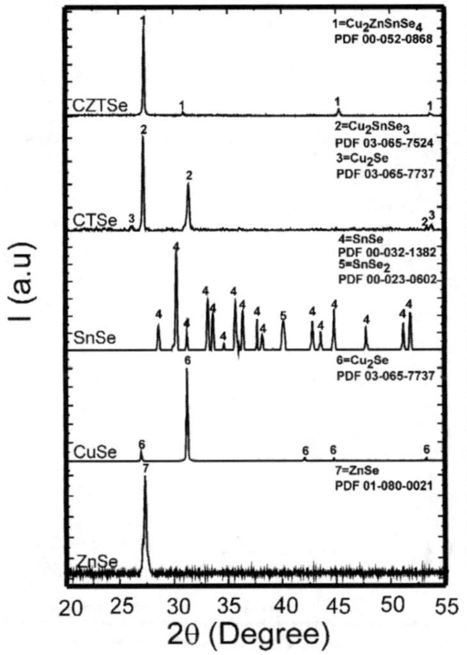

Figure 2. Comparison of the diffractogram of a CZTSe thin film with those of thin films of CuSe, SnSe, ZnSe and Cu_2SnSe_3, deposited under the conditions described previously.

From this study may be established that a possible route for the formation of the Cu2ZnSnSe4 compound is the following:

The sequential evaporation of Cu and Sn at 250°C in presence of elemental selenium followed by annealing at 500°C results in the formation of a mixture of Cu2SnSe3 and Cu2Se . The subsequent evaporation of ZnSe at 250°C and further annealing at 500°C leads to the formation of a single phase Cu2ZnSnSe4 film, according to the following reaction .

3Cu2Se + SnSe+ SnSe2 + Se ⟶ 2Cu2SnSe3 +Cu2Se

Cu2SnSe3 +Cu2Se + ZnSe ⟶ Cu2ZnSnS4 + Se

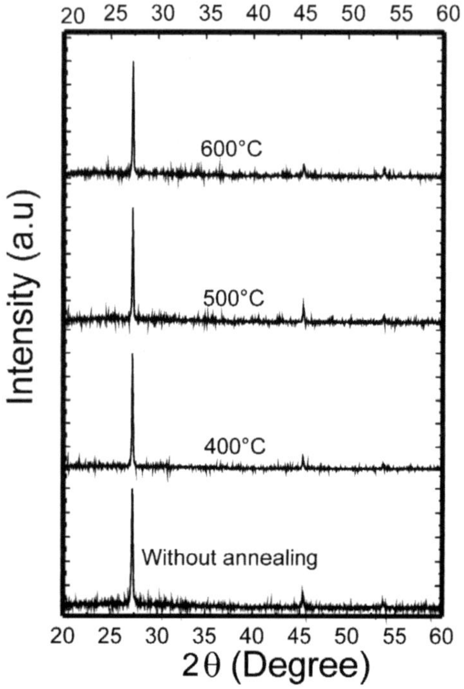

Figure 3. XRD spectra of CZTSe thin film grown varying the annealing temperature (during the 2nd phase of growth) between 400 and 600°C

The effect of annealing temperature of the CuSe/SnS/ZnSe system (during and growth phase), on the structural properties of CZTSe films was also studied. Figure 3 shows XRD spectra corresponding to a CZTSe thin film formed by annealing at temperatures ranging between 400 and 600 °C. This results indicates that in the range studied, the annealing temperature does not affect the phase in which the CZTSe films grow, probably because the thin films obtained at the end of the sequential deposition of precursors at 250°C exhibit a stable Cu2ZnSnS4 phase; the subsequent annealing not affect either the phase or the lattice constants, only affects the density of native and structural defects.

3.2. Optical characterization

In Fig.4 is compared the transmission spectrum of a single phase Cu_2ZnSnS_4 thin film with those of thin films of CuSe, SnSe, ZnSe and ternary chalcogenides of Cu and Sn. The samples were prepared as described previously. It is observed that with the exception of ZnSe films, all other studied samples exhibit low transmittance values, indicating that the CZTSe , SnSe and CuSe films grow with a high density of native defects (vacancies, interstitial and antisite).

These defects generate absorption centers within the energy gap which contribute to the photon absorption. In particular, the transmittance of the CuSe films decreases strongly in the NIR region apparently due to the formation of a very high density of shallow defects. It is also observed that the transmittance curve of the SnSe film has a very small slope. This behavior seems to be caused by absorption of photons in deep centers generated by structural defects.

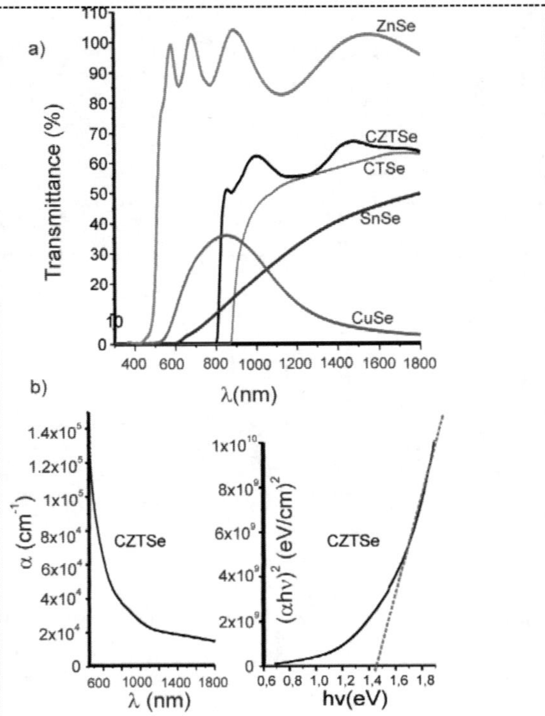

Figure 4. a) Comparison of the transmittance of a typical Cu2ZnSnS4 thin films with those of thin films of CuSe, SnSe, ZnSe and Cu2SnS3 and b) curves of α vs λ and (αhν)2 vs. hν, corresponding to the Cu2ZnSnS4 film.

The optical gap of the CZTSe thin film was determined by extrapolation of the (αhν)2 vs. hν curve with the hν axe. The absorption coefficient α was estimated using the relation T (λ)≈ (1-R(λ)) exp(-αd), where T(λ) is the spectral transmittance, R(λ) the spectral reflectance and d the film thickness. An Eg value of 1.45 eV was found for the CZTSe films.

The effect of annealing temperature during 2nd growth phase, on the optical properties of CZTSe films was also studied through transmittance measurements. Fig. 5 shows transmittance spectra corresponding to a CZTSe thin film grown at annealing at temperatures varying between 400 and 600 °C. This results indicates that in the range studied, the annealing temperature significantly affect the transmittance of the CZTSe films. It is observed that not annealed samples or annealed at temperatures less than 500°C have both, low transmittance as small slope compared with samples annealed at temperatures around 500°C and higher. This behavior indicates that the annealing at high temperatures reduces the density of both, native defects and structural defects. Shifting of cutoff wavelength to lower wavelength values observed in samples annealed at 500°C is apparently due to increased grain size.

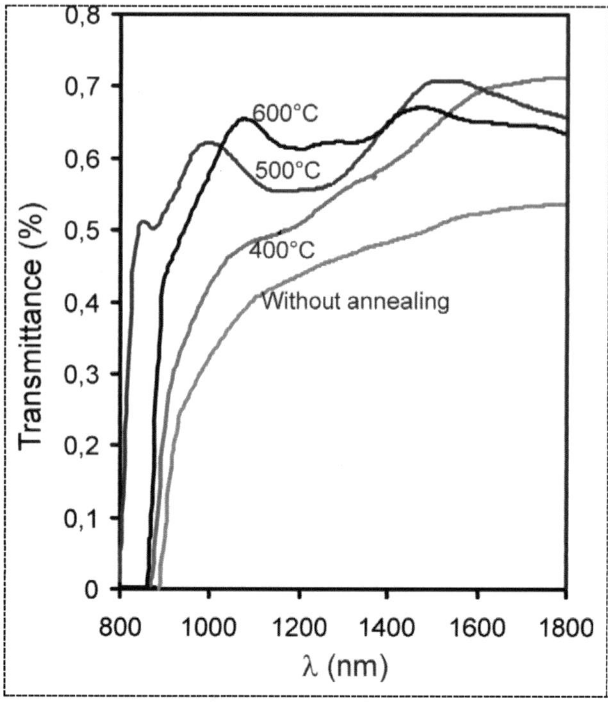

Figure 5: Transmittance spectra of CZTSe thin film grown varying the annealing temperature during the 2nd phase of growth, between 400 and 600°C

3.3. Electrical characterization

Fig. 6 shows curves of both, σ vs T and lnσ vs 1000/T, obtained with a CZTSe film grown in the Cu_2ZnSnS_4 phase as described in section 2.

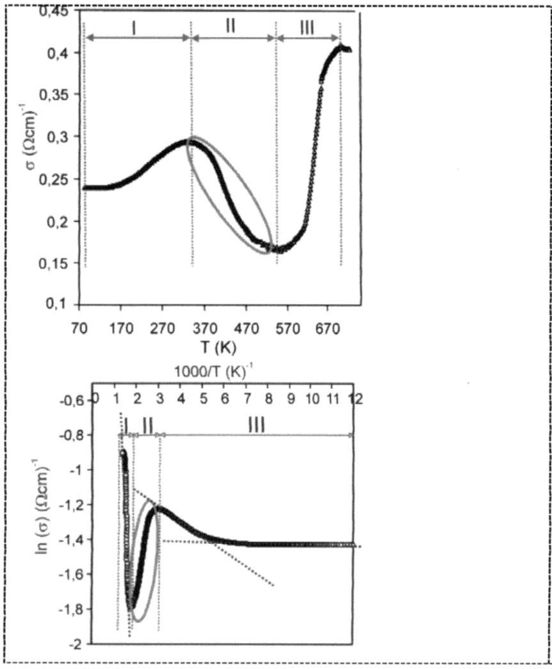

Figure 6: Curves of σ vs. T and ln(σ) vs. 1000/T corresponding to a typical Cu_2ZnSnS_4 thin film

It is observed in Fig.6 that in the regions designated as I and III, the conductivity increases as the temperature increases indicating a typical behaviour of semiconductor materials; however it is observed that in the region II the conductivity decreases with increasing temperature. The abnormal behavior in the region II is caused by a phase change of the CZTS compound induced by the sample heating and not by an effect of temperature on the transport properties; this was verified by XRD measurements in dependence of the temperature, which indicated that these samples grow preferentially in the Cu2ZnSnS4 phase with the presence of Cu and Sn selenides in small proportions; however, when the sample is heated above 350K, begins a transformation of binary phases into the Cu2ZnSnS4 phase. At temperatures of the order of 550K the sample only contains the Cu2ZnSnS4 phase, which is stable as the temperature continues to increase.

The behavior of the lnσ vs. 1000/T curve plotted in Fig. 6b indicates that in the regions I and III, the conductivity of the CZTSe films can be expressed by the relation: $\sigma = \sigma_0 \exp[-(E_C - E_F)/(kT)]$ indicating that the conductivity is predominantly affected by free carrier transport in states of the valence band. The increase in conductivity with increasing temperature is apparently caused by an increase in the density of charge carriers generated by native acceptor impurities. In the range of high temperatures (T>550 K), the increase of σ could be attributed to an increase of the carrier density coming from deep acceptor impurities presents in the Cu2ZnSnS4 compound, whereas the change of σ observed in the range of low temperatures (T<350 K) can be attributed to a change of the carrier density coming from shallow acceptor impurities associated to the binary phases. The two slopes identified in the lnσ vs. 1000/T curve (high temperature region) indicates that the conductivity of the CZTS films is affected by two different types of impurities (probably vacancies and antisite defects). At temperatures greater than 700K, the conductivity begins to decrease with increasing temperature indicating a saturation of the thermal excitation of free carriers from the impurities. The decrease in conductivity with increasing temperature could be then attributed to an increased mobility with temperature. The results also revealed that the Cu2ZnSnS4 films are characterized to get high absorption and an energy band gap of about 1.45 eV, indicating that this compound is suitable to perform as absorbent layer in thin film solar cells.

Conductivity measurements on temperature dependence revealed that conductivity of the CZTSe films is predominantly affected by free carrier transport in states of the valence band. In the range of high temperatures (T>550 K), the increase of σ could be attributed to an increase of the carrier density coming from deep acceptor impurities, whereas the change of σ observed in the range of low temperatures (T<350 K) can be attributed to a change of the carrier density coming from shallow acceptor impurities associated to secondary phases.

ACKNOWLEDGEMENTS: This paper was supported by Colciencias (Cont. #038-2013) and DIB-Universidad National de Colombia.

4. Conclusions

CZTSe thin films with tetragonal-kesterite type structure were grown using a method based on sequential evaporation of thin films of CuSe, SnSe and ZnSe in a two stage process. Characterization performed by XRD gave evidence of the formation of single phase CZTSe films with kesterite structure using optimized growth parameters; however the sequence in which the binary precursors are evaporated and the preparation parameters, significantly affects the phase and morphology, as well as the structural, optical and electrical transport properties of the CZTS films. Optical characterization performed by spectral transmittance measurements revealed that the CZTSe films have low transmittance and poor crystallographic quality, probably associated to structural and native defects, indicating that further studies must be done to improve the properties of the CZTSe films.

5. References

[1] Jackson P, Hariskos D, Lotter E, Paetel S, Wuerz R, Menner R, Wischmann W, Powalla M. New world record efficiency for Cu(In,Ga)Se2 thin-film solar cells beyond 20%. Progress In Photovoltaics: *Research and Applications* 2011; published online DOI: 10.1002/ pip.1078 (Presented at 25th EU PVSEC WCPEC-5, Valencia, Spain, 2010).

Journal of Physics: Conference Series **480** (2014) 012007 doi:10.1088/1742-6596/480/1/012007

[2] H. Katagiri, K. Jimbo, S. Yamada, T. Kamimura, W.S. Maw, T. Fukano, T. Ito, T. Motohiro, Enhanced conversion efficiencies of Cu2ZnSnS4-based thin film solar cells by using preferential etching technique, *Appl. Phys*. Express1 (2008) 41201(2pp).

[3] D. A. R. Barkhouse, O. Gunawan, T. Gokmen, T. K. Todorov and D. B. Mitzi, Device characteristics of a 10.1% hydrazine-processed $Cu_2ZnSn(Se,S)_4$ solar cell,Progr. Photovolt.: *Res. Appl*. 20 (2012) 6-11.

[4] David B. Mitzi , Oki Gunawan, Teodor K. Todorov, Kejia Wang, Supratik Guha, The path towards a high-performance solution-processed kesterite solar cell, *Solar Energy Materials & Solar Cells* 95 (2011) 1421–1436

[5] Todorov TK, Tang J, Bag S, Gunawan O, Gokmen T, Zhu Y, Mitzi DB. Beyond 11% efficiency: Characteristics od state-of-art $Cu_2ZnSn(Se,S)_4$ solar cells. *Advanced Energy Materials* 2012; published online DOI: 10.1002/aenm.201200348

[6] B.A.Schubert, B. Marsen, S. Cinque, T. Unold, R. Klenk, S. Schorr, H.-W. Schock, Prog. Photovoltaics *Res. Appl*. 19 (2011) 93.

[7] I. Repins, C. Beall, N. Vora, C. DeHart, D. Kuciauskas, P. Dippo, B. To, J. Mann, W.-C. Hsu, A. Goodrich, R. Noufi, *Sol. Energy Mater*. Sol. Cells 101 (2012) 154.

[8] G. Zoppi, I. Forbes, R .W. Miles, P. J. Dale, J .J. Scragg, L.M. Peter, $Cu_2ZnSnSe_4$ thin film solar cells produced by selenisation of magnetron sputtered precursors, Prog. Photovolt.:*Res. Appl*. 17 (2009) 315–319.

[9] R. A.Wibowo,W.S. Kim, E.S. Lee, B. Munir, K.H. Kim, Single step preparation of quaternary $Cu_2ZnSnSe_4$ thin films by RFmagnetron sputtering from binary chalcogenide targets, *J. Phys. Chem*. Solids 68(2007)1908–1913.

[10] O.Volobujeva, J. Raudoja ,E.Mellikov, M. Grossberg, S. Bereznev, R. Traksmaa, $Cu_2ZnSnSe_4$ films by selenization of Sn–Zn–Cu sequential films, *J.Phys.Chem*. Solids 70(2009)567–570.

[11] T.K. Todorov, K.B. Reuter, D.B. Mitzi, High-efficiency solar cell with earth- abundant liquid-processed absorber, Adv. Mater. 22 (2010) E156–E159.

[12] M. Ganchev, L. Kaupmees, J. Iliyna, J. Raudoja, O. Volobujeva, H. Dikov, M. Altosaar, E. Melliko, T.Varema, Formation of $Cu_2ZnSnSe_4$ thin films by selenization of electrodeposited stacked binary alloy layers, *Energy Procedia* 2 (2010)65–70.

One-step diffusion membrane assisted CBD synthesis and characterization of Cu_2SnS_3 thin films

R. A. Becerra[1], J. M. Correa[1], H. Suarez[1] and G. Gordillo[2]

[1]Departamento de Química, Universidad Nacional de Colombia, sede Bogotá
[2] Departamento de Física, Universidad Nacional de Colombia, sede Bogotá, Carrera 45 No. 26-85, Edificio Uriel Gutiérrez, Bogotá, Colombia

E-mail: rabecerraa@unal.edu.co

Abstract. This paper present a novel method for growing thin films of Cu_2SnS_3 (CTS) using a solution-based chemical route consisting of simultaneous precipitation of $Cu_{2-x}S$ and SnS_2 performed by diffusion membranes assisted CBD technique. Diffusion membranes are used to optimize the kinetic growth through a moderate control of the releasing metal into the solution. The conditions in terms of concentration of metal species, sulfide anion and temperature required for the precipitation of the Cu_2SnS_3 compound were determined through a study of chemical equilibrium of the system $SnCl_2$, $Na_3C_6H_5O_7{\cdot}2H_2O$, $CuCl_2$ and $Na_2S_2O_3{\cdot}5H_2O$. These conditions were obtained solving the equilibrium equations with the help of the Visual MINTEQ 3.0 package, supported on the program MINTEQA2. X-ray diffraction (XRD) and Raman spectroscopy were used to characterize the structural properties of the CTS films. Optical, morphological and electrical properties were also studied by spectral transmittance, atomic force microscopy (AFM) and resistivity vs temperature measurements. XRD and Raman measurements confirmed the formation of the Cu_2SnS_3 phase

1. Introduction

Thin film photovoltaics is becoming a viable option for large-scale power generation. However, due to the limited availability of In, some concern exists that high material expenses restrict the capacity to lower production costs, especially in view of the desired industrial mass production. Cu_2ZnSnS_4 has been considered as the alternative absorber layer to $Cu(In,Ga)Se_2$ due to its earth abundant and environmental friendly constituents, optimal direct band gap of 1.45eV and high absorption coefficient in the visible range [1,2]. In recent years, great efforts have been focusing on the preparation of CZTS thin films and exploration their potential application in thin film solar cells [2-4].

A variety of routes have been undertaken for thin-film deposition. These include vacuum and solution based deposition approaches. For each of these deposition routes, the compositional control and growth of single phase CZTS films are the main barriers toward a reliable process. These barriers are attributed to the incompletely understood nature of the Cu–Zn–Sn– S phase diagram [5] and to the volatility upon heating of Sn materials [6]. Despite these limitations reasonably successful film deposition and device fabrication has been demonstrated for both vacuum and solution-based deposition approaches. $Cu_2ZnSnSe_4$ and $Cu_2ZnSn(S,Se)_4$-based solar cells using hydrazine-based solution process have already reached an energy conversion efficiency as high as 11.1 % [4], demonstrating the effectiveness of the solution process in CZTSe-based solar cells.

In this work, we propose a novel route for the preparation of Cu_2SnS_3 thin films used as precursor of Cu_2ZnSnS_4 films grown using a solution based method. The Cu_2SnS_3 films are formed by co-precipitation of Cu_2-xS and SnS_2 performed by diffusion membranes assisted CBD technique. Further, results are reported related to structural, optical and morphological properties determined from XRD,

Raman spectroscopy, transmission and AFM measurements. These are compared with those of reference CZTS films prepared by sequential evaporation of metallic precursors in presence of elemental sulfur.

2. Experimental Procedure

The Cu_2SnS_3 compound was prepared using a novel procedure consisted of simultaneous precipitation of $Cu_{2-x}S$ and SnS. The co-precipitation of $Cu_{2-x}S$ and SnS_2 was achieved introducing initially an aqueous solutions of ($SnCl_2.2H_2O$ + $Na_3C_6H_5O_7{\cdot}2H_2O$) and ($CuCl_2{\cdot}2H_2O$) into two spatially separate diffusion membranes and then, both membranes are Introduced into a solution containing ($Na_2S_2O_3{\cdot}5H_2O$ + glucose) dissolved in water. The role of the diffusion membranes is to optimize the kinetic thin film growth through a moderate control of the releasing metal into the solution by osmosis, and glucose is added to promote the reduction of Cu^{2+} to Cu^{+}. The sodium citrate act as complexing agent and as a pH regulator.

A possible reaction mechanism for the formation of the Cu_2SnS_3 compound is the following:

In the reaction medium must exist free species of S^{2-}, Sn^{4+} and Cu^{+} to precipitate sulfides. The S^{2-} ion is generated as free specie in solution due to hydrolysis of sodium thiosulfate in basic medium, according to the following reaction [5]:

$$S_2O_3^{2-} + 2OH^- \rightarrow SO_4^{2-} + S^{2-} + H_2O \quad (1)$$

The cation is obtained from the complex $[SnCit]^-$ content in one of the membranes, which diffuses slowly and in a controlled way toward the solution favoring the ion by ion growth and decreasing the cluster growth, because of the physical barrier provided by the membrane. In solution, the species $[SnCit]^-$ participate in an equilibrium releasing Sn^{2+} in a controlled manner. The precipitation take place in two step, oxidation of Sn^{2+} to Sn^{4+} by thiosulfate, and in the second step the ions in solution (Sn^{4+} and S^{2-}) must exceed the solubility product.

$$Sn^{2+} + Cit^{3-} \rightarrow [SnCit]^- \qquad \beta_{Sn\text{-}Cit} = 2x10^{10} \quad (2)$$

$$4Sn^{2+} + S_2O_3^{2-} + 8H^+ \rightarrow 4Sn^{4+} + 2HS^- + 3H_2O \quad (3)$$

$$Sn^{4+} + 2HS^- \rightarrow 2H^+ + SnS_{2(s)} \qquad K_{sp} = 3.5x10^{-58} \quad (4)$$

At the same time should be given co-precipitation of $Cu_{2-x}S$. For that, Cu^{2+} contained within one membrane diffuses toward the solution containing glucose ($C_6H_{12}0_6$), reducing it to Cu^{+}.

$$C_6H_{12}O_6 + 2Cu^{2+} + 3OH^- \rightarrow C_6H_{11}O_7^- + 2H_2O + 2Cu^+$$

As in the reaction medium there is a high concentration of free thiosulfate ions and Cu^{+} species, the following complexes can be formed:

$$Cu^+ + S_2O_3^{2-} \rightarrow [CuS_2O_3]^- \qquad \beta_{Cu^+\text{-}S_2O_3^{2-}} = 10^{10,3} \quad (5)$$

$$[CuS_2O_3]^- + S_2O_3^{2-} \rightarrow [CuS_4O_6]^{3-} \qquad \beta_{Cu^+\text{-}2S_2O_3^{2-}} = 10^{12,2} \quad (6)$$

$$[CuS_4O_6]^{3-} + S_2O_3^{2-} \rightarrow [CuS_6O_9]^{5-} \qquad \beta_{Cu^+\text{-}3S_2O_3^{2-}} = 10^{13,8} \quad (7)$$

For precipitation to occur, the ionic product $[Cu^+][S^{2-}]$ must exceed the solubility product K_{sp} of Cu_2S.

$$2Cu^+ + HS^- \rightarrow H^+ + Cu_2S_{(s)} \qquad K_{sp} = 1.2\times10^{-35} \quad (8)$$

Subsequent to the formation of sulfides binary, they react in solution to form the ternary compound Cu_2SnS_3, according to the following reaction:

$$Cu_2S_{(s)} + SnS_{2(s)} \rightarrow Cu_2SnS_{3(s)} \quad (9)$$

Journal of Physics: Conference Series **480** (2014) 012008 doi:10.1088/1742-6596/480/1/012008

Parameter values for the reagents constituting the work solution (concentration of metal species, sulfide anion and temperature), required to obtain the Cu_2SnS_3 compound from co-precipitation of Cu_2S and SnS_2 were found by means of a study of the equilibrium conditions of the systems $SnCl_2$ + $Na_3C_6H_5O_7 \cdot 2H_2O$, $CuCl_2$ + $Na_2S_2O_3 \cdot 5H_2O$. The concentrations and equilibrium constants allow building chemical equilibrium equations present in the system which were solved using the Visual MINTEQ 3.0 package, supported on the program MINTEQA2.

Using as input data the values determined through the solution of the chemical equilibrium equation, a parameter study was done with the purpose of finding suitable conditions to grow the Cu_2SnS_3 compound. The following chemical bath composition led to good results: [$CuCl_2$]=50mM, [$SnCl_2$]=100mM, [$Na_2S_2O_3$]=250mM, [glucose]=345mM and [$Na_3C_6H_5O_7$]=237mM. During the deposition the bath temperature was maintained at 70 ◦C and the final solution pH was around 5.

The transmittance measurements were done using a Varian – Cary 5000 spectrophotometer and the film thickness was determined using a Veeco Dektak 150 surface profiler. Further characterization involved X-ray diffraction on a Shimadzu-6000 diffractometer and Raman spectroscopy on a Horiba JobinYvon micro-Raman Spectrometer LabRamHR in backscattering configuration with a laser of 780nm, 20 mW focused with a 50X objective. Morphological studies were also done through AFM measurements carried out with a PSI AP-0100 Scanning Probe Microscope.

3. Results and discussion

3.1. Optical properties

Figure 1 displays transmission and reflection spectra of a typical Cu_2SnS_3 thin film of approximately 450nm thick. This result shows that the sum of the transmittance and reflectance is significantly less than 100 %, indicating that in addition to the fundamental absorption this material contains absorption centers in the gap that significantly contribute to the absorption of photons of lower energies than the corresponding energy gap (Eg). This behavior is apparently caused by the presence of a high density of native defects (vacancies, interstitial and anti-site), which generate absorption centers within the energy gap that contribute to the photon absorption mainly in the VIS and NIR region.

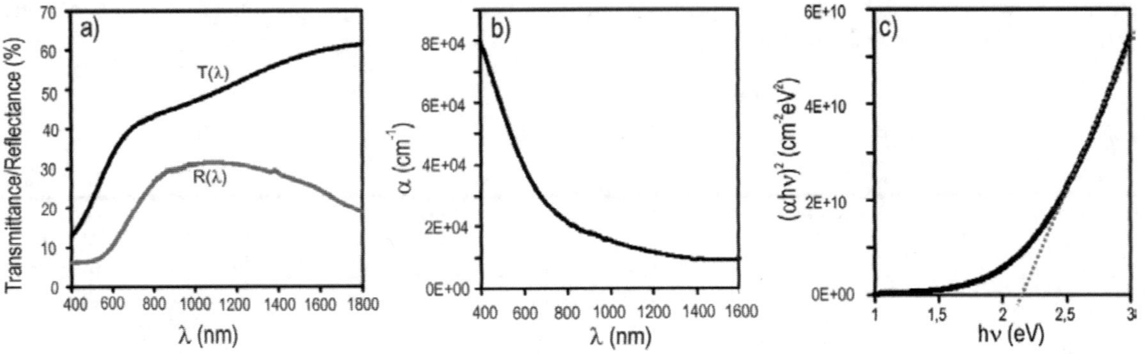

Figure 1. a) Transmission and reflection spectra of a typical Cu_2SnS_3 thin films deposited in one-step using diffusion membrane assisted CBD method b) curve of α vs λ and c) curve of $(\alpha h\nu)^2$ vs hν obtained from transmittance and reflectance measurements.

In Fig. 1 are also displayed curves of α vs λ and $(\alpha h\nu)^2$ vs hν. The absorption coefficient α was determined from measurements of spectral transmittance and reflectance and using the following relationship:

$$T(\lambda) = [1 - R(\lambda)]e^{-\alpha d} \tag{10}$$

Where T(λ) is the spectral transmittance, R(λ) the spectral reflectance and d the sample thickness.

Journal of Physics: Conference Series **480** (2014) 012008 doi:10.1088/1742-6596/480/1/012008

The Cu_2SnS_3 thin films we have grown in this work have a value of Eg = 2.15 eV and an absorption coefficient in the high absorption region (λ<500 nm) greater than 10^4 cm^{-1}. The value of Eg was determined from the intercept with the axis hν of a curve of $(\alpha h\nu)^2$ vs hν.

3.2. Structural characterization

In Fig. 2 is compared the diffractogram of a typical Cu_2SnS_3 thin film prepared by diffusion membrane assisted CBD method with the diffractogram taken to the material precipitated during its growth. It is observed that the CTS thin film present reflections corresponding only to the Cu_2SnS_3 phase (PDF cart #00-019-0412), whereas the diffractogram corresponding to the precipitated powder, show reflexions associated to the Cu_2SnS_3 phase as well as to the Cu_3SnS_4 (PDF cart #00-036-0217)) and CuS (PDF cart #03-065-7111) phases.

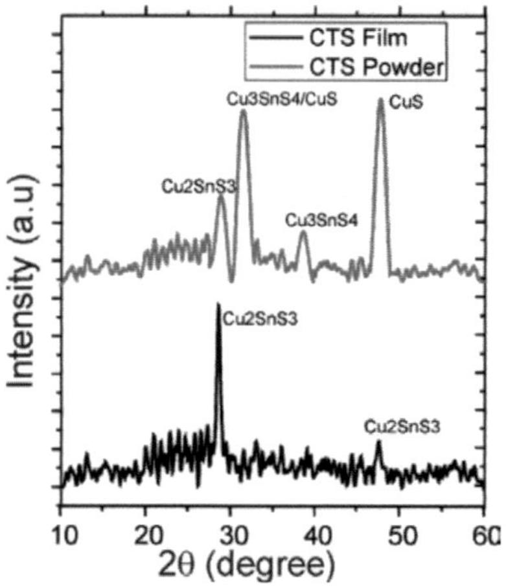

Figure 2. Comparison of XRD pattern of a CTS thin film prepared by One-step diffusion membrane assisted CBD method with that of the material which precipitates during its growth.

Besides X-ray analysis, Raman spectroscopy measurements were done on thin films of Cu_2SnS_3. In Fig. 3 the Raman spectrum of a typical Cu_2SnS_3 film is depicted. Deconvolution of this spectrum, allows identify three peaks around 292, 320 and 352 cm^{-1} respectively.

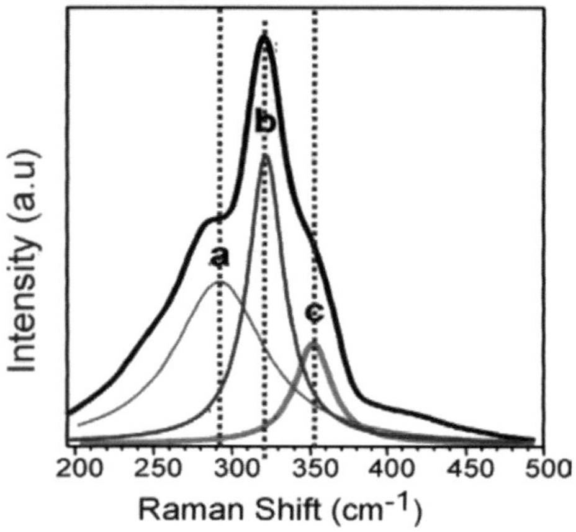

Figure 3. Raman spectrum corresponding to a typical CTS thin film prepared by One-step diffusion membrane assisted CBD method.

The shoulder in the region between 280 cm^{-1} and 300 cm^{-1} with a maximum around 292 cm^{-1} (see deconvolution) has been assigned to the Cu_3SnS_4 compound grown with tetragonal structure [6]. The vibrational mode corresponding to the band b has been attributed to the A′ symmetry of the Orthorhombic Cu_3SnS_4 phase [6] and the peak at 352 cm^{-1} to the A′ symmetry vibrational modes from the monoclinic Cu_2SnS_3 phase [7].

3.3. Electrical characterization

Electric conductivity measurements carried out as a function of temperature in a range of -80 to 700 K were used to identify scattering processes affecting the transport mechanisms of the CTS films. The conductivity (σ) was measured using electrical contacts of molybdenum deposited by sputtering, arranged to get the resistance through the standard four-stripe method; the film thickness was obtained using an α-step profiler. Fig. 4 shows a typical curve of σ vs T, obtained with a CTS films prepared by CBD.

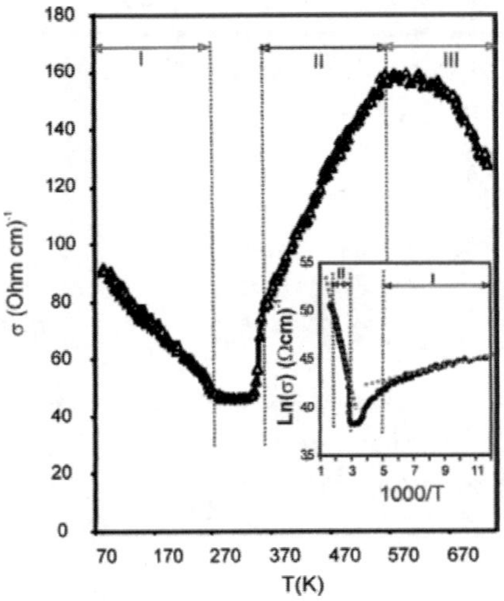

Figure 4. Curve of σ vs T corresponding to a CTS film deposited by CBD.

The σ vs T curves show two negative slopes and one positive, indicating that the conductivity is affected by different transport mechanisms, each one predominating in a different temperature range.

The results of Fig. 4 were analyzed assuming predominance of one determined transport mechanism in the temperature ranges denoted as I, II and III respectively. The curve of Ln σ vs 1000/T displayed in inset of Fig. 4 shows that in the region II (between 350 and 570K) the conductivity can be expressed by the relation: $\sigma = \sigma_0 \exp[-(E_C - E_F)/(kT)]$ indicating that in this range of temperature the conductivity is predominantly affected by free carrier transport in extended states of the conduction band. Since the prefactor σ_0 varies slowly with temperature and the term exp[-(E_C-E_F/kT] depends strongly on T, σ_0 can be assumed as approximately constant. Under this approximation, the slope of the graph lnσ vs 1/T gives the activation energy $E_\sigma = E_C - E_F$.

In the range of low temperatures (region I), the curve of σ vs T can be expressed by a relation of the form: $\sigma = \sigma_0 \exp[(Q\varphi)/(kT)]$. Since in this temperature range the free carrier density is not significantly affected by the temperature, the conductivity in the region I seems mainly affected by the carrier mobility, which in his turn is affected by the height of the potential barrier ϕ that the carrier must overcome in the grain boundary regions; the high barrier ϕ can be estimated from the slope of the σ vs. 1000/T curve.

In the highest temperature region (region III), the conductivity decreases as the temperature increases. This behavior could be caused by phase change of the CTS compound (verified through XRD measurements) or mobility decrease induced by interaction of free carriers with lattice phonons.

3.4. Morphological characterization

Fig. 4 shows the AFM image of a Cu_2SnS_3 thin films prepared by One-step diffusion membrane assisted CBD method. It is observed that these type of samples have an unusual morphology characterized by sheet-shaped grains grouped together in big clusters with sizes varying in the range of

1X0.4 μm and 2.5x1.4 μm. Similar results were observed with Cu_2SnS_3 films prepared by co-evaporation of elemental precursors at 250°C followed by annealing at 550°C in sulfur atmosphere.

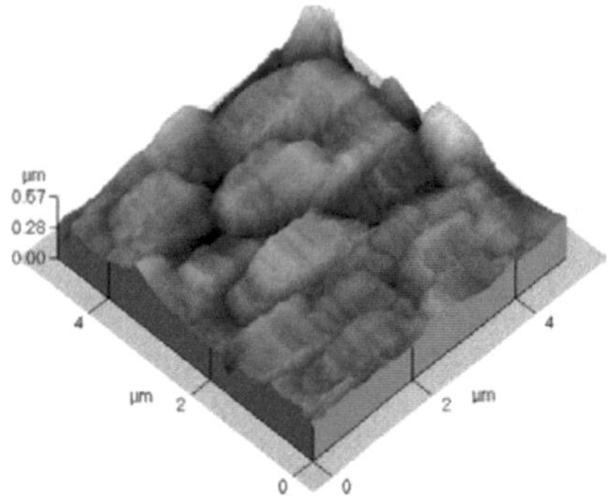

Figure 5. AFM image of a Typical Cu_2SnS_3 thin film.

4. Conclusions

Cu_2SnS_3 thin films were grown using a novel chemical route consisting of simultaneous precipitation of $Cu_{2-x}S$ and SnS_2 performed by diffusion membranes assisted CBD technique. The conditions in terms of concentration of metal species, sulphide anion and temperature required for the precipitation of the Cu_2SnS_3 compound were determined through a simulation of the chemical equilibrium of the systems ($SnCl_2$ + $Na_3C_6H_5O_7 \cdot 2H_2O$) , $CuCl_2$ and $Na_2S_2O_3 \cdot 5H_2O$ performed with the help of the Visual MINTEQ 3.0 package, supported on the program MINTEQA2.

XRD measurements gave evidence of the formation of single phase Cu_2SnS_3 films grown with tetragonal structure. Raman spectroscopy confirmed this result and additionally revealed the presence of the Cu_3SnS_4 phase grown with both tetragonal and orthorhombic structure. Transmittance and reflectance measurements indicated that in addition to the fundamental absorption the CTS films contains absorption centers in the gap that significantly contribute to the absorption of photons of lower energies than the corresponding energy gap, mainly in the VIS and NIR region. AFM studies revealed that the morphology of CTS thin films prepared by One-step diffusion membrane assisted CBD method have an unusual morphology characterized by sheet-shaped grains grouped together in big clusters.

ACKNOWLEDGEMENTS: This paper was supported by Colciencias (Cont. #038-2013, DIB-Universidad Nacional de Colombia and FODEIN-Universidad Santo Tomás

5. References

[1] Wang K, Gunawan O, Todorov T, Shin B, Chey S J, Bojarczuk N A, Mitzi D, Guha S, *Appl. Phys. Lett.* **97** (2010) 143508 (3pp).

[2] Katagiri H, Jimbo K, Yamada S, Kamimura T, Maw W S, Fukano T, Ito T, Motohiro T, *Appl. Phys. Express 1* (2008) 41201(2pp).

[3] Barkhouse D A R, Gunawan O, Gokmen T, Todorov T K and Mitzi D B, *Res. Appl.* **20** (2012) 6-11.

[4] Todorov T K, Tang J, Bag S, Gunawan O, Gokmen T, Zhu Y, Mitzi D B, *Advanced Energy Materials* 2012; published online DOI: 10.1002/aenm.201200348

[5] Ionkin A S, Fish B M, Marshall W J & Senigo R H, *Solar Energy Materials and Solar Cells* **104**, 23–31 (2012).

[6] Fernandes P A, Salomé P MP, da Cunha A F, *J. of Alloys Comp.* , **509** (2011) 76000-76606.

[7] Berg D M, Djemour R, Gütay L, Siebentritt S, Dale Ph J, Fontane X, Izquierdo-Roca V, and Pérez-Rodriguez A, *Appl. Phys. Lett.* **100**, 192103 (2012); http://dx.doi.org/10.1063/1.4712623 (*4 pages*).

Journal of Physics: Conference Series **480** (2014) 012009 doi:10.1088/1742-6596/480/1/012009

Effect of thermal treatment on the CO and H_2O sensing properties of MoO_3 thin films

M Torres-Luengo, H M Martínez, J Torres and L D López-Carreño
Grupo de Materiales con Aplicaciones Tecnológicas, GMAT. Departamento de Física, Universidad Nacional de Colombia, Bogotá, Colombia

E-mail: mtorresl@unal.edu.co

Abstract. MoO_3 thin films were prepared on Corning glass substrates using the chemical spray pyrolysis technique. A 0.1 M solution of ammonium molybdate tetrahydrate was used as precursor one. 5ml and 20 ml of the precursor solution was sprayed with the substrate temperature maintained at 623 K. Thermal treatment involved drying at 393 K for 8 h with continuous N_2 flow, followed by a vacuum annealing at 473 K for 2 h in a residual inert atmosphere. XRD indicates that the crystallographic structure corresponded to the orthorhombic α-MoO_3 phase. Electrical characterization was carried out in a system operating under high vacuum conditions. The samples could be cooled down to LN_2 temperature and heated in a controlled way up to 473 K. To elucidate the electrical response of the films to CO and H_2O exposure, the I-V characteristic curve was measured over the whole temperature range. The electrical resistance of the films decreased with increasing temperature. In 5 ml films, the sensitivity to both gases increased which thermal treatment, reaching values between 40% and 60% at room temperature. On the contrary, the 20 ml films' sensitivity decreased almost half of their original values after thermal treatment.

1. Introduction

Because of its technological importance, scientific research in the thin films-based gas sensor field has shown a significant increase over the last years. From a technological point of view, efficiency and low cost of production are some of the most important features in the development of these devices [1–4]. The operating principle of these devices is based on the changes in the sensor's electrical resistance induced by gas-surface electrostatic interactions. Semiconductor transition metal oxides have been widely used as a detection material. Among these, molybdenum trioxide has received special attention. Thin films have been prepared using different techniques, and their properties depend on the method as well as the parameters used during the preparation process. The physical-chemical properties of MoO_3 thin films prepared through spray pyrolysis and their potential application as a gas sensor have been studied, and the effect of the deposition parameters on the crystallographic structure, morphology, and opto-electronic properties have also been widely reported [5–10].

2. Experimental procedure

MoO_3 thin films were deposited on Corning glass substrates using the spray pyrolysis technique. A 0.1 M solution of ammonium heptamolybdate tetrahydrate ($(NH_3)_6Mo_7O_{24}{\cdot}4H_2O$) was used as a precursor. Specific details of the spray pyrolysis equipment and the substrate's cleaning procedure have already been described [11]. Regarding the present article, two types of samples were prepared: 5 and 20 ml of the precursor solution was sprayed onto substrates maintained at 623 K, respectively. The thermal treatment involves two steps: a drying process at 393 K in atmospheric pressure under continuous N_2 flow for 8 hours and an annealing at 473 K in a vacuum with a residual gas mixture of N_2, Ar and H_2 for 2 hours.

The films' crystallographic structure was studied using an XPert PRO PANAnalytical X-ray diffractometer. X-ray diffractograms were taken in a 2θ range 10° - 70°, with a step of $\Delta 2\theta = 0.02°$ and an exposure time of 2s per step using Cu-Kα radiation (λ=1.5406Å) and Bragg -Brentano geometry.

The films' I-V characteristic curves were measured via the two-point technique on a separate high-vacuum system that operates at a base pressure of about 10^{-5} mbar. In this system, the samples' temperature can be varied in a controlled way in the range 133-473 K. The samples' temperature was measured with a K-type thermocouple attached to the sample.

In order to measure the I-V curves, two comb-shaped aluminum electrodes were deposited through thermal evaporation in a high vacuum (10^{-5} mbar base pressure). Cu wires were attached to the aluminum electrodes using silver staining (GA18573 Leitsilver BioChemika) [12]. An AC-voltage 20 V peak-to-peak and 0.05 Hz signal provided by a digital PeakTech 4070 signal generator was used as excitation source.

The current through the sample was measured using a Keithley 617 digital high-impedance electrometer. I-V curves were recorded using an automatic data acquisition system.

The interaction of CO and H_2O molecules with the MoO_3 thin films' surface was studied by measuring the relative change of the films' electrical resistance, which is obtained from the measured I-V curves. Before exposing the sample to a CO or H_2O atmosphere, the film was heated to 473 K at a 10^{-5} mbar residual gas pressure in order to clean it through desorption the sample's surface. After that, CO or H_2O was injected into the HV-chamber until a 10^{-2} mbar pressure was reached. Then, the film was cooled down to 133 K using LN_2.

3. Results and discussion

3.1. X-ray diffraction (XRD)

It is well known that MoO_3 can exist in two crystallographic forms: the α-phase, which presents an orthorhombic structure, and the β-phase, with a monoclinic structure. Figure 1 shows the XRD spectra of the samples deposited with a 5 ml solution. According to crystallographic card JCPDS-ICDD 00-005-0508, the prepared samples' structure corresponds to the orthorhombic α-MoO_3 phase. Analysis of the diffraction peak's integrated intensity suggests that MoO_3 films grow with a preferred orientation along the [0k0] direction. This characteristic behavior has already been reported [11].

From the X-ray spectra, the lattice constants were also calculated. For thermally untreated films, values of a=3.96 Å, b=13.86 Å and c=3.69 Å were determined. These values are consistent with those reported in the crystallographic card for MoO_3-powder samples. In all cases, the maximum discrepancy does not exceed 0.2%. As can be seen in Figure 1 (lower spectrum), thermal treatment has no significant effect on the MoO_3 films' crystallographic structure.

For the films deposited with 20 ml of solution, the X-ray spectra analysis (figure 2) shows that the MoO_3 films also have an orthorhombic-type crystalline structure. Diffraction patterns were also indexed by comparing them with JCPDS-ICDD card 00-005-0508, and the presence of a single α-MoO_3 phase was confirmed. Lattice parameters obtained from the diffraction peaks' positions are also in agreement with those obtained for the MoO_3 films deposited with 5 ml of solution. A preferred growth along the [0k0] direction was also observed. As in the previous case, no effect on the crystallographic structure can be observed as a result of thermal treatment.

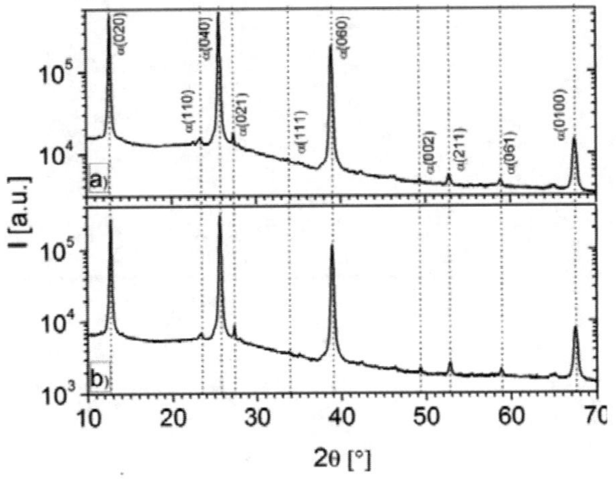

Figure 1. 5 ml α-MoO_3 XRD patterns of a) non-thermally treated films and b) thermally treated films.

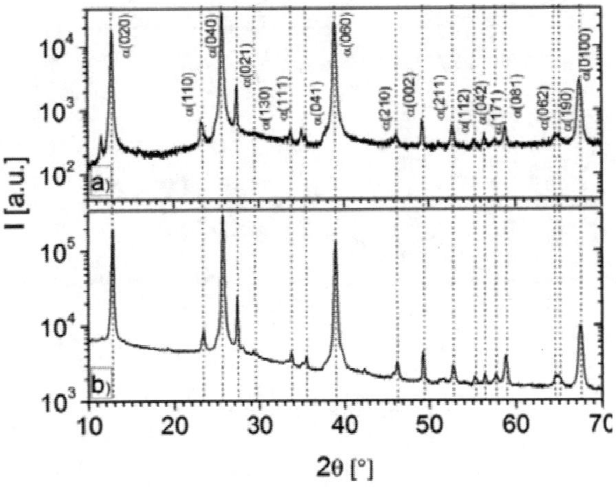

Figure 2. 20 ml α-MoO_3 XRD patterns of a) non-thermally treated films and b) thermally treated films.

Although the MoO_3 films' crystallographic structure analysis does not suggest significant effects resulting from the thermal treatment, effects are more evident for the MoO_3 films' electrical properties, as is discussed in the next section.

3.2. Electrical Properties

Sensitivity was used as a figure of merit in order to quantify the effect of CO or H_2O adsorption on the MoO_3 films' electrical properties. In this case, the sensitivity S is defined as:

$$S=100\cdot(R_a-R_g)/R_a \quad (1)$$

where R_a and R_g are the film's electrical resistance measured in a vacuum and in the corresponding gas atmosphere, respectively.

In all cases, the I-V characteristic curve of the α-MoO_3 thin films deposited with 5ml of volume exhibits a linear behavior over the whole temperature range studied. A similar behavior was found after CO and H_2O adsorption. Regardless of thermal treatment, the MoO_3 films' electrical resistance decreases with increasing temperature of the films. On the contrary, prior to thermal treatment, the I-V characteristic curve of films deposited with 20 ml of volume exhibits a non-linear behavior, which is more evident for temperatures lower than 253 K. Ohmic behavior is restored throughout the whole operating temperature range after thermal treatment.

In high vacuum conditions, the 5 ml MoO_3 films' electrical resistance varies from 100 KΩ to 500 MΩ, depending on the film's temperature. After thermal treatment, the variation range of the electrical resistance increases by at least an order of magnitude. In the case of the MoO_3 films deposited with 20 ml of volume, in vacuum conditions the electrical resistance varies between MΩ and 2000 GΩ. However, after thermal treatment, values of the MoO_3 films' electrical resistance in the range 70 KΩ to 10 GΩ were found. In this case, the films' electrical resistance decreases by two orders of magnitude after thermal treatment.

The effect of thermal treatment on the CO and H_2O sensitivity for MoO_3 films deposited with 5ml of volume is shown in Figure 3. For both gases, and prior to thermal treatment (left graph), the percentage relative change at the lowest temperature of the electrical resistance is close to 20%. For these films, the CO and H_2O sensitivity gradually increases as the operating temperature is raised. For a film's temperature close to room temperature, the sensitivity tends to stabilize, reaching values of around 33% and 38% for CO and H_2O, respectively.

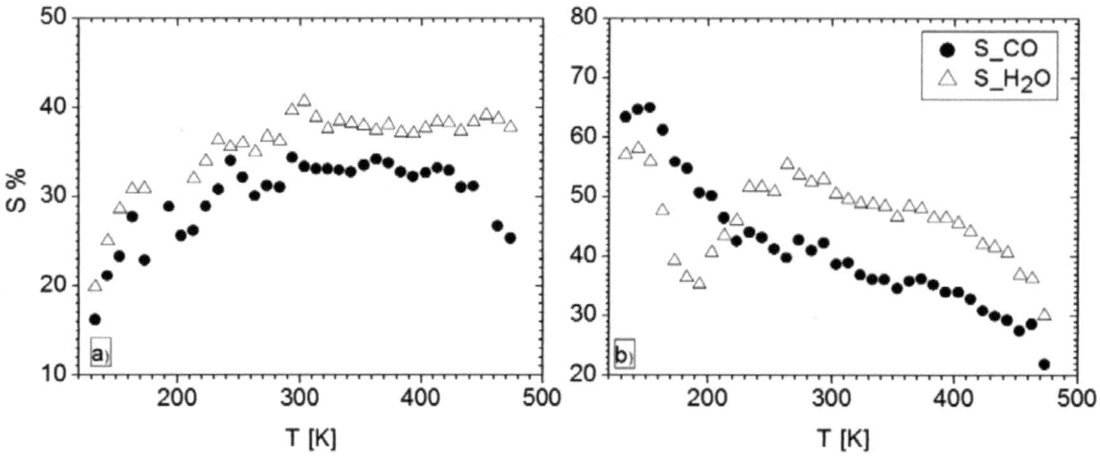

Figure 3. 5ml α-MoO_3 sensitivity as a function of temperature for a) non thermally treated films and b) thermally treated films.

Conversely, after thermal treatment, the CO and H_2O sensitivity reaches values close to 60% at the films' lowest measured temperature. When the operating temperature increases, the sensitivity for CO and H_2O gradually decreases to 20% and 30%, respectively. The H_2O sensitivity curve shows an untypical behavior characterized by the presence of a local minimum located near 190 K. At this temperature, a significant increase in the total pressure of the system there was also observed. These correlated events could be attributed to H_2O desorption from the MoO_3 film's surface.

The CO and H_2O sensitivity curves for the films deposited with 20 ml of volume are shown in Figure 4. In this case, regardless of the thermal treatment, the percentage relative change of the electrical resistance due to gas adsorption diminishes with increasing temperature of the films. For thermally untreated films, it was not possible to obtain a reference curve in the lower temperature range measured because of the nonlinear I-V characteristic curve measured in this region. Despite that, for temperatures above 253 K, the H_2O sensitivity remains practically constant, with values close to 100%, showing a slight decrease around 390 K, where it reaches a value of 90%. The CO sensitivity varies from 88% at the lowest measured temperature up to about 75% around 400K. After thermal treatment, the sensitivity for both gases remains constant along the whole measured temperature range, with values near to 50% and 40% for H_2O and CO, respectively.

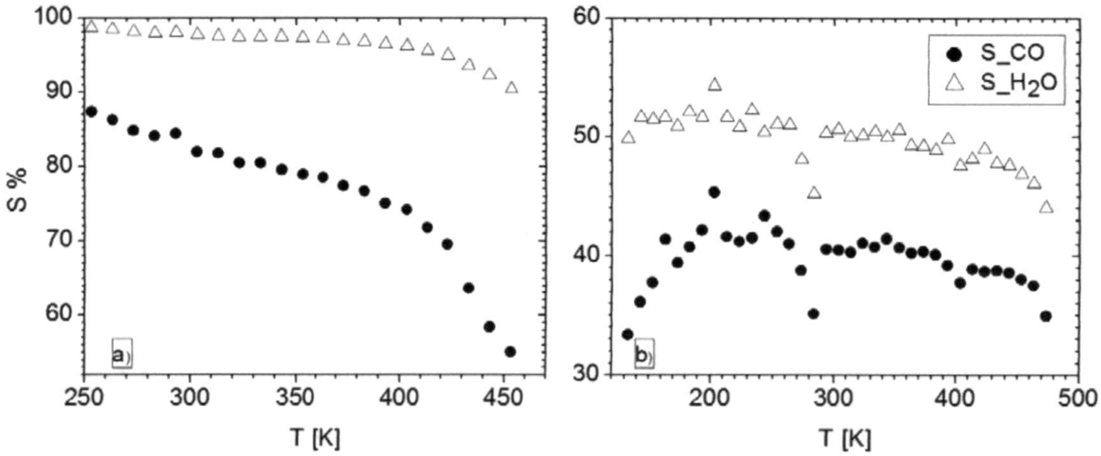

Figure 4. 20ml α-MoO_3 sensitivity as a function of temperature for a) non thermally treated films and b) thermally treated films.

Journal of Physics: Conference Series **480** (2014) 012009 doi:10.1088/1742-6596/480/1/012009

4. Conclusions

In this paper, we compared the structural, electrical, and sensing properties of MoO_3 thin films. The effects of thermal treatment on these properties were also studied.

Results obtained from XRD revealed a single orthorhombic phase of α-MoO_3 and showed that an increase of the deposited volume of the solution also increases the number of diffraction peaks observed, but they do not vary significantly with thermal treatment.

However, thermal treatment increases the percentage relative change of the electrical resistance, due to the interaction of CO and H_2O molecules with the MoO_3 films deposited with 5 ml of volume. When the volume of the deposited precursor solution is increased, thermal treatment has an opposite effect, and it also eliminates the nonlinear behavior in the I-V characteristic at low operating temperatures. Despite this, both gases have reducing effects on MoO_3 thin film's sensitivity properties.

References

[1] Patil P S 1999 *Mater. Chem. Phys.* **59** 185.

[2] Garzella C, Comini E, Tempesti E, Frigeri C and Sberveglieri G 2000 *Sensors Actuators B Chem.* **68** 189.

[3] Galatsis K, Li Y X, Wlodarski W, Comini E, Sberveglieri G, Cantalini C, Santucci S and Passacantando M 2002 *Sensors Actuators B Chem.* **83** 276.

[4] Itoh T, Matsubara I, Shin W, Izu N and Nishibori M 2008 *Sensors Actuators B Chem.* **128** 512.

[5] Martínez H M, Torres J, López-Carreño L D and Rodríguez-García M E 2013 *J. Supercond. Nov. Magn.* **26** 2485.

[6] Martínez H M, Torres J, Rodríguez-García M E and López Carreño L D 2012 *Phys. B Condens. Matter.* **407** 3199.

[7] Imawan C, Steffes H, Solzbacher F and Obermeier E 2001 *Sensors Actuators B Chem.* **78** 119.

[8] Hussain O and Rao K 2003 *Mater. Chem. Phys.* **80** 638.

[9] Barazzouk S, Tandon R P and Hotchandani S 2006 *Sensors Actuators B Chem.* **119** 691.

[10] Prasad A K, Kubinski D J and Gouma P I 2003 *Sensors Actuators B Chem.* **93** 25.

[11] Martínez H M, Torres J, López-Carreño L D and Rodríguez-García M E 2013 *Mater. Charact.* **75** 184.

[12] Martínez H M, Torres J, López-Carreño L D, Alfonso J E, Moreno L C and Pardo A 2011 *Ing. Cienc.* **7** 105.

First-principles study of the (0001)-MgB_2 surface finished in Mg and B

Sully Segura[1], Jairo Arbey Rodríguez Martínez[2] and María Guadalupe Moreno-Armenta[3]

[1]Escuela de Física, Universidad Pedagógica y Tecnológica, Tunja, Colombia and Universidad Santo Tomás, Tunja, Colombia.
[2]GEMA: Grupo de Estudio de Materiales, Dept. of Physics, Universidad Nacional, Bogotá, Colombia.
[3]Centro de Nanociencias y Nanotecnología - CNyN, UNAM, Ensenada, BC, México.
Emails: SS: ssegura@ustatunja.edu.co, JARM: jarodriguez@unal.edu.co, MGMA: moreno@cnyn.una.mx

Received 4 October 2013, in final form XX Novembrer 2013

Abstract. We present a study based on Density Functional Theory (DFT) of the volume and two surfaces (0001) of MgB_2, one of them terminated in Mg and the other one terminated in B. Each one of the surface was relaxed and their electronic properties were determined. From calculation of the enthalpy of formation we found that the Mg-terminated surface is energetically favored. The bands seem to present a formation similar to the Dirac's cone as that are presented in graphene, but in MgB_2 is above of the Fermi level. In the three cases, volume and the two surfaces, the behaviour is boron-metallic, because there are strong presence of B orbital's in the neighborhood of the Ferm level.

Keywords
MgB2 surfaces, DFT, wien2k, superconductor, surface electronic states, apparent Dirac's cones

1. Introduction

The main application of the MgB_2, from the discovery of its superconducting properties [1], is its facility to manufacture superconducting wires and tapes. In nanotechnology, is used to construct nanotubes, nanowires, and nanoparticles [2-9]. Additionally in its normal state, is a resistive electrical conductor, which is an important difference with superconducting ceramics and can be an important advantage for design of nano-devices. MgB_2 is also significantly cheaper than other superconducting materials. Josephson junctions based on MgB_2, are only reproducible if surfaces are smooth and high-quality. Experimentally, high-quality surfaces were obtained [10-13], and theoretically, calculations of the (0001) surfaces were carried out in [14, 15]. This work is devoted to present a comparison of the electronic properties in volume and surfaces.

2. Calculation details

Calculations were carried out by means or the code wien2k [16], which is based on DFT [17, 18]. We used the Generalized Gradient Approximation in the parametrization of PBE [19]. The muffin-tin radii were 2.15 and 1.52 u.a. for Mg and respectively. Energy separation between core and valence states was 6 Ry. Convergence of energy was 0.0001Ry. R_{mt} *K_{max} and l_{max} were 7 and 10 respectively and these parameters control the size of the basis-set to expand the solutions of the Kohn-Sham equations [18], according to LAPW method. The surfaces are simulated by means of the slab method, where each slab has 5 bilayers of MgB_2, with a vacuum space between slabs of ~12Å, which guaranteed that there are no-interactions between surfaces. The bulk cell parameters obtained are $a = 3.52$Å and $c = 3.52$Å, wich are in a good agreement with other authors [14,15]. Figure 1a shows the hexagonal volumetric cell used.

3. Structural properties

Let us to cut a bulk of a material (Gibbs potential G_0) in two pieces (G_1 and G_2) and two surfaces (G_{surf}). Then $G_0=G_1+G_2+G_{surf}$, with $G_{surf}=(U+PV-TS)_{surf} = U_{surf} - TS_{surf}$ (V=0 for a surface). By application the first law of thermodynamics to an open system at equilibrium:

$$U_{surf} = TS_{surf} - PV_{surf} + 2\gamma A + \sum \mu_i n_i \tag{1}$$

where $2\gamma A$ is the energy of formation of the two surfaces ($2\gamma A = E_{surface}$, A is the area of each surface), μ_i is the chemical potential of the atomic specie i and n_i is the number of particles. Therefore, $G_{surf} = 2\gamma A + \sum n_i \mu_i$. At P=0 and T=0 we can approximate $G_{surf} \approx E_{DFT-slab}$ where $E_{DFT-slab}$ is the total energy of the slab calculated by wien2k.

If we take as area unit the area of and 1x1-cell, then

$$E_{surface}\,(1x1\ cell) = \frac{1}{2}(E_{DFT-slab} - \sum n_i\mu_i)_{1x1cell} \tag{2}$$

Particularly,

$$E_{surface\ Mg-terminated}\,(1x1cell) = \frac{1}{2}(E_{DFT-slab-\ Mg-\ terminated} - n_B\mu_B - n_{Mg}\mu_{Mg})_{1x1cell} \tag{3}$$

and similarly for $E_{surface-\ B-terminated}$ (1x1cell). Using $\mu_{Mg} - \mu_{Mg-Bulk}$ as the independent variable, and after an algebraic job, we obtained

$$E_{surface}^{(Mg/B)-terminated} = \frac{1}{2}\left[\left(\frac{1}{2}n_B - n_{Mg}\right)\left(\mu_{Mg} - \mu_{Mg}^{Bulk}\right) + E_{slab}^{(Mg/B)-terminated} + \frac{1}{2}n_B\left(\mu_{Mg}^{Bulk} - \mu_{MgB_2}^{Bulk}\right) - n_{Mg}\mu_{Mg}^{Bulk}\right] \tag{4}$$

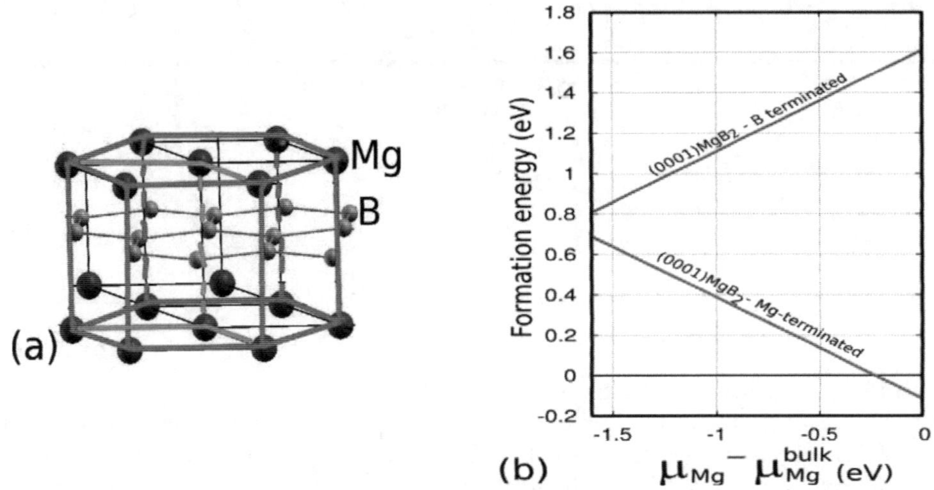

Figure 1. a) The cell used in calculations. The hexagonal symmetries are shown. The unit cell is marked with green lines. b)Energy of formation of the surfaces of MgB_2 as a function of the chemical potential of Mg.

where μ_{Bulk} is obtained from $E_{DFT-bulk}$ by each particle, and $-\Delta H_F \leq \mu_{Mg} - \mu_{Mg}^{Bulk} \leq 0$ and ΔH_F is the enthalpy of formation of the MgB_2 [15]. We obtained, $\Delta H_F = 1.59eV$ (discrepancy ~5.5% with [15]), and $\mu_{Mg-Bulk} = -5451.38eV$, $\mu_{MgB_2-Bulk} = -6804.15eV$, and for the surface Mg-terminated, we have $n_{Mg} = 6$ and $n_B = 10$ and for the surface B-terminated, $n_{Mg} = 5$ and $n_B = 12$. Additionally, $E_{DFT-slab-Mg-terminated}(1x1) = -39472.358eV$ and $E_{DFT-slab-B-terminated}(1x1) = -35370.303eV$. Therefore,

$$E_{surface}^{Mg-terminated} = \frac{-1}{2}\left[\left(\mu_{Mg} - \mu_{Mg}^{bulk}\right) - 0.1eV\right]; E_{slab}^{B-terminated} = \frac{1}{2}\left(\mu_{Mg} - \mu_{Mg}^{bulk}\right) + 0.16eV \quad (5)$$

Figure 1b shows the plot of the formation energy for the two types of surfaces of MgB_2 as a function of the chemical potential using as reference the chemical potential of the Mg-bulk (Eq.5). We can conclude that the Mg-terminated surface is more favorable than the other one as have been informed by other authors [15], in all the range of the chemical potential of the Mg.

4. Electronic properties

Figure 2 shows the bands of the surfaces and the bulk. The material presents a conductor behavior. It is important to note that the bulk bands of this material at K-point, above the Fermi level, present a formation like as the Dirac's cones of graphene (highlighted with red circle in Figure 2 c). And these apparent Dirac's cones are also present in the surfaces bands for the two types of surface. Composition of bands, around the Fermi level, are due to the main contributions are due to B orbitals. p-B orbitals are situated in the region between -6.0eV up to Fermi level. There are orbitals of Mg in the conduction band and there also a bit at the bottom of the valence band. One can see that the surface band are similar to those volume, but the lines look parallel and repeated 5 times. Mg orbitals are mainly located above of the Fermi level. In the most stable surface, around the Fermi level, the presence of p-B orbitals is notorious and the Mg orbitals are majority above the Fermi level.

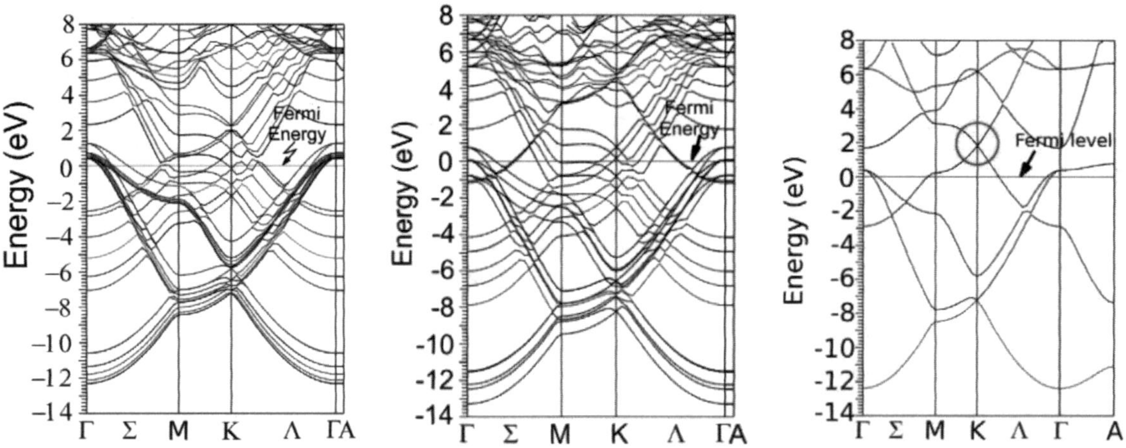

Figure 2 Band structures of MgB_2. (a) (0001) B-terminated. (b) (0001) Mg-terminated. (c) Bulk

5. Sumary and Conclusions

We presented calculations about the volume and (0001) surfaces of the superconductor MgB_2. The results we obtained are in overall agreement with other authors. Taking in account the energy of formation, the Mg-terminated surface is more favorable than the other one. The material is a good electrical conductor in volume and any of the two surfaces, and the conduction of electricity is mainly carried out by p-B orbitals. Other observation is that the bands seem to present a formation very similar to the Dirac's cones as those are presented in graphene, but they are above of the Fermi level.

6. Acknowledgments

Calculations were carried out in the DGCTIC UNAM Supercomputing Center in México and at the cluster of the GEMA Group at the Physics Dept. of Universidad Nacional, Bogotá Colombia. SS was partially supported by the M. Sc. Program of the Physics Dept. at the Universidad Pedagógica y Tecnológica de Colombia, Tunja, Colombia.

7. References

[1] J Nagamatsu, N Nakagawa, T Muranaka, Y Zenitari, and J Akimitsu. 2001 *Superconductivity at 39K in magnesium diboride.* Nature, **410** 63

[2] N Sano, O Kawanami, H Tamon. 2011 *Journal Applied Physics,* Vol **109** 034302 – 034302-6

[3] X. Zhang, Y Ma, Z Gao, D Wang, L Wang, W Liu, and C Wang. 2008 *J. Appl. Phys.*, Vol. **103** 103915

[4] E W Collings, M D Sumption, M Bhatia, M Susner, and S D Bohnenstiehl. 2008 *Supercond. Sci. Technol.* Vol **21** 103001

[5] S Zhou, P Wang, B Zhang, H Gong, X Zhang. 2009 *Materials Letters,* **63**(20) 1680-1682

[6] Shekhar C, Giri R, Malik SK, Siravastra ON. 2007 *J. Nanosciencie and Nanotechnology* **7**, (6) 1804-1809(6)

[7] M Dias Lima, J Bykova, A. Howard, Ray, A Zakhidov. 2012 *American Physical Society*

[8] R Ma, Y Bando, T Mori and D Golber. 2003 *National Institute for materials Science* **15** (16) 3194-3197

[9] N Novosel, S Galić, D Pajić, Ź Skoko, I Lončarek, M Mustapić, K Sadro and E Babić. *Superconductor Science and Technology* **26**(6)

[10] M Xu, Y Takano, T Hatano, T Kimura, D Fujita. 2003 *Applied Surface Science 205.* 225-230

[11] V Despoja, D J Mowbray and V M Silking. 2011 *Phys. Rev.* B **84** 104514

[12] D R Lamborn, R H T Wilke, Q Li, X X Xi, D W Snyder, and J M Redwing. 2007 *Transactions on Applied Superconductivity* **17**(2) 2862-2866

[13] P Orgiani, Y Cuil, J Chen, V Ferrando, A V Pogrebnyakov, J M Redwing, Q Li and X X Xi. 2006 *Advances in Science and Technology,* **47** 55-62

[14] J Kortus, I I Mazin, K D Belashchenko, V P Antropov and L L Boyer. 2001 *Phys Rev* B. **86** (20)4656-4659

[15] Z Li, J Yang, J G Hou, Q Zhu. 2002 Phys Rev B. **65** 100507-1

[16] P Blaha, K Schwarz, G K H Madsen, D Kvasnicka and J Luitz. 2001 *WIEN2k, An Augmented Plane Wave + Local orbitals Program Oro Calculating Cristal Properties (Karlheinz Schwarz, Techn. Universität Wien, Austria)* ISBN 3-9501031-1-2

[17] P Hohenberg and W Kohn. 1964 *Phys Rev.* **136** B864

[18] W Khon and L J Sham. 1965 *Phys Rev* **140** A1133

[19] J P Perdew, K Burke, M Ernzerhof, 1996 *Phys Rev* **77**(18) 3865-3868

Journal of Physics: Conference Series **480** (2014) 012011 doi:10.1088/1742-6596/480/1/012011

Fractal structures in casting films from chlorophyll

G C Pedro, F D S Gorza, N C de Souza and J R Silva[1]
Grupo de Materiais Nanoestruturados, Campus Universitário do Araguaia, Universidade Federal de Mato Grosso, MT, Brazil

E-mail: josmarysilva@ufmt.br

Abstract. Chlorophyll (Chl) molecules are important because they can act as natural light-harvesting devices during the photosynthesis. In addition, they have potential for application as component of solar cell. In this work, we have prepared casting films from chlorophyll (Chl) and demonstrated the occurrence of fractal structures when the films were submitted to different concentrations. By using optical microscopy and the box-count method, we have found that the fractal dimension is $D_f = 1.55$. This value is close to predicted by the diffusion-limited aggregation (DLA) model. This suggests that the major mechanism – which determines the growth of the fractal structures from Chl molecules – is the molecular diffusion. Since the efficiencies of solar cells depend on the morphology of their interfaces, these finds can be useful to improve this kind of device.

1. Introduction

Chlorophyll molecules (Chls) are important because they can act as a natural light-harvesting devices occurring during the photosynthesis [1], and also as component of dye-sensitized solar cells - DSSCs [2]. There has been shown that the efficiency of solar cell depends on their surface morphologies [3]. Therefore, fractal structures – which determine different morphological structures of surfaces – can play an important role on conversion efficiency. A fractal structure is that that exhibits a fractal dimension, which exceeds its topological dimension assuming noninteger values [4]. Fractal structures are well described by the so-called Diffusion Limited Aggregation (DLA) model [5]. The DLA model account for that each immobile cluster grows by diffusion of particles with a unitary probability of sticking. A lot of experimental systems have been successfully described by using this model [6]. Fractal structures in organic films are usually studied by using analyzes of optical microscopy [7], scanning electron microscopy - SEM [8], and atomic force microscopy – AFM [6]. In particular, films from Chl were already prepared by using techniques including Langmuir-Blodgett [9], self-assembly monolayer [10], and casting [11]. In these systems, properties such as electrical, photoelectrical, as well as photogeration of charges were examined. However, while properties of Chl films have been widely investigated, studies of surface morphology focused on fractal structures are lacking for this kind of films.

In this work, we report on investigation of the formation of fractal structures in Chl films prepared by casting technique. Two different concentrations of Chl solution were used and the fractal dimension of the structures was determined by using box count method. The results obtained were discussed on basis of DLA model.

[1] josmarysilva@ufmt.br

Journal of Physics: Conference Series **480** (2014) 012011 doi:10.1088/1742-6596/480/1/012011

2. Materials and methods

Chlorophyll (Chl) was purchased from MP Biomedicals, LLC, and used as receipted. Chl solutions with different concentrations (0.50 and 0.12 g/L) have been prepared by dissolving Chl in purified water as a solvent. The solution pH was adjusted to 8 by adding appropriate amounts of ammonium hydroxide. Thin films were obtained by casting the solutions onto BK7 slices and drying in an air atmosphere at room temperature (23 ^{0}C). Surface morphologies of films were examined by using an optical microscope (Nikon, Eclipse Ci/L, USA). The fractal analyzes were performed by the box counting method [4] using the freeware software ImageJ [12]. In this method, the fractal dimension, D_f, is determined as the slope of the regression line for log–log plots of the scale r and count $N(r)$, i.e. $D_f = -\lim(\log N(r)/\log(r))$.

3. Results and discussion

It is well known that several systems in Nature exhibit fractal structures and they present a dependence on the temperature and concentration of solution used [13]. In order to examine the possibility of occurrence of fractal structures in Chl films, we have performed a study of dependence of film topography on the concentration. Figure 1 shows images of optical microscopy for Chl films as a function of concentration at room temperature. We can note a well-known fractal form of aggregates, as showed in Figures 1A and 1B.

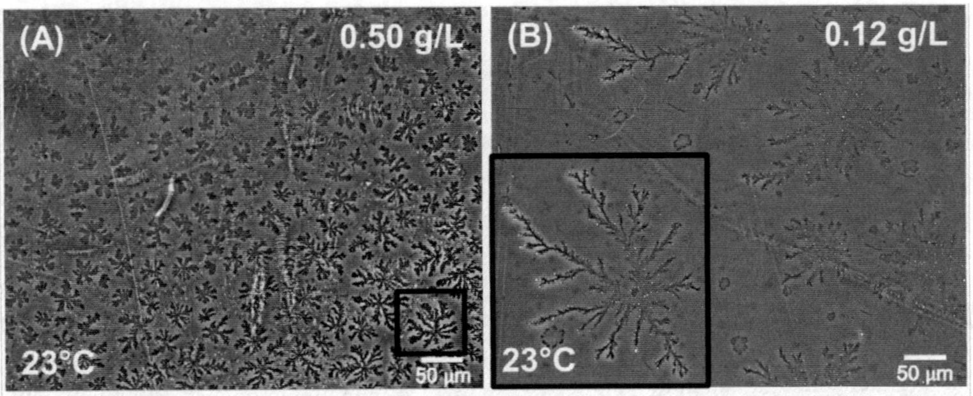

[P1]

Figure 1. Optical microscopy images of Chlorophyll films at concentrations of (A) 0.50 g/L and (B) 0.12 g/L. Two selected fractal structures are indicated.

Figures 2 and 3 exhibits the processed images obtained from Figure 1 by using outline

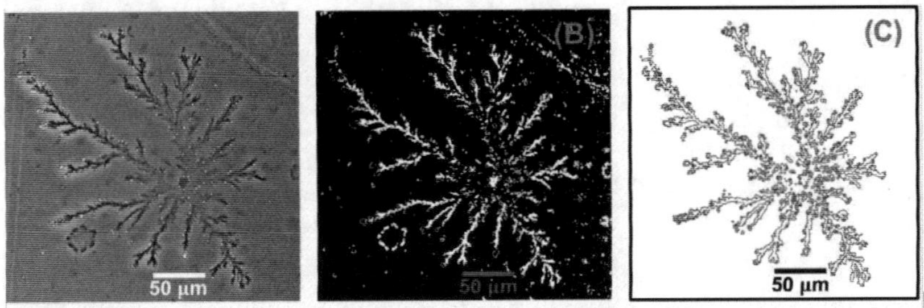

[P2]

Figure 2. Enlarged image of selected area in Figure 1B before (A) and after application of binarization (B), outline and noise reduction methods (C), respectively.

approach implemented in ImageJ software. Here, we can see clearly the fractal structure formed by

Chl molecules. The fractal form observed in Figures 2C and 3C is also found in several kinds of

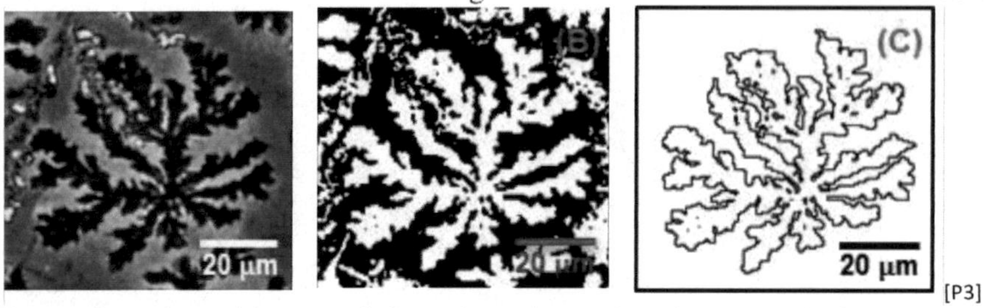

Figure 3. Enlarged image of the selected area in Figure 1A before (A) and after application of binarization (B), outline and noise reduction methods (C), respectively.

materials, including DNA [14], germanium nanoparticles and nanowires [15], as well as poly(ethylene oxide) crystals [16].

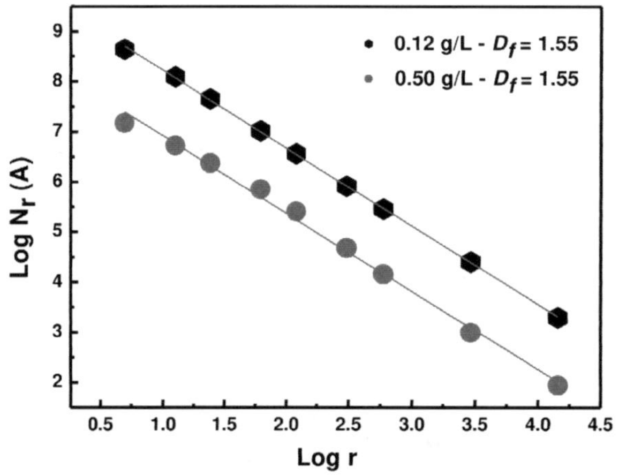

Figure 4. Box counting analysis for the fractal structures showed in Figure 2C and Figure 3C. r is the scale (box size/image size) and $N(r)$ is the count of boxes of size r containing foreground pixels.

In order to gain an insight in formation mechanism of the fractal structures, we have determined the fractal dimension of the structures showed in Figures 2C and 3C using the box count method [17]. Figure 4 displays a log-log plot of number of boxes as a function of scale for fractal structures. As expect due to similar form of the fractals in Figures 2C and 3C, the value found is similar, $D_f = 1.55$. This value is close to $D_f = 1.71$, predicted for a typical case of diffusion-limited aggregation (DLA) model [18]. According to this model, the lower value found in our study suggests a faster aggregation of Chl molecules. Interestingly, the concentration did not affect the fractal dimension. This can be caused by the variation of concentrations used, which may not have been enough to change the dynamics of the formation of the fractal.

4. Conclusion

In summary, we have prepared casting films from chlorophyll molecules and investigated the formation of fractal structures at two different concentrations. The fractal dimension value of the structures was found be $D_f = 1.55$, which is close to that predicted by DLA model. These finds can provide insight into the understanding and improvement of devices which use Chl such as photovoltaic cells. In addition, future works can be focused on relationships between the surface morphological structure and the photovoltaic efficiency.

References

[1] Renger G 2008 *Primary Processes of Photosynthesis: Principles and Apparatus* (Cambridge: RSC Publishing).
[2] Senthil T S, Muthukumarasamy N, Velauthapillai D, Agilan S, Thambidurai M and Balasundaraprabhu R 2011 *Renewable Energy* **36** 2484.
[3] Grätzel M 2004 *Journal of Photochemistry and Photobiology A: Chemistry* **164** 3.
[4] Mandelbrot B B 1983 *The Fractal Geometry of Nature* (New-York: W. H. Freeman).
[5] Sander L M and Somfai E 2005 *Chaos* **15** 026109.
[6] Haberko J, Bernasik A, Łużny W, Raczkowska J, Rysz J and Budkowski A 2010 *Synthetic Metals* **160** 2459.
[7] Zhang G, Weeks B, Gee R and Maiti A 2009 *Appl. Phys. Lett.* **95** 204101-1.
[8] Venkatesh B, Chen D L and Bhole S D 2008 *Scripta Materialia* **59** 391.
[9] Kassi H, Barazzouk S, Brullemans M, Leblanc R M and Hotchandani S 2010 *Thin Solid Films* **518** 5345.
[10] Lee N-S, Shin H-K, Kwon Y-S and Lee B-J 2010 *Ultramicroscopy* **110** 650.
[11] Radwan R M, Lotfy S and Desouky O S 2008 *Nuclear Instruments and Methods in Physics Research B* **266** 3953.
[12] Schneider C A, Rasband W S and Eliceiri K W 2012 *Nat. Methods* **9** 671.
[13] Barabási A L and Stanley H E 1995 *Fractal Concepts in Surface Growth* (Cambridge: Cambridge University Press).
[14] Dai S, Zhang X, Du Z and Dang H 2005 *Materials Letters* **59** 423.
[15] Ottaviano L, Parisse P, Grossi V and Passacantando M 2010 *J. of Non-Cryst. Solids* **356**, 2076.
[16] Ma Z, Zhang G, Zhai X, Jin L, Tang X, Yang M, Zheng P and Wang W 2008 *Polymer* **49** 1629-1634.
[17] Enculescu M, Enculescu I and Topa V 2006 *J. of Opt. and Adv. Materials*, **8(3)** 1230.
[18] Witten T A and Sander 1981 *Phys. Rev. Lett.* **47** 1400.

Journal of Physics: Conference Series **480** (2014) 012012 doi:10.1088/1742-6596/480/1/012012

Preparation and characterization of $Fe_{50}Co_{50}$ nanostructured alloy

N Yepes[1], J Orozco[1], Z Caamaño[1], J Mass[2] and G Pérez[3]

[1]Science and Characterization of Materials Group, Universidad Del Atlántico, Km 7 Antigua Vía Puerto Colombia, A.A. 1890, Barranquilla, Colombia.
[2]Applied Physics Group, Universidad Del Norte, Km 5 Vía Puerto Colombia, A. A. 1560, Barranquilla, Colombia.
[3]Physical Metallurgy and Phase Transitions Theory Group, Universidad Del Valle, Ciudad Universitaria Meléndez, Calle 13 No 100-00, A. A. 25360, Cali, Colombia.

Email: zuliacaamano@mail.uniatlantico.edu.co

Abstract. Nanostructured $Fe_{50}Co_{50}$ alloy was prepared by mechanical alloying of Fe and Co powders in a planetary high energy ball milling. The microstructure and structural evolution of the alloy have been investigated as a function of milling time (0 h, 8 h, 20 h and 35 h) by scanning electron microscopy (SEM) and X-Ray diffraction (XRD) characterization techniques. SEM micrographs showed different powder particles morphologies during the mechanical alloying stages. By XRD analysis it could be identified the structural phases of the alloy and the crystallite size was calculated as a function of the milling time.

1. Introduction

FeCo-based soft magnetic materials have the largest saturation magnetization of all known magnetic alloys. These alloys exhibit high saturation magnetization and high Curie temperatures [1], that make them potential candidates for the high temperatures applications such as new generation turbine engines as well as in recording media [2]. In particular the $Fe_{50}Co_{50}$ alloy has a considerably large permeability, low magnetocrystalline anisotropy and low electrical resistivity [3]. All these properties are shown to be superior at the nanoscale. That's why there has been a great interest in researching on processing and characterization of nanostructured magnetic alloys.

Mechanical Alloying (MA) is one of the most important methods for preparation of nanostructured magnetic materials, such as equiatomic iron-cobalt alloy [4]. This method is an intensive energy process of mechanical grinding for the preparation of alloyed powders. It involves a repeated fracturing and welding of particles, leading to size reduction of nanometer scale [3].

Nanostructured $Fe_{50}Co_{50}$ alloys prepared by high energy mechanical alloying method at different milling times and characterized by SEM and XRD have been reported in the literature [1-6].

In this work, a study of morphological and structural characterization of nanostructured $Fe_{50}Co_{50}$ alloy prepared by high energy mechanical alloying (MA) is presented.

2. Experimental procedure

Fe (99.5 %, 6-9 μm) and Co (99.8 %, 2 μm) powders supplied by Sigma Aldrich, were separately weighted and mixed to get the desired composition. The mechanical alloying process was performed

in Fritsch Pulverizette 5 (P5) planetary high energy ball mill, belonging to Physical Metallurgy and Phase Transitions Theory Group of the Universidad Del Valle (Cali, Colombia). The mixed powders were introduced in stainless steels containers of 250 ml with balls of the same material, making previously vacuum. The ball to powder mass ratio was about 20:1 and the milling intensity was 305 rpm. Un-milled (0 hours) and milled samples during 8, 20 and 35 hours were prepared. To avoid excessive heating during milling each milling time was followed by a pause of one hour.
For morfological characterization of samples by SEM a Jeol JSM - 5600 microscope from the Universidad Del Norte (Barranquilla, Colombia) was used. SEM micrographs for un-milled powders (0 hours) and milled during 8, 20 and 35 hours were taken.
Structural characterization was performed on a PANalytical X′Pert diffractometer, from the Universidad Del Valle (Cali, Colombia) XRD laboratory. Measurements were carried out at room temperature; using Co radiation over 40-130° 2Θ and 0.54°/2.22s step. The mean crystallite size was estimated by the Scherrer formula [7], taking into account the diffraction peak integral width calculated from the diffraction peaks refinement software GSAS.

3. Results and Discussion

3.1. Morphological characterization by SEM.

Figure 1 shows the SEM micrographs of the $Fe_{50}Co_{50}$ alloy at different milling times. Figure 1 (a) displays the un-milled powder, the different shapes of the alloying elements particles can be seen. The Co particles have spherical shape while the Fe particles have cylindrical shape. Similar results were found by Akkouche et al. [3] for nanocrystalline powders of the same composition prepared by mechanical alloying.
Figure 1 (b) shows the micrograph obtained from the $Fe_{50}Co_{50}$ milled during 8 h. It is observed the agglomeration of powder particles, as consequence of powders that are repeatedly flattened, cold weld, fractured and rewelded, what is typical of mechanical alloying process for ductile elements as Fe and Co, as it is reported in the literature [1-6]. The powder size distribution and the particles shape are not uniform, as indicated by N. Yepes [8].
SEM micrograph for the powders milled during 20 h is shown in Figure 1 (c). It can be seen the flattening and welding of powders, resulting in a narrower distribution of particle size, also reported by Chermahini et al. [5] for powders of similar compositions.
The cold welding and fracture events continue to take place for milled powders during 35 h, leading to microstructural refinement, as it is shown in Figure 1 (d), which is a characteristic of ductile-ductile systems.

3.2. Structural characterization by XRD.

3.2.1. Phase identification.
Figure 2 shows the evolution of the XRD patterns for $Fe_{50}Co_{50}$ alloy as a function of the milling time. For the initial powders mixture the XRD pattern (0 h) confirms the presence of body-centred cubic Fe (bcc), hexagonal closed-packed (hcp) Co and face-centred cubic (fcc) Co phases. Also, it is observed a slight broadening of the bcc Fe-Co diffraction peaks between 8 and 35 hours of milling, due to the deformation processes and internal stresses occurring at macroscopic level during the milling process [3, 5]. After 8 hours of milling, the hcp and fcc Co diffraction peaks disappear, the Co atoms diffuse into the Fe bcc lattice leading to the onset of alloy formation with iron. Analogous results were found by Chermahini et al. [5] for powders of similar compositions milled during 45 h. These results differ from those found by Akkouche et al. [3] for alloys of the same composition, where a bcc Fe (Co) solid solution is formed after 100 h of milling. The formation of a crystalline solid solution is typical for all FeCo alloys prepared by MA [1-5]; unlike FeNi alloys in which an amorphous structure is formed by MA [1-2].

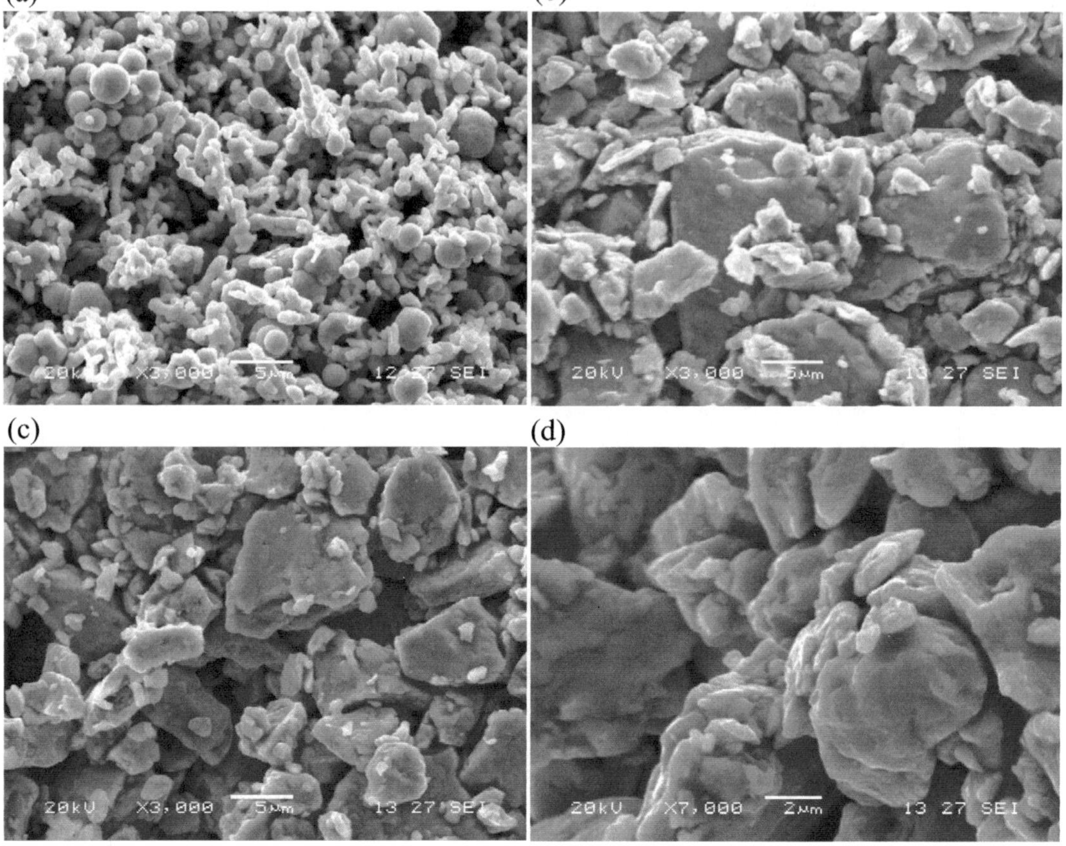

Figure 1. SEM micrographs of $Fe_{50}Co_{50}$ powders for various milling times: a) 0 h (un-milled), b) 8 h, c) 20 h and d) 35 h.

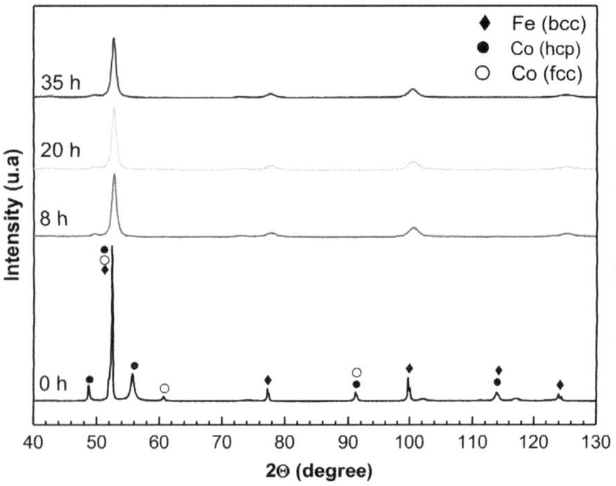

Figure 2. XRD of mechanically alloyed $Fe_{50}Co_{50}$ for various milling times.

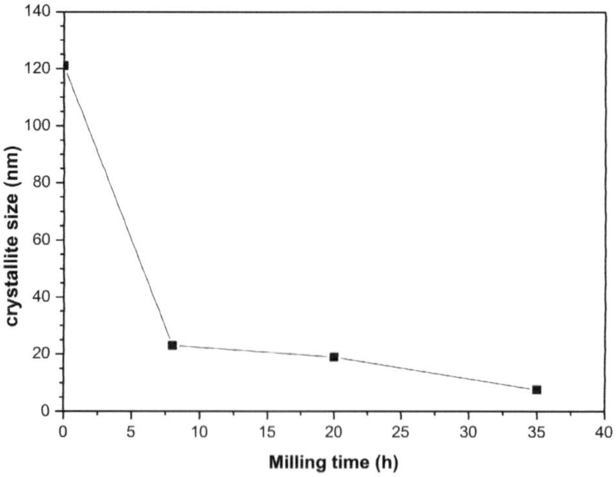

Figure 3. Evolution of crystallite size as a function of milling time.

3.2.2. Crystallite size.

Figure 3 shows the evolution of the crystallite size as a function of the milling time. It is observed that the crystallite size decreases with increasing milling time, from a value of 121 nm for un-milled powders (0 h, corresponding to Fe powders) to a value of 7, 5 nm for milled powders during 35 h. It is also noticed that the crystallite size decreases rapidly from 121 nm (0 h) to 23 nm (8 h) followed by a relatively slower one at 20 h of milling. These results are comparable with those obtained by Chermahini et al. [5] for nanostructured $Fe_{1-x}Co_x$ (x between 0, 2 and 0,7 at. % of Co) alloys prepared by MA using very similar milling times. Nevertheless, the crystallite size between 20 and 35 h changed slowlier, reaching a value of 12 nm. Different results were obtained by Akkouche et al. [3], for alloys of the same composition prepared by high energy MA, whose average crystallite size values were 150 ± 2 nm for un-milled powders (0 h) and 14 ± 2 nm for milled samples during 200 h. These differences are due to the used methods and the conditions for preparing the alloy.

The decreasing crystallite size with milling time behavior is attributed to dislocation generation caused by severe plastic deformation [4, 5].

4. Conclusions

Nanostructured alloys of $Fe_{50}Co_{50}$ in powder form were prepared by mechanical alloying method by means of a high energy planetary ball milling P5, in dried milling, making previously vacuum and with a milling intensity of 305 rpm. The morphology powders evolution with milling time was revealed by SEM. The formation of nanostructured alloy $Fe_{50}Co_{50}$ was confirmed by XRD. The crystallite size decreased with the increasing milling time; the lowest value in the crystallite size (7, 5 nm) was obtained at 35 hours of milling. The evolution of powders morphology as well as the crystallite size is attributed to deformation processes that powders undergo when they are mechanically alloyed within a high energy ball milling.

Acknowledgments

To Natalia Martelo who helped me with grammatical correcting in this article.

References

[1] Zeng Q, Baker I, McCreary V and Yan Z 2007 *J. Magn. Magn. Mater* **318** 28–38.
[2] Shokrollahi H 2009 *Mater. Des* **30** 3374–3387.
[3] Akkouche K, Guittoum A, Boukherroub N and Souami N 2011 *J. Magn. Magn. Mater* **323** 2542–2548.
[4] Chitsazan B, Shokrollahi H, Behvandi A and Mirzaee O 2011 *Powder Technol* **214** 105–110.
[5] Chermahini M D, Sharafi S, Shafyei A, Zandrahimi M and Shokrollahi H 2009 *J. Alloys. Compd* **484** 54-58.
[6] Poudyal N, Rong C, Zhang Y, Wang D, Kramer M J, Hebert R J and Liu J P 2012 *J. Alloys. Compd* **521** 55-59.
[7] Cullity B D 1978 *Elements of X-Ray Diffraction* United States of America (Addison-Wesley Pub. Co).
[8] Yepes N A June 2013 *Preparation and morphological characterization of FeCo magnetic nanoparticles* (Monograph dissertation, undergraduate physics program, faculty of basic sciences, Universidad del Atlántico, Barranquilla, Colombia,) 41–51.

Journal of Physics: Conference Series **480** (2014) 012013 doi:10.1088/1742-6596/480/1/012013

Overscreened Kondo physics in four-fold symmetric Au systems doped with Co

P. Roura-Bas[1], S. Di Napoli[1] and A. A. Aligia[2]
[1]Dpto de Física, Centro Atómico Constituyentes, Comisión Nacional de Energía Atómica, Buenos Aires, Argentina
[2]Centro Atómico Bariloche and Instituto Balseiro, Comisión Nacional de Energía Atómica, 8400 Bariloche, Argentina

E-mail: aligia@cab.cnea.gov.ar

Abstract. We analyze the relation between a model recently proposed to describe Co doped Au chains with four-fold symmetric leads, and the generalized SU(2)×SU(2) Anderson model. We compare the structure of the Kondo limit in the general case. We calculate the spectral density and the conductance as a function of temperature in both cases, showing that both models display similar non-Fermi liquid behavior.

1. Introduction

The Kondo effect is one of the paradigms in modern solid state theory. In its simplest version, the spin 1/2 of a magnetic impurity is *screened* by the spin of a conduction band (also called conduction channel) below a characteristic temperature T_K, resulting in a singlet ground state [1]. In nanoscience, ideal systems with a single localized spin 1/2 were studied and confirmed predictions of the SU(2) Kondo impurity model, or the more general SU(2) impurity Anderson model [2, 3, 4]. More recently, nanoscopic systems were studied [5, 6] in which a spin 1 was *underscreened* by one conduction channel, leaving a doublet ground state, and having a quantum phase transition to a singlet ground state [6], as predicted in an Anderson model which mixes a singlet and a triplet with a doublet through one conduction channel [7].

The physical realization of nanostructures with *overscreened* spins is more difficult, due to the sensibility of the physical properties to asymmetry of the channels or inter-channel charge transfer [8] which spoil the observation of the expected non-Fermi liquid behavior [8, 9, 10]. Recently, it has been shown that if in Co doped Au chains, the axial symmetry along the chain is broken by a sufficiently strong crystal field of four-fold symmetry, which splits the $3d$ x^2-y^2 and xy levels of Co, then the system displays clear signatures of non-Fermi liquid behavior, like an impurity contribution to the entropy $\ln(2)/2$ and a conductance per channel at low T of the form $G(T) \simeq G_0/2 - a\sqrt{T}$, where G_0 is the conductance at zero temperature in the one-channel case [11]. A similar behavior is expected for Co doped Cu chains attached to a four-fold symmetric lead [12].

A model studied before which displays non-Fermi liquid behavior is the SU(N)×SU(M) generalization of the Anderson model [13, 14], where M identical conduction channels screen the N degrees of freedom of the impurity. It has been shown that for the multichannel non-Fermi-liquid case M≥N, the non-crossing approximation (NCA) reproduces the exact power-law

Journal of Physics: Conference Series **480** (2014) 012013 doi:10.1088/1742-6596/480/1/012013

behavior of the main physical properties, including the spectral density of states [13, 14].

In this work we show that the model mentioned above for Co doped nanoscopic systems, can be considered as a generalization of the SU(2)×SU(2) Anderson model. We also calculate the spectral density and the conductance for both models using the NCA. We show that rescaling the Kondo temperature, the low-temperature properties of both models become very similar.

2. Model

First principles calculations [11] show that the ground state of the Co ion is a quadruplet of the $3d^7$ configuration with the xz and yz orbitals occupied by only one electron, and also either the xy or x^2-y^2 orbital is singly occupied, depending on the (irrelevant) sign of the crystal field. We denote the components of this quadruplet with three holes in the $3d$ shell by $|M_3\rangle$ where M_3 is the spin projection. This configuration is hybridized via the hopping to the xz and yz $5d$ orbitals of Au with *two* $3d^8$ triplets ($|\alpha M_2\rangle$) in which either the xz or the yz orbital of Co is doubly occupied. The effective Hamiltonian which describes this physics is [11]

$$H = \sum_{M_3}(E_3 + \frac{D}{2}M_3^2)|M_3\rangle\langle M_3| + (E_2 + D'M_2^2)\sum_{\alpha M_2}|\alpha M_2\rangle\langle\alpha M_2| + \sum_{\nu k\alpha\sigma}\epsilon_{\nu k}c^\dagger_{\nu k\alpha\sigma}c_{\nu k\alpha\sigma}$$
$$+ \sum_{\alpha M_2 M_3}\sum_{\nu k\sigma}V_\nu\langle 1\frac{1}{2}M_2\sigma|\frac{3}{2}M_3\rangle(c^\dagger_{\nu k\alpha\sigma}|\alpha M_2\rangle\langle M_3| + \text{H.c.}), \tag{1}$$

where $c^\dagger_{\nu k\alpha\sigma}$ creates a hole in the 5d band of Au states with symmetry α at the left ($\nu = L$) or right ($\nu = R$) of the Co site. A crucial role is played by the splitting D between the quadruplet states with spin projection $\pm 3/2$ and those with $M_3 = \pm 1/2$. Instead, D' which splits the excited states with $M_2 = \pm 1$ from those with $M_2 = 0$, has a negligible effect for realistic values [11]. Both terms appear as a consequence of the spin-orbit coupling $\hat{H}_{\text{SOC}} = \lambda\sum_i \hat{\mathbf{l}}_i\cdot\hat{\mathbf{s}}_i$. In spite of the irrelevance of D' for the system studied, it is interesting to note that for $D, D' \to +\infty$, the model becomes equivalent to the SU(2)×SU(2) Anderson model with a hybridization V_ν reduced by a factor $(2/3)^{1/2}$, which is the value of the Clebsch-Gordan coefficients $\langle 1\frac{1}{2}0\sigma|\frac{3}{2}\sigma\rangle$ in Eq. (1). This is easily seen using two pseudofermions to represent the states $|M_3\rangle$ with $M_3 = \pm 1/2$ and two pseudobosons for $|\alpha 0\rangle$. In the general case we need four pseudofermions and six pseudobosons to represent all states for the NCA treatment. The formalism is similar to that used by two of us for the singlet-triplet Anderson model [15].

Further insight into this equivalence is given by a Schrieffer-Wolff canonical transformation that eliminates the hybridization term (the last one) in Eq. (1) for infinite D (actually $D \gg T_K$ is enough, where T_K is the Kondo temperature [11]), and leads to an effective exchange interaction between an effective spin 1/2 ($\hat{S}$) for the states $|M_3\rangle$ with $M_3 = \pm 1/2$ and the spin of the extended states at the impurity site ($\hat{s}^\alpha$). The resulting Kondo model has the form

$$\hat{H}_{\text{2CK}} = \frac{2}{3}\frac{V_L^2 + V_R^2}{\delta}\sum_\alpha(\frac{2D' + \delta}{D' + \delta}\hat{S}_z\hat{s}^\alpha_z + 2\hat{S}_x\hat{s}^\alpha_x + 2\hat{S}_y\hat{s}^\alpha_y) + \sum_{\nu k\alpha\sigma}\epsilon_{\nu k}c^\dagger_{\nu k\alpha\sigma}c_{\nu k\alpha\sigma}, \tag{2}$$

where $\delta = E_2 - E_3 - D/8$ is the charge-transfer energy (we set the Fermi energy $\epsilon_F = 0$). For $D' \to +\infty$, the interaction is the usual one for the isotropic two-channel Kondo model (2CKM), except for the prefactor 2/3 which comes from th above mentioned Clebsch-Gordan coefficients. For $D' = 0$ the coefficient of the $\hat{S}_z\hat{s}^\alpha_z$ term is exactly half the other two, and the model reduces to an anisotropic 2CKM, which also has non-Fermi liquid properties.

3. Effect of spin-orbit coupling

To calculate D we have solved exactly the Hamiltonian of the 3d^7 configuration [16] including $\hat{H}_{SOC}$. We show in Fig. 1 how D depends on the splitting ΔE between 3d $x^2 - y^2$ and xy orbitals on D. We took the values of the energies of the different $3d$ orbitals from *ab initio* results [11]: $E_{3z^2-r^2} = -0.1$ eV, $E_{xz} = E_{yz} = 0.6$ eV, and assume $-E_{x^2-y^2} = E_{xy} = \Delta E/2$. For the Coulomb integrals F_i [16] and λ, we have taken the values $F_2 = 0.16$ eV, $F_4 = 0.011$ eV, $\lambda = 0.08$ eV, obtained from a fit of the low energy spectra of late transition metal atoms. From the figure, one sees that for $\Delta E = 0$, $D = -0.0426$ eV is negative (80% of the value $-2\lambda/3$ expected from first order perturbation theory in $\hat{H}_{SOC}$) indicating that the Co spin is oriented along the chain. As ΔE increases, D also increases and changes sign for $\Delta E \approx 0.8$ eV. For the value $\Delta E = 1.2$ eV estimated from *ab initio* calculations when a lead with four-fold symmetry is attached to the Co atom [11], we obtain $D = 0.0017$ eV.

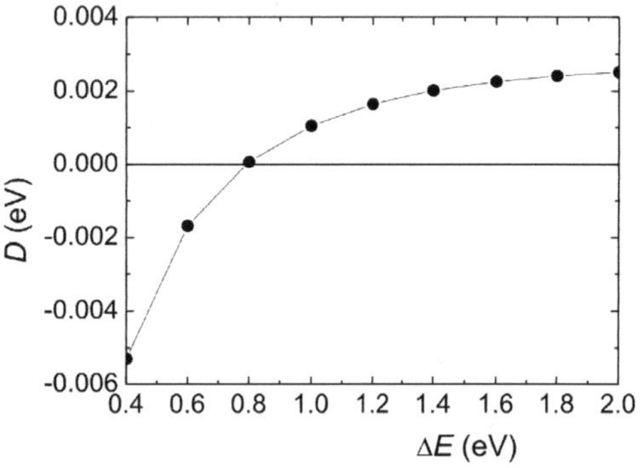

Figure 1. Splitting between states with $S_z = \pm 3/2$ and $S_z = \pm 1/2$ of the isolated Co atom in a Au chain, as a function of the splitting between Co xy and $x^2 - y^2$ orbitals

Stretching the device, the interatomic positions increase, reducing the effect of the crystal field and therefore decreasing D. This renders in principle possible to tune the system from the *overscreened* regime which takes place for $D > 0$ to the non Kondo regime for $D < 0$ through the *underscreened* regime for $D = 0$.

4. Spectral density and conductance

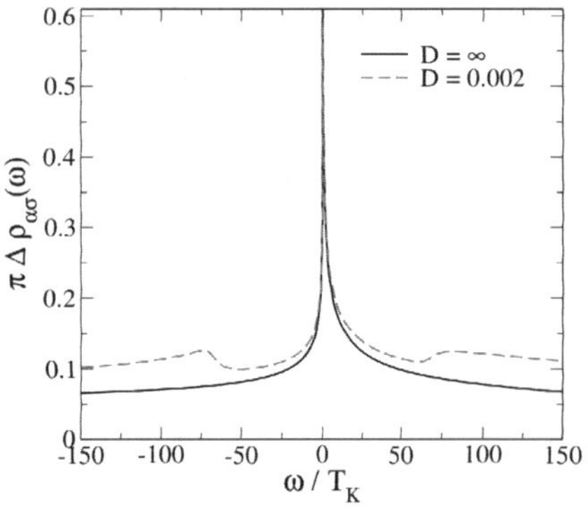

Figure 2. Spectral density per channel and spin for $D, D' \to +\infty$ (full line) and $D = 0.002$, $D' = 0$. Other parameters are $W = 1$, $\delta = 0.67$, $\Delta = 0.225$, and $T = 0.005T_K$.

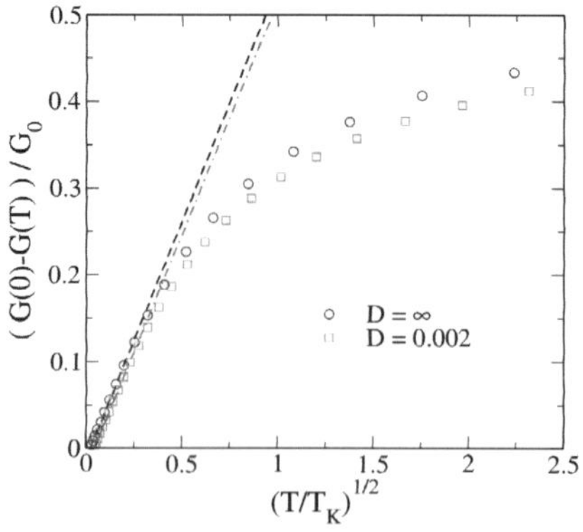

Figure 3. Conductance as a function of temperature for the same parameters as in Fig. 2. $G(0) = 1.2G_0$, where $G_0 = 2e^2/h$.

For the NCA calculations, we have taken the half band width of the conduction band $W = 1$ as the unit of energy and the resonant level width $\Delta = \pi\rho_0(V_L^2 + V_R^2) = 0.225$, where $\rho_0 = 1/(2W)$ is the density of states of the free conduction band. In Fig. 2 we compare the spectral density $\rho_{\alpha\sigma}(\omega)$ of the SU(2)×SU(2) Anderson model with that for more realistic values of the anisotropies. The main difference is that in the latter case, the Kondo temperature T_K defined from the half width at half maximum of the Kondo peak, decreases by a factor near 4, due to the decrease of the z component of the effective exchange interaction [see Eq. (2)] and two shoulders appear near energies $\pm D$ due to the effect of the $|M_3\rangle$ states with $M_3 = \pm 3/2$, which are neglected in the SU(2)×SU(2) model. For $D, D' \to +\infty$, we obtain $T_K = 2 \times 10^{-4}$, while for $D = 0.002$, $D' = 0$, $T_K = 5 \times 10^{-5}$. Note that in contrast to Fermi liquid systems, for which $\pi\rho_{\alpha\sigma}(0)\Delta = 1$ in the Kondo limit, here the correct value is 1/2. The NCA gives a value slightly larger.

In Fig. 3 we show the dependence of the conductance $G(T)$ on temperature. As before, the value $G(0) = 1.2G_0$ is nearly half the value $2G_0$ expected for a Fermi liquid with two channels. The deviation of 20% is due to shortcomings of the NCA. Moreover, the $\sqrt{T}$ dependence that $G(T)$ shows for $T < 0.2T_K$ is also characteristic of non-Fermi liquid behavior. The slope is very similar for both sets of parameters when it is scaled by the corresponding T_K

5. Discussion

We have shown that the model Eq. (1) contains the SU(2)×SU(2) Anderson model as a special case. Using the NCA we show that the non-Fermi liquid structure of the former is kept for more realistic parameters. Since the NCA can be extended to calculate the conductance out of equilibrium, this opens the possibility to study to what extent non-Fermi liquid behavior can be seen for a finite applied bias voltages V_b. In principle one expects that $G(V_b)$ has a dependence of the type $\sqrt{V_b}$ for eV_b smaller than a fraction of $k_B T_K$.

Acknowledgments

We are partially supported by CONICET, Argentina. This work was sponsored by PIP 112-200801-01821, PIP 00258 and PIP 00273 of CONICET, and PICT 2010-1060 and 2011-1187 of the ANPCyT-Argentina.

References

[1] Hewson A C, 1993 *The Kondo Problem to Heavy Fermions*, Cambridge University Press, Cambridge.
[2] Goldhaber-Gordon D, Shtrikman H, Mahalu D., Abusch-Magder D, Meirav U and Kastner M A, 1998 *Nature* (London) **391**, 156
[3] Cronenwet S M, Oosterkamp T H and Kouwenhoven L P, 1998 *Science* **281**, 540
[4] van der Wiel W G, de Franceschi S, Fujisawa T, Elzerman J M, Tarucha S and Kowenhoven L P, 2000 *Science* **289**, 2105
[5] Parks J J *et al* 2010 *Science* **328** 1370
[6] Florens S, Freyn A, Roch N, Wernsdorfer W, Balestro F, Roura-Bas P and Aligia A A 2011 *J. Phys.: Condens. Matter* **23** 243202; references therein.
[7] Allub R and Aligia A A, 1995 *Phys. Rev. B* **52**, 7987
[8] Mitchell A K, Sela E and Logan D E, *Phys. Rev. Lett.* 2012 **108**, 086405
[9] Zaránd G, Chung C-H, Simon P and Vojta M, 2006 *Phys. Rev. Lett.* **97**, 166802
[10] Bork J, Zhang Y, Diekhöner L, Borda L, Simon P, Kroha J, Wahl P and Kern K, *Nature Phys.* 2011 **7**, 901
[11] Di Napoli S, Weichselbaum A, Roura-Bas P, Aligia A A, Mokrousov Y, and Blügel S, 2013 *Phys. Rev. Lett.* **110**, 196402
[12] Di Napoli S, Barral M A, Roura-Bas P, Aligia A A, Mokrousov Y and Llois A M, 2013 *IEEE T. Magn.* **49**, 4683
[13] Cox D L and Ruckenstein A E, 1993 *Phys. Rev. Lett.* **71**, 1613
[14] Hettler M H, Kroha J and Hershfield S, 1998 *Phys. Rev. B* **58** 5649
[15] Roura-Bas P and Aligia A A 2010 *J. Phys.: Condens. Matter* **22** 025602
[16] Aligia A A and Kroll T, 2010 *Phys. Rev. B* **81**, 195113

Journal of Physics: Conference Series **480** (2014) 012014 doi:10.1088/1742-6596/480/1/012014

Synthesis and study of the optical properties of dielectric Bragg reflectors infiltrated with 6G-Rhodamine

E Gómez-Barojas[1], V Aca-López[1], E Sánchez-Mora[2], R Silva-González[2], and J A Luna-López[1]

[1]CIDS-IC, Benemérita Universidad Autónoma de Puebla, Apdo. Postal 196, Puebla, Pue. 72000. México
[2]Instituto de Física, Benemérita Universidad Autónoma de Puebla, Apdo. Postal J-48, Puebla, Pue. 72570 México

e-mail: egomez@ifuap.buap.mx

Abstract. We report the study of the optical properties of 6G-Rhodamine (Rhd) infiltrated porous silicon dielectric Bragg reflectors (DBRs) with 31 constituent periods. The DBRs were obtained by an electrochemical anodizing process of Si in a two electrodes Teflon cell. The porosity was determined by gravimetric measurements on single Porous silicon (PSi) layers. Based on the characterization results of single layers the DBRs were synthesized. After anodizing, the DBRs were silanized with a 3-mercaptopropyltrimethoxysilane solution and functionalized with Rhd solutions at different concentrations. Cross section scanning electron micrographs show that the DBRs synthesis was successful. After each preparation step, Reflectance and Fluorescence (FL) spectra were recorded. These spectra show that as the Rhd concentration in solution is increased the stop band intensity as well as the FL intensity are enhanced due to constructive interference effects.

1. Introduction

Bragg mirrors, also called distributed Bragg reflectors formed with PSi multilayers have been already suggested and demonstrated [1-3]. The most important property that the DBRs present is the high reflectivity region or photonic stop band and its wavelength range selectivity. The stop band is formed by the effect of multiple constructive interferences when a light beam is reflected by many dielectric interfaces. Infiltrating the PSi-DBRs pores with a fluorophore and tuning the DBR stop band with the fluorophore FL emission band it is possible to enhance the fluorophore FL emission intensity.

In this work we investigate the effect of infiltrating the DBRs pores with different concentrations of Rhd on the optical properties of the DBRs. These properties were studied with reflectivity- and fluorescence- spectroscopy.

2. Experimental procedure

This investigation was carried out in 2 parts. In the first part a set of single PSi layers were synthesized and characterized. In the second part, based on the experimental results obtained in the first part, a set of DBRs were synthesized by electrochemical etching. As substrate it was used polished silicon p-type wafers (1,0,0) orientation with resistivity of 0.01- 0.02 Ω·cm.

The electrochemical anodization setup is formed by an electrochemical cell and a programmable BK Precision 9130 used as a current source. The electrochemical cell consisted of: a Teflon base, a Si wafer (1.5 cm×1.5 cm) that acted as the anode, a neoprene O-ring of internal diameter of 0.91 cm, a small Teflon cell, and a Tungsten filament in spiral form that acted as the cathode. The etching solution contained Ethanol (purity: 99.5%), Hydrofluoric acid (45%) and Glycerol (purity: 99%). The PSi samples have been prepared by keeping the Ethanol:HF:Glycerol ratio at 60:30:10. A set of 7 PSi single layers were prepared in a Teflon two electrode electrochemical cell with anodizing current density (J) of 20, 45, 70, 95 and 120 mA/cm2 for 20 min.

The equipment used in this investigation was: Profilemeter Veeco model DEKTAK 150, an analytical balance AND HR-200, a Field Emission SEM, JEOL JSM-7800 F, a Spectrometer Varian Cary 100 Scan and a Varian Fluorescence Spectrometer Cary Eclipse model.

3. Results and discussion

The porosity of the set of PSi layers was determined by the gravimetric technique [3]. Then, a plot porosity vs anodizing current density of the set of single PSi layers was generated. The thicknesses of the PSi layers were calculated using the formula given by Vinegoni et al. [3], then, a plot of thickness data vs anodizing current density was done. With the thicknesses data of the samples, the etching rate defined as the thickness of the PSi layer divided by the anodizing process time, was determined. Then, the refraction indices of the set of PSi layers were determined using the equation given by Pap et al. [4].

The dielectric Bragg reflectors are narrow–band dielectric mirrors obtained by stacking periodically two layers of high refractive index (layer H) and low refractive index (layer L) whose thickness is $\lambda/4$. This means that the Bragg's relation $n_H d_H = n_L d_L = \lambda/4$ was used to design the DBRs where λ= 570 nm, the wavelength at which Rhd fluoresces [5] was chosen. Taking into account the experimental results obtained for the PSi single layers set, DBRs with 31 periods were fabricated. The chosen parameters are: the layer-H with n_H=1.81, P=61% and thickness of 0.78 μm. The layer-L with n_L=1.29, P=81% and thickness of 0.11 μm. Care was taken to form reflecting interfaces of H-layers with the adjoining media so that the reflecting beams interfere constructively. Due to the short anodizing process times a computer program to control the current source was used.

The as anodized (0.0 Rhd) DBRs were stabilized by immersion into a 2.5 % v/v a solution of 2-Propanol with 5% v/v of 3-Mercaptopropyltrimetoxysilane (MPTS) for 15 min. Then, they were functionalized by dipping them into a Rhd solution. The Rhd concentration was varied from 1.2 to 1.8 M in increments of 0.2 M also the 0.6 M concentration was included.

In figure 1 SEM micrographs of the surface morphology and of the cross section of an as anodized DBR formed with 31 periods by electrochemical etching are shown. The fabrication of this kind of multilayer structures is possible because the silicon anodization only occurs at the pore tips that is, at the interface between the Si substrate and the electrolyte so that the porous layer already formed is not affected by the subsequent conditions of fabrication.

Journal of Physics: Conference Series **480** (2014) 012014 doi:10.1088/1742-6596/480/1/012014

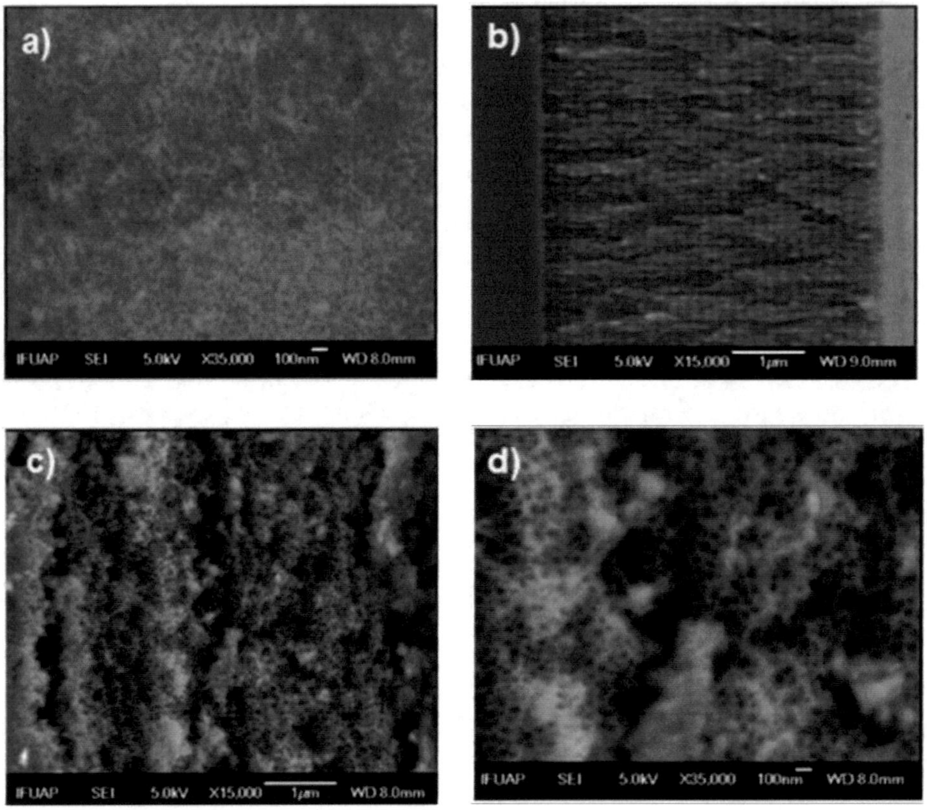

Figure 1. SEM micrographs of an as grown DBR formed with 31 periods by electrochemical etching. (a) image at x35000 of the surface morphology of the uppermost layer (b) Cross section image at x35000, it is seen clearly the constituent layers of the DBR, (c) Top down view (x15000) of a free standing piece of cross section layer of the DBR. It is seen that the fringes frontiers are not abrupt and (d) an image of the region shown in (c) at x35000 where the deep pores were focused so that they can be seen clearly in a nanometric scale.

The set of DBRs were characterized by 45° incidence reflectance measured at room temperature with an Halogen-W lamp and normalized with respect to the reflectance of an Al coated mirror. All DBRs present a stop band whose peaks are at slightly different wavelengths in the range from 478-520 nm and the amplitude of the stop bands are increased when the Rhd concentration is varied from 1.2 M to 1.8 M, see figure 2a.

Room temperature Fluorescence spectra were recorded using the λ=330 nm line of a Xe lamp as excitation source The spectrometer entrance and output slits were 10 nm wide. All DBRs present a very wide FL emission spectrum (400-800 nm) whose peaks are located at 569 nm, also it is seen that amplitude of the spectra increases as the Rhd concentration in the functionalization solution is increased, see figure 2b. The DBR functionalized with 0.6 M Rhd does not follow the tendency; the reason is that when the DBR stop band coincides with λ=570 then the FL amplitude increases due to a constructive interference otherwise a destructive interference is caused.

Journal of Physics: Conference Series **480** (2014) 012014 doi:10.1088/1742-6596/480/1/012014

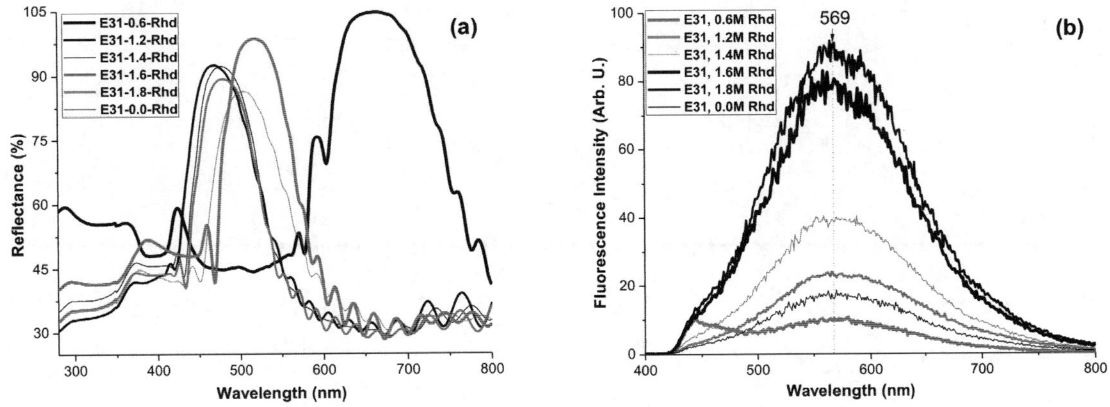

Figure 2. (a) Specular Reflectance spectra and (b) FL spectra recorded from the set of DBRs formed with 31 periods and functionalized with different concentrations of Rhd. The FL spectrum of the as anodized DBR is included for comparison.

In summary the SEM micrographs are the evidence that the BRDs were successfully synthesized by an electrochemical anodizing process, they show also interesting details of the forming layers, of their interfaces, and of the pores size. The FL spectra show that the DBRs functionalization with Rhd was successful too, since the amplitude of these spectra is increased with respect to the one of the as anodized DBR, as the Rhd concentration is increased.

4. Acknowledgments

Work partially supported by Benemérita Universidad de Puebla, México. Project No. GOBE-EXC/13-I and by SEP-CA190.

References

[1] Vincent G 1994 Optical properties of porous silicon superlattices J. *Appl. Phys. Lett.* **18** 2367
[2] Mazzoleni C and Pavesi L 1995 Application to optical components of dielectric porous silicon mutilayers *J. Appl. Phys. Lett.* **20** 2983
[3] Pavesi L 1997 Porous silicon dielectric multilayers and microcavities *Revista del Nuovo Cimento* **10** 1-76
[4] Pap A E, Kordás K, Vähäkangas J, Uusimäki A, Leppävuori S, Pilon L and Szatmári S J 2006 *Opt. Mat.* **28** 506-513
[5] Lakowicz J R 1999 *Principles of Fluorescence Spectroscopy* 2nd Ed (Kluwer Academic/Plenum Publisher, New York)

Journal of Physics: Conference Series **480** (2014) 012015 doi:10.1088/1742-6596/480/1/012015

Effect of the repulsion interaction on the ground-state of the Kondo lattice model with a superlattice potential

J Silva-Valencia[1], R Franco[1], M S Figueira[2]

[1]Departamento de Física, Universidad Nacional de Colombia, Bogotá, Colombia

[2]Instituto de Física, Universidade Federal Fluminense, Av. Litorânea s/n, 24210-346 Niterói, Rio de Janeiro, Brazil

E-mail: `jsilvav@unal.edu.co`

Abstract. We investigate the ground-state of a new Kondo lattice model, where the free carriers interact repulsively between them and undergo an external superlattice potential. This model can be simulated with ^{171}Yb atoms confined in optical lattices. We use the density matrix renormalization group method to evaluate the charge and spin gaps, and the structure factors. We found that the ground-state evolves from a Kondo spin liquid state to a charge-gapped antiferromagnetic state with zero spin gap, when the antiferromagnetic exchange increases. Also, we verify that the quantum critical point varies linearly with the repulsion and the exchange.

1. Introduction

The cold atoms community has been observed that the alkaline-earth atoms are usefull to study the interplay of spin and charge degrees of freedom, therefore emulate some Hamiltonians that describe the physical properties of heavy fermions materials [1]. The first degenerate gas of alkaline-earth atoms was achieved by Fukuhara *et al.* [2]. After this, it was possible to observe the superfluid-insulator quantum phase transition with bosonic isotopes, and a Mott-insulator state with fermionic isotopes.

Daley *et al.* [3] recently proposed that the atoms in $^1\mathrm{S}_0$ (g) and $^3\mathrm{P}_0$ (e) states can be confined independently in two different optical lattice potentials with the same periodicity. In this system, the collisions do not cause spin changes, because at low temperatures the scattering lengths are independent of the nuclear spins. Based on the above proposal, and encouraged by the possibility of studying the interplay between orbital and spin degrees of freedom, it was shown that using alkaline-earth atoms confined in an optical lattice it is possible to simulate the Kondo lattice model [4, 5], which can be obtained when the strong repulsion between e atoms is taken into account through a unit-filling constraint such that they are localized.

Using the mean field theory with a local density approximation, Foss-Feig *et al.* [6, 7] consider the free g atoms confined in a harmonic potential and interacting with localized e atoms in one and two dimensions. They modeled the system with the Kondo lattice model plus a quadratic potential term and observed that metallic regions and Kondo insulator domains with one g atom at each site coexist. A beyond mean field study of the 1D case was made using DMRG [8].

Journal of Physics: Conference Series **480** (2014) 012015 doi:10.1088/1742-6596/480/1/012015

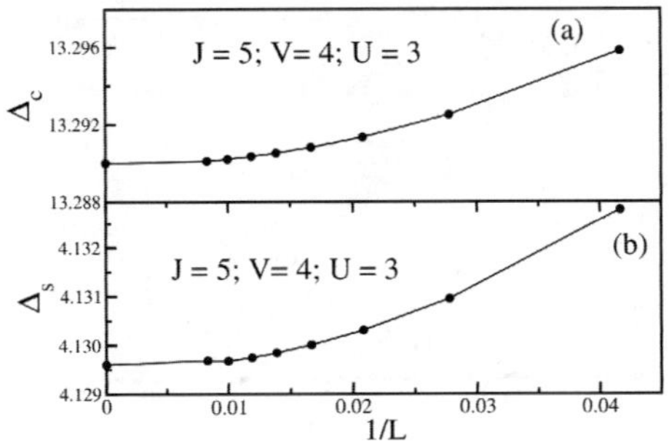

Figure 1. Charge (a) and spin (b) gaps for different lattice sizes. Here $V = 4$, $U = 3$ and $J = 5$.

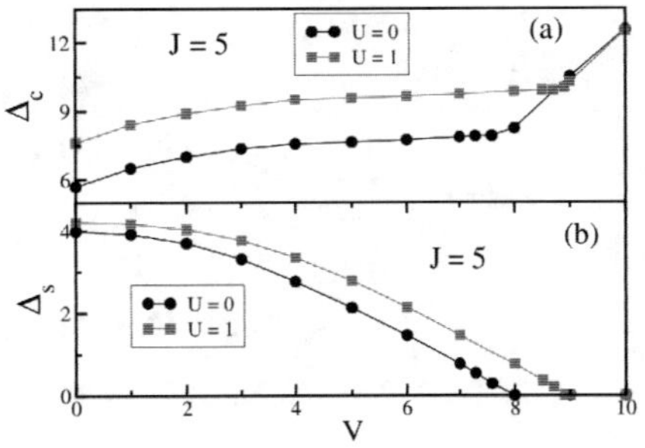

Figure 2. Thermodynamic limit of charge (a) and spin (b) gaps as a function of the superlattice potential strength.

The progress in the trapping technique allows the experimentalist to confine bosonic and fermionic atoms in different kinds of potentials: harmonic, superlattice, and anharmonic [9, 10, 11].

Here, we take into account the repulsive interaction between the g atoms, which experience a superlattice potential, while they interact locally with the e atoms through a Heisenberg type term.

2. Model and results

Using the ^{171}Yb atoms, we can emulate the Kondo lattice model plus a external potential, which Hamiltonian is given by

$$H = -t \sum_{i\sigma} \left(c^{\dagger}_{i,\sigma} c_{i+1,\sigma} + \text{h.c.} \right) + \sum_{i} \left(U n_{i,\uparrow} n_{i,\downarrow} + J \mathbf{S}_{ic} \cdot \mathbf{S}_{if} + V_i \left(n_{i,\uparrow} + n_{i,\downarrow} \right) \right), \qquad (1)$$

where, $c^{\dagger}_{i\sigma}$ creates an atom at site i in the electronic state $^1S_0$, and (nuclear) spin state $\sigma =\uparrow$ or $\downarrow$. $n_{i\sigma}$ is the local density operator of delocalized atoms with spin σ, and the spin operator of localized (delocalized) atoms is $\mathbf{S}_{if}$ ($\mathbf{S}_{ic}$). Finally, we consider a superlattice confinement potential, which is zero for even sites and nonzero (V) for odd sites of the lattice. This potential can be created through the superposition of two optical lattices of different frequencies, with the frequency of one optical lattice being double that of the other.

We set the hopping integral $t = 1$ and adopt the density matrix renormalization group method [12] maintaining $m = 400$ states per block.

In this paper, we consider a local antiferromagnetic coupling between the localized and delocalized atoms, and we analyze the case when the number of delocalized atoms matches the lattice size. Without superlattice potential ($V_i = 0 \ \forall \ i$), this system has a finite charge and spin gaps for different values of the exchange (J) and repulsion (U). Calculating the the ground state energy for the S^z sector of a lattice with L sites, we obtain the spin gap $\Delta_s = E_g(S^z = 1, N = L) - E_g(S^z = 0, N = L)$. The charge gap is given by $\Delta_c = E_g(S^z = 0, N = L+2) - E_g(S^z = 0, N = L)$. To characterize the ground-state, we calculate the charge structure factor $N(q) = \frac{1}{L}\sum_{j,k} e^{iq(j-k)} \langle n_j n_k \rangle$, and also the spin structure factor $S(q) = \frac{1}{L}\sum_{j,k} e^{iq(j-k)} \langle \mathbf{S}^T_j \cdot \mathbf{S}^T_k \rangle$, where $\mathbf{S}^T_j = \mathbf{S}_{jc} + \mathbf{S}_{jf}$ is the total spin. Without superlattice potential, the spin structure factor is nonzero and constant indicating that the ground-state is composed by a local singlet in each site, whereas the charge structure factor has a strong peak at $q = 0$ indicating the homogeneous distribution of charge in the lattice.

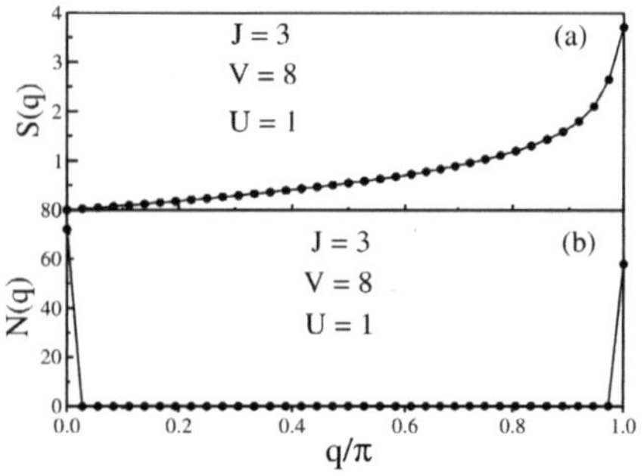

Figure 3. The spin (a) and charge (b) structure factors in the charge gapped antiferromagnetic state.

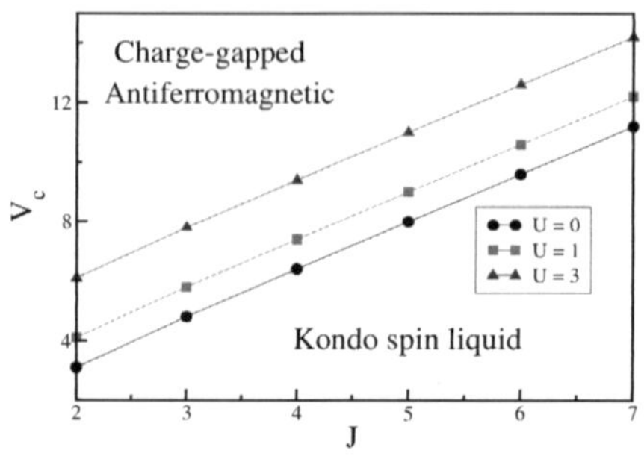

Figure 4. Phase diagram of ^{171}Yb atoms in a superlattice potential with constant repulsion interaction.

We show in Fig. 1 the spin and charge gaps for different lattice sizes of ^{171}Yb atoms in a superlattice potential of strength $V = 4$. The antiferromagnetic coupling is $J = 5$ and the repulsion between the g atoms is $U = 3$. We observed that the charge gap decreases as the lattice size increases; however, for larger sizes this behavior is slow, indicating that at the thermodynamic limit this quantity can be finite (see Fig. 1(a)). To determine the thermodynamic limit of the charge gap, we fit the curve to the following expression $\Delta(L) = \Delta(\infty) + \frac{\alpha}{L} + \frac{\beta}{L^2}$, where α and β are constants [13]. By adjusting the curve in Fig. 1(a) to the above expression, we obtain that the charge gap in the thermodynamic limit is $\Delta_c(J = 5, V = 4, U = 3) = 13.2900$, which is larger than the value without repulsion interaction $\Delta_c(J = 5, V = 4, U = 0) = 7.5624$.

In Fig. 1(b), we can see the spin gap as a function of the lattice size. Using the expression for the gap, we obtained that the spin gap at the thermodynamic limit is $\Delta_s(J = 5, V = 4, U = 3) = 4.1296$, again this value is larger than the value without repulsion interaction. Note that the spin and charge gaps increase with the repulsion when we consider the ^{171}Yb atoms confined in a superlattice potential, a similar result was found by Shibata *et al.* [13] for the homogeneous Kondo lattice model with a Coulomb interaction between the conduction electrons, however for our system the effect of the repulsion on the charge gap is stronger than on the spin gap.

For the parameters of Fig. 1, we obtained a ground-state with finite charge and spin gaps, hence this state is a Kondo spin liquid state, i. e. despite that the charge redistribution generates by the superlattice potential, the local repulsion and the strong exchange guarantee the formation of a singlet in each site of the lattice.

We are interested in finding new ground-states, i. e. to identify quantum phase transitions. For this reason, we evaluate the thermodynamic limit of the spin and charge gaps as a function of the strength of the optical superlattice potential. In Fig. 2, we show our results when the exchange and the repulsion are fixed to $J = 5$ and $U = 0, 1$ respectively. We observe that the spin gap decreases as the strength of the optical potential increases and vanishes at different values depending of the repulsion parameter(see Fig. 2(b)). Regardless the repulsion value, the system goes from a spin gap state to a spin gapless one, the critical point for $U = 1$ ($U = 0$) is $V_c(U = 1) = 9.0 \pm 0.1$ ($V_c(U = 0) = 8.0 \pm 0.1$). In Fig. 2(a), we note that the charge gap change his behavior at the critical point, after this, the charge increases linearly with the potential strength. Before the critical point the charge gap increases slowly and the diference between the curves with and without repulsion is almost the same, on the contrary to the spin gap, where the diference increases with the strength of the superlattice potential.

To understand the above results, we must to consider that the superlattice potential generates a redistribution of charge in the lattice, the even (odd) sites tends to be doubly occupied (empty) for a critical value and the localized atoms order antiferromagnetic, under this conditions the effective local coupling vanishes and the delocalized and localized chains of atoms are disconnected. The repulsion effect prevent the double occupation and we have to increase the charge redistribution to overcome the repulsion barrier.

We conclude that for any repulsion value, the system suffers a quantum phase transition at the critical point: the system goes from a Kondo insulator phase with finite charge and spin gaps to a antiferromagnetic insulator phase with only a charge gap. The latter phase is characterized by non-local coupling and a superlattice ordering of the charge, with a site doubly occupied and the neighbor site empty when the strength is that high.

In Fig. 3, we show the charge and spin structure factors for ^{171}Yb atoms in a superlattice potential in the charge-gapped antiferromagnetic state ($J = 3$, $V = 8$, $U = 1$). We observe that the spin structure factor $S(q)$ is an increasing function with a pronounced maximum at $q = \pi$, which is consistent with the fact that the spin gap is zero and the localized atoms are in antiferromagnetic ordering (see Fig. 3(a)). The charge structure factor [Fig. 3(b)] vanishes for all values of q except at $q = 0$ and $q = \pi$, indicating that we have a superlattice charge ordering, characterized by empty and fully occupied sites.

3. Conclusions

We studied the ground-state of ^{171}Yb atoms confined in a one-dimensional optical superlattice, such that it simulates a Kondo lattice model with a superlattice potential. We found a quantum phase transition when the number of delocalized atoms matches the lattice size. This transition take place from the Kondo insulator state characterized by local singlets to a charge-gapped antiferromagnetic state without singlets. The latter state is characterized by an antiferromagnetic ordering of localized atoms, non-local effective coupling and a superlattice ordering of the charge, with a site doubly occupied and the neighbor site empty. We obtained that for any value of the repulsion the dependence of the critical strength potential V_c versus J is linear (see Fig. 4). Given a repulsion value, the line divides the phase diagram into two parts: the bottom, in which the system is in the well-known Kondo insulator phase, and the top, where the system is in the charge-gapped antiferromagnetic state. The evolution of the critical points is consistent with the expresion $V_c(J, U) = 3J/2 + U$.

Acknowledgments

The authors are thankful for the support of Universidad Nacional de Colombia (under grant DIB No. 15309) and the Brazilian agency CNPq.

References

[1] Gorshkov A V *et al* 2010 *Nature Phys.* **6** 289
[2] Fukuhara T, Takasu Y, Kumakura M and Takahashi Y 2007 *Phys. Rev. Lett.* **98** 030401
[3] Daley A J, Boyd M M, Ye J and Zoller P 2008 *Phys. Rev. Lett.* **101** 170504
[4] Tsunetsugu H, Sigrist M and Ueda K 1997 *Rev. Mod. Phys.* **69** 809
[5] Shibata N and Ueda K 1999 *J. Phys.: Condens. Matter* **11** R1
[6] Foss-Feig M, Hermele M and Rey A M 2010 *Phys. Rev.* A **81** 051603
[7] Foss-Feig M, Hermele M, Gurarie V, Rey A M 2010 *Phys. Rev.* A **82** 053624
[8] Silva-Valencia J and Souza A M C 2012 *Eur. Phys. J.* B **85** 5
[9] Snoek M, Titvinidnze I, Bloch I and Hoffstetter W 2011 *Phys. Rev. Lett.* **106** 155301
[10] Nascimbene S, *et al* 2012 *Phys. Rev. Lett.* **108** 205301
[11] Jauregui R, Poli N, Roati G and Modugno G 2001 *Phys. Rev.* A **64** 033403
[12] White S R, 1992 *Phys. Rev. Lett.* **69** 2863
[13] Shibata N, Nishino T, Ueda K and Ishii C 1996 *Phys. Rev.* B **53** R8828

Journal of Physics: Conference Series **480** (2014) 012016 doi:10.1088/1742-6596/480/1/012016

Magnetotransport properties and Seebeck effect in the superconductor $FeSe_{0.5}Te_{0.5}$

J.L. Pimentel Jr.[1,2], P. Pureur[1], M.A. vila[3], R.A. Ribeiro[3]

[1]Instituto de Fsica, Universidade Federal do Rio Grande do Sul, 91501-970, Porto Alegre, RS, Brazil

[2]Instituto de Matemtica, Estatstica e Fsica, Universidade do Rio Grande (FURG), 96201-900 Rio Grande, RS, Brazil

[3]Universidade Federal do ABC, CCNH, Av. dos Estados, 5001, 09210-580 Santo Andr, SP, Brazil

E-mail: J.L. Pimentel Jr., jlpimenteljr@gmail.com

Abstract. We carried out measurements of the electrical resistivity, magnetoresistance, Hall resistivity and Seebeck effect in a highly oriented sample of the Fe-based $FeSe_{0.5}Te_{0.5}$superconductor. Complementary structural and magnetic characterizations were also performed. Our sample do not show long-range magnetic order down to 4.2 K. Superconductivity occurs with critical temperature $T_c \simeq 15$ K. In the normal phase, the resistivity versus temperature behavior mimics that of a Kondo-lattice system. The magnetoresistance, Hall coefficient and Seebeck coefficient show sign reversals. These results are discussed with basis on the combined effects from two-band conduction and weak magnetic fluctuations. Effects from superconducting fluctuations are also observed near T_c.

Keywords: Fe-based superconductors; magneto-transport properties; Seebeck coefficient
PACS: 72.15.Eb; 72.15.Gd; 74.25.fc; 74.40.Xa

1. Introduction

The discovery of the Fe-based superconductors [1] attracted considerable interest since most of the compounds having high Fe content present strong magnetic properties rather than superconductivity. Moreover, the Fe-based superconductors have some similarities with the high-T_c cuprates, mostly concerning the strong planar anisotropy, doping allowance and the occurrence of spin fluctuations. In the cuprates, these magnetic excitations lead to the pseudogap phenomenon [2]. In non-doped Fe-based parent compounds, enhanced spin fluctuations lead to spin density wave (SDW) ordering [3]. In general, doping either with donors or acceptors disrupts this long-range magnetic ordering, so that a superconducting ground state may emerge [3,4]. In many cases, however, coexistence between magnetism and superconductivity has been reported [5].

In the present communication, we report on magneto-transport and Seebeck thermal transport experiments on a highly oriented sample of $FeSe_{0.5}Te_{0.5}$. This compound does not order magnetically [6]. However, our results on resistivity, longitudinal magnetoresistance, Hall resistivity and Seebeck coefficient as functions of the temperature consistently show effects from spin fluctuations. In particular, the Hall and Seebeck coefficients reveal an energy scale for these fluctuations that we interpret as related with the pseudogap phenomenon.

2. Experimental

Highly c-axis oriented samples with nominal composition $FeSe_{0.5}Te_{0.5}$ were prepared using the self-flux method [7]. The obtained samples were characterized by x-ray diffraction and scanning electron microscopy. Both techniques confirm the highly textured microstructure of the samples, where large grains having 50 μm thickness are staked perpendicularly to the c-axis of the PbO-type tetragonal structure. Auxiliary magnetization measurements were performed in several applied fields as functions of the temperature. A weak Pauli-type susceptibility is observed above $T_c^{(mag)}$= 15.1 K, where the onset of a strong diamagnetic contribution to the magnetization is located. The resistivity, magnetoresistance and Hall effect experiments were performed on one of the samples using a conventional Physical Properties Measuring System (PPMS)$^{@}$manufactured by Quantum Design, Inc. Magnetic fields up to 9 T were applied in the orientation parallel to c-axis. The measurements of the thermoelectric Seebeck effect were carried out with a cryogen-free PPMS$^{@}$.

3. Results

The inset in Fig. 1 shows resistivity measurements as a function of the temperature for a typical sample of $FeSe_{0.5}Te_{0.5}$. At high temperatures, the resistivity $\rho(T)$ is semiconducting-like, then goes through a broad maximum centered around 150 K before decreasing strongly in temperatures below 100 K, approximately. The superconducting transition given by the maximum of $d\rho/dT$ is T_c =14.5 K. The overall behavior of $\rho(T)$ reminds that of a Kondo-lattice system, where coherent scattering by Kondo resonances produces a strong resistivity decrease in temperatures below the crossover between the single-impurity and the coherent regimes.

The magneto resistance is small and approximately linear with the field in the whole normal phase, except near T_c, where the suppression of superconducting fluctuations produces strong positive magnetoresistance. Figure 1 shows the amplitude of the magnetoresistance measured at H = 9 T as a function of the temperature. Results are normalized with respect to the resistivity at zero applied field. The magnetoresistance is positive at high temperatures, changes sign to negative at $T \sim 240$ K, then changes back to positive below $T \sim 50$ K.

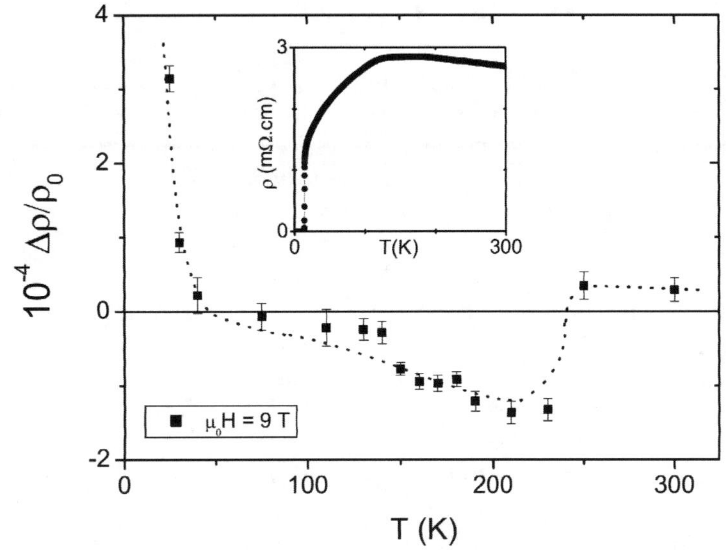

Figure 1. Intensity of the magnetoresistance, $[\rho(H) - \rho(0)]/\rho(0)$, measured at H = 9 T as a function of temperature for $FeSe_{0.5}Te_{0.5}$. The dotted line is a guide for the eyes. The inset shows resistivity versus temperature results for the same sample.

The Hall coefficient, defined as $R_H = \rho_{xy}/B$, where ρ_{xy} is the Hall resistivity and B is the magnetic induction, is plotted in Fig. 2 as a function of the temperature. Three sign reversals are observed. As for the magnetoresistance, at high temperatures R_H is positive. Decreasing the temperature, $R_H(T)$ becomes negative at T ~240 K, then changes back to positive around T ~50 K. The Hall coefficient shows a sharp maximum just above T_c. Then, upon further decreasing the temperature R_H abruptly reverses sign once more and goes through a sharp negative peak before becoming zero in the superconducting phase. Excepting for this low temperature feature, results in Fig. 2 are qualitatively similar to results previously reported for thin film samples of $FeSe_{0.5}Te_{0.5}$ [8].

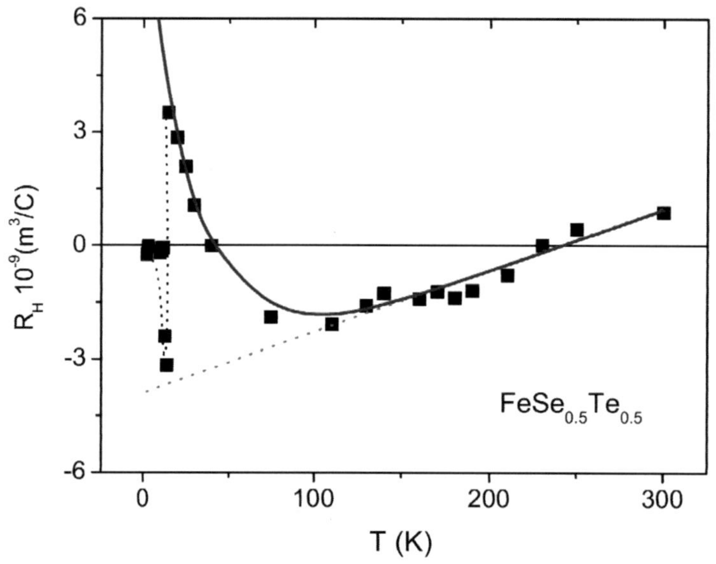

Figure 2. Hall coefficient for $FeSe_{0.5}Te_{0.5}$as a function of the temperature. The solid line is a fit based on Eq. (1). The dotted line represents the term linear in temperature in Eq. (1).

Figure 3 shows the Seebeck coefficient, defined as $S = E/\nabla T$, where E is the electrical field, as a function of the temperature for the studied system. This thermoelectric property also presents a sign reversal. In high temperatures, $S(T)$ is small and positive. Upon decreasing the temperature, S changes sign at T ~160 K, reaches a negative maximum with significant amplitude at T ~20 K , then decreases steeply to zero at T_c.

4. Discussion

The results in Figs. 1-3 strongly suggest that the electrical magneto-transport properties and the electronic contribution to thermal transport in $FeSe_{0.5}Te_{0.5}$ may be described by mechanisms combining two-current conduction with magnetic and superconducting fluctuations. The conduction by electron-like and hole-like carriers is expected with basis on the known band structure for the Fe-based superconductors [9]. Near room temperature, the conduction is dominated by the hole's mobility, as obtained from the positive R_H. Since the magnetoresistance is also positive in this region, one deduces that the Lorentz force mechanism is the most important in high temperatures. When the temperature is decreased below T ~240 K, both the Hall coefficient and the magnetoresistance become negative. This means that the suppression of spin disorder is appreciable in the intermediate and low temperature regimes. In the same temperature regions $\rho(T)$ goes through a maximum before strongly decreasing with negative curvature when temperature approaches T_c from above. We propose that spin resonances

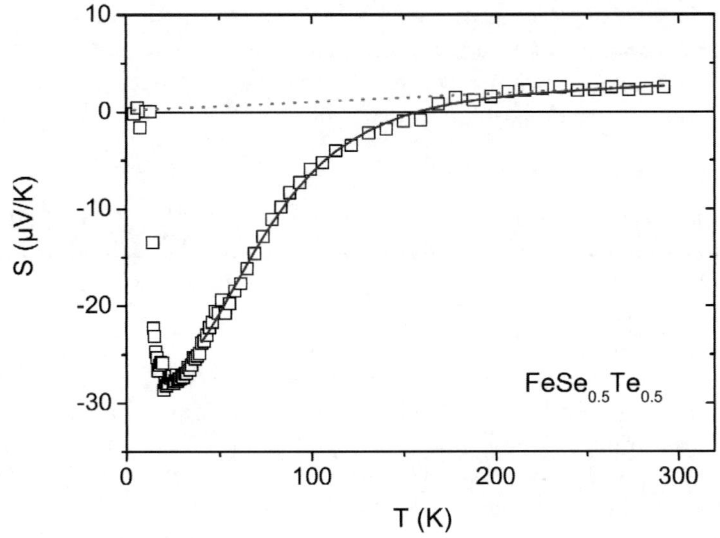

Figure 3. Seebeck coefficient versus temperature for $FeSe_{0.5}Te_{0.5}$. The continuous line correspond to a fit with Eq. (2). The dotted line is the extrapolation of the linearly T-dependent high temperature behavior.

start to built up localized magnetic momenta in the Fe sites in the intermediate temperature range. These resonances scatter carriers coherently in the low temperature regime leading to a significant drop in resistivity, as much as in Kondo-lattice systems. The same magnetic fluctuation term influences the Hall resistivity. Indeed, as shown in Fig. 2, we could describe the Hall coefficient phenomenologically as

$$R_H = a\,(T - T^*) + b\,e^{-\frac{T}{\Delta}}, \tag{1}$$

where a = 1.6 10^{-11}m^3/C·K , b = 1.3 10^{-8}m^{-3}/C, T^*= 240 K, and Δ = 32 K. We interpret the first term in Eq. (1) as resulting from the interplay between Lorentz-force-driven hole and electron mobilities [10]. The second term in Eq. (1) accounts for the fast decay of spin correlations with increasing temperature [11] or, alternatively, to progressive localization of holes in the Fe sites when the temperature is decreased. These processes are related to the coherent scattering observed in the resistivity experiment at low temperatures. The exponential term in Eq. (1) leads to a sign reversal in $R_H(T)$, from negative to positive, at T ~50 K. Near the superconducting transition, the sign reversal and deep minimum observed in R_H have been reported to occur in other superconductors [12]. The almost coincident values of the intensity of this peak and the extrapolation of the normal contribution to R_H (see Fig. 1) suggest that this controversial feature of the Hall effect might be related to remaining electron-like quasiparticles just below T_c in the case of $FeSe_{0.5}Te_{0.5}$.

The Seebeck coefficient shown in Fig. 3 may be fitted to the expression

$$S = cT + gT\,e^{-\frac{T}{\Delta}}, \tag{2}$$

where c = 8.6 $10^{-3}\mu$V/K^2, g = 2.48 μV/K^{2}and Δ = 32 K. The value for Δ is identical, within the experimental uncertainties, to the one obtained from the fit of $R_H(T)$ to Eq. (1). Using the nearly-free-electron model [10], one may show that $S \propto T\,R_H$. In view of Eq. (2), this result is consistent with the assumption that both the Hall and Seebeck effects in $FeSe_{0.5}Te_{0.5}$are related to the same damped spin fluctuations in the temperature range where $\rho(T)$ exhibits a behavior seemly consistent with the scattering of carriers by localized magnetic resonances. The cutoff energy Δ leading to the suppression of the magnetic excitations at high temperatures might be reminiscent of the pseudogap phenomenon, mostly studied in the cuprates context [2] but also expected to be relevant in the Fe-based superconductors [13].

5. Conclusions

Measurements of the electrical resistivity, magnetoresistance, Hall effect and Seebeck effect in melt-textured samples of $FeSe_{0.5}Te_{0.5}$ indicate that two-band conduction must be taken into consideration to describe the electronic transport in this material. The obtained results also show that scattering due to correlated spin fluctuations play an important role in the electrical and thermal transport properties of this Fe-based superconductor, mostly in the temperature range between $T_c$and room temperature. In particular, the observed exponential temperature dependence of both Hall and Seebeck coefficients consistently furnishes the characteristic energy scale related to the damping of these magnetic excitations. Tentatively, we propose that the relevant spin excitations in $FeSe_{0.5}Te_{0.5}$ are short-range spin density waves, having extension larger than the electron mean-free-path and life-time longer than the relaxation time of the normal electronic transport in the studied compound. Close to the critical temperature, effects from superconducting fluctuations are observed in all studied properties.

Acknowledgments

This work was partially financed by the Brazilian Agencies FAPERGS and CNPq under the Grant PRONEX 10/0009-2.

References

[1] Y. Kamihara, T. Watanabe, M. Hirano, H. Hosono, J. Am. Chem. Soc. 130 (2008) 3296.
[2] B. Fauqu et al., Phys. Rev. Lett. 96 (2006) 197001.
[3] Y. Liu, D.L. Sun, J.L. Park, C.T. Lin, Physica C 470 (2010) S513.
[4] K. Ishida, Y. Nakai, H. Hosono, J. Phys. Soc. Jpn. 78 (2009) 062001.
[5] E. Wiesenmayer et. al., Phys. Rev. Lett. 107 (2011) 237001.
[6] C. Dong et al., Phys. Rev. B 84 (2011) 224506.
[7] J.L. Pimentel Jr. et al., J. Appl. Phys. 111 (2012) 033908.
[8] I. Tsukada et al., Phys. Rev. B 81 (2010) 054515.
[9] K. Nakaima et al., Phys. Rev. Lett. 105 (2010) 197001.
[10] J.M. Ziman, in *Principles of the Theory of Solids*, 2nd. edition, Cambridge University Press, UK, 1972.
[11] L. Fanfarillo, E. Cappelluti, C. Castellani, L. Benfatto, Phys. Rev. Lett. 109 (2012) 096402.
[12] M. Galffy and E. Zirngiebl, Solid State Commun. 68 (1988) 929.
[13] H. Ikeda, J. Phys. Soc. Jpn. 77 (2008) 123707.

Microstructure AFM study and raman spectra of In-GaAs bilayers prepared by R.F. magnetron sputtering on Si(100) substrates

M. Venegas[3], R. Bernal [1, 2], M. López [3] and A. Pulzara[1]

[1]Laboratorio de Nanoestructuras Semiconductoras, Universidad Nacional de Colombia, sede Manizales. A.A. 127, Colombia.
[2]Escuela de Materiales, Universidad Nacional de Colombia, sede Medellín, Colombia.
[3]Grupo de Estado Sólido, Departamento de Física, Centro de Investigación y de Estudios Avanzados del I.P.N. Av. Instituto Politécnico Nacional No. 2508, Apartado Postal 14-740, 07000, México D.F, México.

Received 15 September 2013, in final form 18 Novembrer 2013

Abstract. Preliminary results of In-GaAs bilayers prepared by RF Magnetron Sputtering on Si(100) substrates are presented. The growth temperatures were 300 and 580 °C for the high purity targets of In and GaAs, respectively. Under an Ar atmosphere, three samples were prepared, each one with a GaAs buffer layer deposited for 60 minutes, then two cycles were prepared as follows: In layers from 5, 10, and 15 minutes with a subsequent GaAs layer deposited for 20 minutes. The morphological and optical studies of the samples were made by means of Amplitude Modulation Atomic Force Microscopy (AM-AFM) and Micro Raman Spectroscopy (μRAMAN). In order to analyze and correlate the surface morphology and alloy composition, the samples were cleaved along the [001] direction and subsequently characterized by AM-AFM and μRAMAN. From the topographic images, a statistical study of the roughness and grain size is made. Additionally cross sectional images were performed for each sample, from where the phase channel, which is sensitive to the material properties was of particular interest. The Raman spectra showed the two vibrational phonon modes related to the InAs and GaAs.

[1] Introduction

The Indium Gallium Arsenide (InGaAs) ternary alloy, often regarded as GaAs and InAs binary alloys, is a III-V semiconductor compound with an increasing demand for the micro/opto -electronic industry [1,2]. Among the several processes used to growth InGaAs, the Magnetron Radio Frequency (RF) Sputtering offers a relative low cost fabrication technique. Understanding, controlling and characterizing the growing process is necessary in order to obtain the required film that best meets the specifications. Between the several characterization techniques, AM-AFM is a powerful, flexible and economic tool that enable us to study the surface and the heterostructure morphology [3,4]. Certainly with AM-AFM in phase contrast mode it is possible to study material properties [5]. Therefore, cross-sectional analysis in semiconductor heterostructures by AM-AFM allows us to study the inner features

[6]. On the other hand, μRAMAN spectra can give us the opportunity to study and understand interesting semiconductor materials properties, like crystal orientations, carrier concentrations and mixed crystals composition [10]. For all this, we are presenting our preliminary results of the GaAs/In/GaAs/In/GaAs/Si(100) heterostructure, grown by Magnetron RF Sputtering. Our interest to fabricate In/GaAs bilayers was to obtain the InGaAs ternary allow at the interface. Having the correct In content, a required band gap for solar cells applications can be obtained. For this reason, three samples were fabricated with different In content. The surface morphology was described in terms of the statistical quantities (Maximum Height value and RMS roughness) and Grain analysis, by using the watershed algorithm [4] (Grain Population Density, Mean Grain Size and Total Grain Volume). The hetero-structure features were studied by cross-sectional imaging, obtaining the topography and the phase images. Then, the μRAMAN spectra were obtained on the same area by using 532 nm and 785 nm laser lines respectively, detecting the phonon compositional related modes to GaAs and InAs.

[2] Experimental Procedure

High purity (95.5 %) GaAs(100) and In targets where used. The Si(100) surfaces were firstly degreased with acetone and methanol. After that, they were cleaned with hydrofluoric acid at 2 %, rinsed with deionized water, dried with nitrogen and introduced into the sputtering chamber. The growth chamber was evacuated to 1.2 x 10^{-6} Torr, then back-filled with high purity Argon gas to 1.2 x 10^{-3} Torr. For each sample, with the temperature of the Si(100) substrate at 580 °C, a GaAs buffer layer was deposited for 60 minutes, then the heterostructures were prepared as follows: with a temperature of 300°C , In deposition at 5, 10 and 15 min for Sample 1 (S1), Sample 2 (S2) and Sample 3 (S3), respectively, was performed. Then, another GaAs layer was deposited during 20 minutes at 580 °C. This process is then repeated once again for each sample, in order to deposit a total of two In-GaAs films onto a GaAs buffer layer with different In content. The AM-AFM and μRAMAN characterizations where performed at ambient conditions with the N8 NEOS SENTERRA system from Bruker. This equipment combines an optics SENTERRA Raman spectrometer with the Nano's N8 NEOS AFM, to allow the morphological, structural and chemical analysis of the same area. The cantilevers used were PPP-NCLR from Nanosensors, with 190 Khz resonance frequencies, and force constants around 48 N/m. The AFM was setup to work in AM-AFM contrast phase mode, with 195 nm Free oscillation Amplitudes (FA), Set Points (SP) of 50 %, 65 % and 80 %, Oscillation Frequencies (fosc) around 169 Khz, Scanning Speeds (SS) of 0.7 l/s, 0.5 l/s and 0.2 l/s, and the Free oscillation Phase (FP) was 90° for the cantilever free oscillation. The software used to treat the AFM images were Gwyddion [7] and NanoScope Analysis [8]. μRAMAN spectra were obtained with, a green laser (532 nm at 20 mW, 350 nm spot diameter), and a red one, (785 nm at 100 mW and 550 nm spot diameter).

[3] Results and Discussions

After optical imaging inspection, it was selected an area for each of the samples, from where μRAMAN and large scales AFM images were obtained. Then, 10 x10 μm^2 areas were imaged (figure 1) to perform the grain discretization by the Gwyddion watershed algorithm [4]. The statistical quantities and grain statistics resulted are presented in table 1. The cross-section topographic and phase images, figures 2(a,g), 2(b,h) and 2(c,i), for S1, S2 and S3, respectively, are presented with their height and phase profiles. At this particular locality they presented a film thickness of about 2.16 μm for S1, 3.85 μm for S2, and 6.15 μm for S3. Note that these values are slightly larger than those observed by the Maximum heights values presented in table 1. The phase signal presented a different behavior when the tip was interacting with the Si surface (left part of the image), with the film (middle and/or right), and with the free space, after the specimen presence (right part).

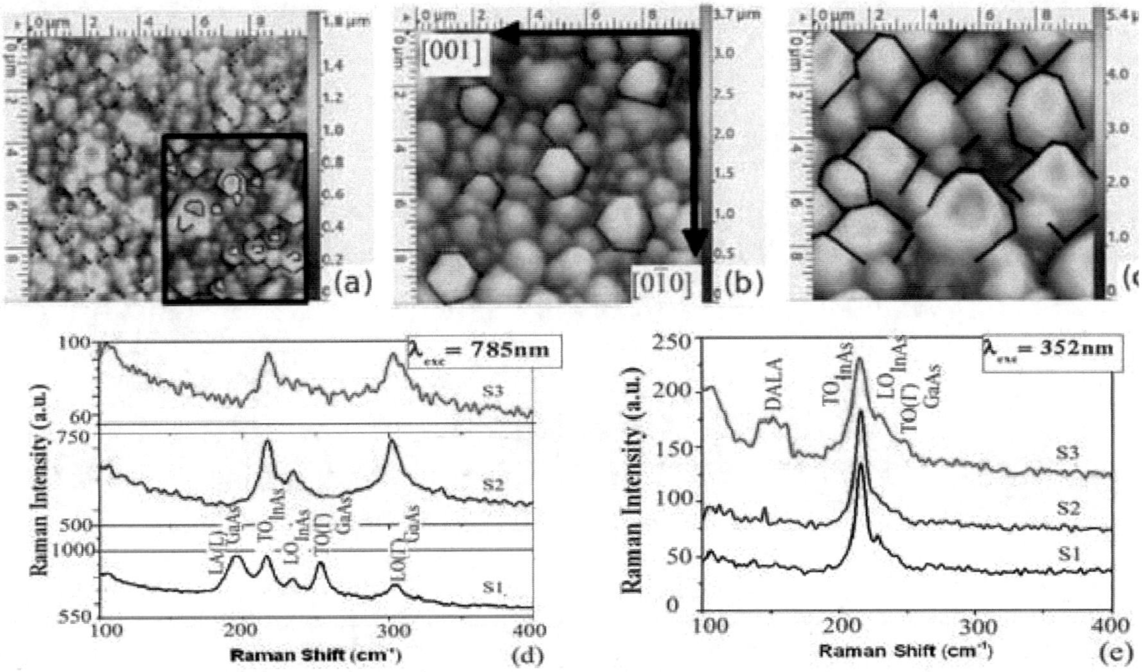

Figure 1: AFM topography images for S1, (a), SP = 50 % and SS = 0.5 l/s, The inset is an amplification of the zone below. For S2, (b), SP = 65 % and, SS = 0.6 l/s. And for S3, (c), SP = 50 %, and SS = 0.5 l/s. These images presented linear features (marked with black lines). The corresponding μRAMAN for all the samples are resumed in figures (d) and (e), for the 785 nm and 532 nm laser lines respectively.

Table 1. Statistical quantities and Grain statistics obtained from the AFM images of figure 1.

	Statistical Quantities		*Grain Statistics*		
Sample	**Maximum height [μm]**	**RMS Roughness [μm]**	**Grain population density [μm^{-2}]**	**Mean grain size [μm]**	**Total grain volume [μm^3]**
1	1.79	0.24	11.2	0.18	48
2	3.78	0.51	1.75	0.56	121
3	5.44	1.01	0.73	0.82	227

From figure 1(d), all the μRAMAM presented the two-phonon modes behaviour of first-order Raman scattering, TO and LO, of both InAs and GaAs. These spectra presented a shift of 5 cm^{-1} to lower frequencies with respect to the bulk, we are considering that these shifts are caused by the In content, due to the change of the integrated intensity, and the FWHM observed in InAs and GaAs Raman peaks. The average residual strain in the polycrystalline samples disappeared, because it is randomly distributed over the samples and then relaxed at the grain boundaries [9]. From cross-sectional SEM images (not presented here) we did not observe cracking in our samples. Additionally to the InAs and GaAs LO and TO phonon modes, the LA(A) mode centered at 195cm^{-1} is evident for S1. When the In-GaAs/In layers were excited with the 532 nm laser, figure 1(e), the GaAs-like LO phonon mode is not evident any more. However a weak signal of TO GaAs-mode can be identified, as well as the strong signal centered at 215 cm^{-1}, corresponding to a InAs TO phonon mode, with two small shoulders to the right, related to LO InAs and TO GaAs phonon modes. Finally, the band presented at 150 cm^{-1} (possibly produced by a Disorder Activated Longitudinal Acoustic (DALA) effect) increased in intensity while the In layer thickness increased.

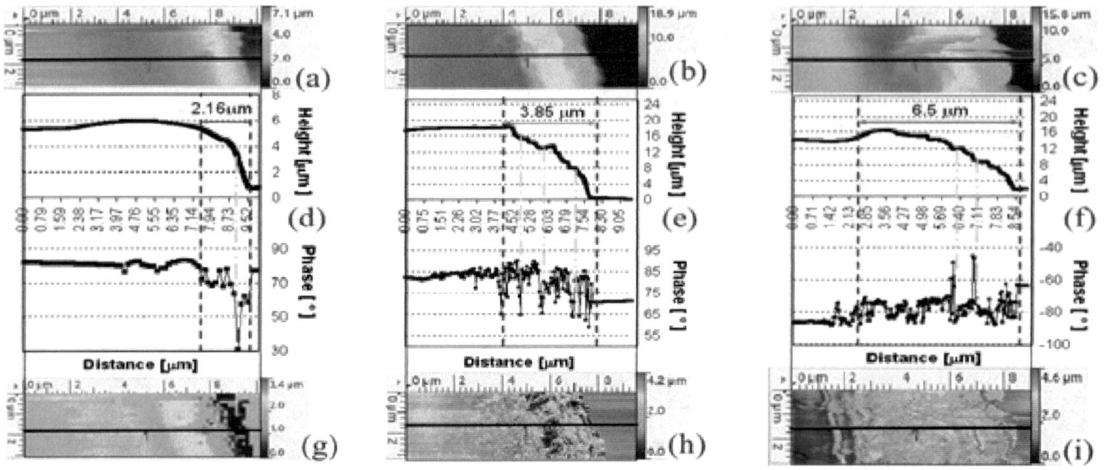

Figure 2: AFM Topography and Phase cross-sectional qualitative comparative analysis.

In summary, grain lateral dimensions and heights were increased upon larger In deposition times. Linear features with some order in certain crystallographic orientations were formed. A topography V.S. phase qualitative analysis of the specimens cross section, presented different phases values between the Si(100), the GaAs/In/GaAs/In/GaAs films and the free space. While the topography profile decreased logarithmically, the tendency of the phase signal did not present this behavior for all the samples. Raman spectra and composition analysis, presented the phonon related modes to GaAs and InAs. This Mean that the ending semiconductor compound was presented in the fabricated heterostructures. However, for the moment it is very premature to give a conclusion regarding the correct In content for a suitable InGaAs ternary alloy at the interface.

[4] Acknowledgements
We appreciate the "Instituto de Ciencia y Tecnología del Distrito Federal, (IcyTDF)" now "Secretaría de Ciencia, Tecnología e Innovación del Distrito Federal" (SECITI), for financial support. Sincere gratitude to Prof. José Trujillo and M.S. Marving Soriano. R. Bernal appreciate very much to "Colciencias – Colombia" for the financial support ("Francisco José de Caldas -2011" scholarships).

References

[1] P Borri, S Schneider, W Langbein and D Bimberg 2006 *J. Opt. A: Pure Appl. Opt.* **8** S33
[2] A J Zilkie, J Meier, M Mojahedi, P J Poole, P Barrios, D Poitras, T J Rotter, Y Chi, A Stintz, K J Malloy, P W E Smith, J S Aitchison, 2007 *IEEE J. Quantum Electron.* **43** 1873
[3] D Franta, I Ohlídal, P Klapetek, A Montaigne-Ramil, A Bonanni, D Stifter and H Sitter 2002 *J. Appl. Phys.* **92** 1873
[4] P Klapetek, I Ohlídal, D Franta, A Montaigne-Ramil, A Bonanni, D Stifter and H Sitter 2003 *Acta Phys. Slovaca* **53** 223
[5] S N Magonov, V Elings and M H Whangbo 1997 *Surf. Sci.* **375** L385
[6] C Jenkins, D I Westwood, M Elliott, J E Macdonald, C Meaton and S Bland 2001 *Mater. Sci. Eng.* **80** 138
[7] http://gwyddion.net/
[8] http://nanoscaleworld.bruker-axs.com/nanoscaleworld/media/p/2740.aspx
[9] A M Mintairov, P A Blagnov, V G Melehin, N N Faleev, J L Merz, Y Qiu, S A Nikishin, and H Temkin 1997 *Phys. Rev. B* **56** 15836
[10] G Abstreiter, E Bauser, A Fischer, and K Ploog 1978 *App. Phys.* 16 345

Journal of Physics: Conference Series **480** (2014) 012018 doi:10.1088/1742-6596/480/1/012018

Phospho-olivines as cathode materials for lithium ion battery: trends on microwave synthesis

C E Alarcón-Suesca, T Nilges

Fachgebiet Synthese und Charakterisierung Innovatiert Materialien, Chemistry Department, Technische Universität München, Lichtenbergstrasse D-85748, Garching, Germany

Email: tom.nilges@lrz.tu-muenchen.de

Abstract. In recently years, scientific research has been focused on finding new sources of efficient energy storage and lithium batteries are one of the most outstanding alternatives. Nowadays, investigation on cathode materials constitute one of the key factors of lithium rechargeable batteries studies. Recently, oxides with olivine structures has triggered the battery field, due to its excellent electrochemical properties, crystal stability, energy density, power capacity and are not expensive. However, the demand and development of more enhanced cathode materials can be achieved by the use of sophisticated synthetic methods. The purpose of this article is to provide a perspective to the materials community on the opportunities and limitations of nanostructured materials by highlighting examples in synthesis, fundamental studies and applications on microwave synthesis as novel route to obtain high voltage phospho-olivines.

1. Introduction

Since their commercialization in the early 1990´s by Sony®, Li-ion batteries (LIB) are considered an important element inside the so-called "modern electricity-based society". So far, LIB constituted the most appropriate method of energy storage: they show long cycle life, wide range of operating temperatures, high energy density, low cost and safety [1]. LIB have prompted the progress of electronic devices commercialization (mobile phones, laptops, tablets), internal combustion engines (ICE), electric vehicles (EVs) and support the storage of energy produced by renewable energy sources (RES) [2]. In currently LIB technology, the electrochemical performance (cell voltage, capacity, Li^+ transportation rate) is determined mainly by the cathode material and this justifies the intense research in this field [3]. Since report of electrochemical activity of Li*M*PO_4 (*M*=Fe, Mn) by Goodenough [4], lithium transition metals phosphates with ordered olivine structure (Phospho-olivines) has emerged as a promising cathode material due to its excellent electrochemical properties, good thermal stability, low cost and environmentally friendly [5]. Crystalline Li*M*PO_4, has an orthorhombic unit cell (D_{2h}^{16} -space group *Pmnb*), where oxygen ions form strong covalent bonds with P^{5+} to form PO_4^{3-} tetrahedral units stabilizing the entire network assuring a stable operation and high temperatures; present the so-called "inductive effect" due to the P_{tet}-O-M_{oct} linkage allowing an interaction that tunes the M^{3+}/M^{2+} redox energy to useful levels (3.4 to 5.0 V) and present high theoretical capacity (up to 170 mAh/g) that provides high energy density compared with other cathode materials such as spinel-oxides or layered oxides [6,7]. The insertion and extraction of Li^+ ions into the stable crystal lattice of the electrode is the key factor inside LIB operation. One of the main causes of poor performance of the battery is the instability of electrode materials, due to the existence of secondary phases, inhomogeneity in the compound formation, stoichiometry, crystal structure and morphology of the active material [8]. These factors are closely related and the selection of an appropriate synthesis route is the key factor to enhance the battery working. Following the report of Higuchi [9], microwave-assisted route is considered a novel processing technique to obtain nanoscale cathode materials with excellent electrochemical properties. The purpose of this article is not provide a comprehensive review; rather, this paper offers a perspective to the materials community on the microwave (mw) synthesis by highlighting examples and propose some promising developmental directions on assisted-microwave route for the synthesis of phospho-olivines for LIB cathode-materials.

2. Microwave assisted-solid state synthesis

Microwave processing method is based on a self-heating process that occurs by the absorption of electromagnetic energy; due to the use of dielectric microwave (which reduce the thermal gradients inside the reactor), a very fast and uniform heating of the total volume of reactants is obtained within a short time, resulting in very homogenous reaction conditions. Moreover, it allows the possibility to synthesized cathode materials at temperatures lower than those required in solid state synthesis using an oven [10]. The first attempts of direct microwave synthesis evolved from preliminary steps of mixing precursors, pelletizing and the use of domestic microwave ovens; in the case of $LiFePO_4$ a capacity value of 125 mAh/g was reported [9] and then improved until 161mAh/g at C/10 mixing stoichiometric amounts of Li_3PO_4 and $Fe_3(PO_4)_2{\cdot}8H_2O$ in a ball-milling jar with 5 wt.% acetylene black, pressed in a pellet and then irradiated for several minutes (2-5) at 750 W [11]. Recently reports shows alternative procedure using $FePO_4.4H_2O$ and Li_2CO_3 mixed with glucose and milling in ball-mill for 5 min at 400 rpm. Mixture is transferred in a crucible and alternatively heated at a procedure of 20 seconds with 10 seconds pause for 4 times at 1000W and 2.45 GHz without use of inert gas; the reported capacity reach 162 mAh/g at C/10 which is one of the highest discharge specific capacity reported so far [12]. The use of carbonaceous substance (such as acetylene and glucose) is a good strategy to include in mw synthesis. The carbon source acts as a room temperature susceptor that reaches a critical temperature to couple with the mw radiation; besides, carbon supplies thermal energy by heat transfer by surrounding and immersing the sample in both thermal and electric fields [8] and finally, carbon covers the precursor particles and create a layer around them to facilitate the transport of electrons between the current collector and the particle surface [13].

On the other hand, reports on formation of $LiCoPO_4/C$ nanocomposites, mixing CH_3COOLi, $(CH_3COO)_2Co$, $(NH_4)_2HPO_4$ with acetylene black in alcohol media, followed by decomposition under Ar atmosphere for 2 h at 350°C and mw irradiation (2.45 GHz, 700W) for 11 min; shows particle sizes of 150 nm and capacities of 144 mAh/g at C/10 [14]. Develop of nanosized electrode materials is an essential strategy to achieve short diffusion path lengths from the particle-core to the surface, faster ion diffusion and higher rate capability in LIB´s [15]. Synthesis of LiMnPO4 nanocrystallites deposited in nanopores of porous carbon was made using mw irradiation process. Here, LiH_2PO_4 and $Mn(COOCH_3)_2{\cdot}4H_2O$ were dissolved with a Li/Mn/P molar ratio of 1/1/1 in diethylene glycol and mixed with porous carbon with an average pore diameter of 110 nm (PC110) afterward, subjected to a mw irradiation (2.45 GHz, 400 W) for 5 and 8 min at 200 °C and 250 °C respectively and finally dried at 110 °C in air for 12 h. . Although capacity of the samples is not high so far (≈ 104 mAg/h at C/10), the formation $LiMnPO_4$ nanocrystallites-embedded porous carbons exhibited a retention capacity of 87% at 0.5 C against 0.1 C, which was comparable to that reported for $LiMnPO_4$-C coated [16].

3. Microwave assisted-solvo/hydrothermal synthesis

The requirement of uniformity conditions during the mw irradiation is fundamental to assure good electrochemical activity in the final material. The use of an adequate hydro/solvothermal settings, facilitates the synthesis of polyanionic compounds with well controlled morphology and fine crystal particles; tuned by factors such as temperature, pH, concentration of species and kinetic parameters (speed stirring) [17]. Murugan [18], present a combination of mw and solvothermal process using as starting materials LiOH and $(CH_3COO)_2Fe$ solved in tetraethyleneglycol (TEG) and mixed with H_3PO_4. The homogenous gel is placed in an Anton Paar microwave synthesis system (2.45 GHz, 400 W) machine and heated for 5 min at 300 °C. In addition, produced $LiFePO_4$ particles were coated with *p*-toluene sulfonic acid (*p*-TSA) doped poly(3,4-ethylenedioxythiophene) (PEDOT) in ethanol. TEM analysis reveals well-defined crystalline nanorod morphology of 25 ± 6 nm and a length of up to 100 nm. This material exhibits high discharge capacity (166 mAh/g) at C/10. PEDOT is used due to its redox behavior (over a wide range of potentials), relative high environmental and thermal stability and high electronic conductivity in its doped state compared to other commonly available conducting polymers. A key factor in hybrid mw-solvothermal synthesis is the use of an organic solvent; indeed polyalcohols (ethylene glycol, tetraethylene glycol) acts first as a solvent in the process and then as a stabilizer to allow limit the particle growth and prohibiting agglomeration [19]. Versatility of the method includes recent contributions on synthesis of doped samples. Goodenough [20] use mw-solvothermal method to produce $LiFe_{1-3x/2}V_{x-x/2}PO_4$ ($x \leq 0.2$) doped samples by mixing $LiOH.H_2O$, H_3PO_4, $(CH_3COO)_2Fe$ in TEG and adding $OV(OCH(CH_3)_2)_3$ at 300 °C for 30 mins. First charge–discharge curves of the samples shown values around 160 mAh/g in the range of 3.8-4.5 V. This synthetic approach demonstrated a maximum V doping of 10% without the formation of impurity; in fact, the method avoid to use conventional high-temperature routes. Finally, formation of vacancies on Fe site provide an additional conduction pathway for Li^+ ions to transfer between neighbouring 1D conduction channels along the crystallographic b axis.

Microwave-hydrothermal have gained reception and made remarkable progress due to increase in kinetic rates of reactions by one or two orders of magnitude, reduce reactions times, offers large scale batch reactions and saves energy [21]. Furthermore, control of the morphology shows an improvement in electrochemical performance as published by Ji [22]. In this study, $LiMnPO_4$ crystals were obtained using $MnSO_4 \cdot H_2O$, H_3PO_4, LiOH as starting materials in the stoichiometric ratio 1:1:3. Sodium dodecyl benzene sulfonate (SDBS) and citric acid (H_3cit) were used to affect the morphology of the products and 5 min mw irradiation at 180 °C – 300 °C and 300 W conditions were used. SEM analysis shows the formation of nanoplatelets (150 nm) orientated in the ac plane whose thickness is controlled mainly by the concentration of precursors. The plate-like morphology could be useful for fast charge–discharge performances as it offers short lengths for Li^+ ion insertion/exertion as comment previously. Addition of organic additives to limit the size shown and improve of capacity value from 44.1 to 89.0 mAh/g. On the other hand, the mw-hydrothermal was used in synthesis of high homogeneity $LiFePO_4$ with elongated parallelepiped shape in the range 0.5–1.0 mm for suitable studies onto the intrinsic defectivity, chiefly the so-called lithium iron *anti-site*, considered one of the most critical issues when envisaging electrochemical applications [23].

4. Microwave assisted-sol-gel synthesis

From the point of view of yields and use of novel strategies such as carbon-nanotubes to increase the conductivity, new mw-sol-gel synthesis offer better characteristics rather than conventional sol-gel method [24]. Gyroscope-like lithium iron phosphate/multiwalled carbon nanotubes composites has been synthesized with initial discharge capacities of 153.3, 149.6, 140.7, 135.3 and 130.1 mAh/g at 0.1, 0.2, 0.5, 0.8 and 1 C rate, respectively. Multi-wall carbon nanotubes (MWCNTs) are considered ideal materials due to hollowfiber structure and ability to be inserted in the centre of $LiFePO_4$, thus forming a conductive network, which not only provide pathways for electron transference but also lead to interparticle electronic connection [25]. Not only carbon compounds are suitable to use to make coating in phospho-olivines. Zhang [26] use mw-sol-gel method to prepare $LiFePO_4$-Li_4SiO_4 cathode material. The Li_4SiO_4 coating prevent the direct contact between $LiFePO_4$ and the electrolyte solution. The discharge capacity of the uncoated $LiFePO_4$ decayed with the increase of the discharge rates, whereas Li_4SiO_4-coated $LiFePO_4$ shown a very steady cycling behaviour. Moreover, from electrochemical impedance spectroscopy coated cathode exhibits a smaller charge-transfer the resistance (135.1 Ω) than that of the uncoated $LiFePO_4$ cathode (305.6 Ω).

5. Conclusion

Microwave processes constitute a powerful tool to synthesize phospho-olivines with a strong reduced power consumption. Microwave assisted process enables us to produce crystalline cathode particles with different shapes and sizes and has the advantage that the solvent used in the synthesis can often be converted into the carbon source upon heating. Finally it is a simple method for mass production at low cost and short time of synthesis.

6. Acknowledgment

One of the authors (Alarcón-Suesca) wish thanks to Colciencias and DAAD for the grant founding.

References

[1] Tao Z F H, Liu H, Kan X and Chen P 2011 *Open Mater. Sci. J* **5** 204
[2] Cairns E J and Albertus P 2010 *Annu. Rev. Chem. Biomol. Eng.* **1**, 299
[3] Xu B, Qian D,Wang Z and Meng J S 2012 *Mat. Sci. Eng. R* **73** 51
[4] Padhi A K, Nanjundaswamy K S and Goodenough J B 1997 *J. Electrochem. Soc* **144** 1188
[5] Zaghib K, Mauger A and Julien C M 2012 *J. Solid State Electr.* **16** 835
[6] Yamadaa A, Hosoyaa M, Chunga Sai-H, Kudob Y, Hinokumaa K, Liub K Y and Nishi Y 2003 *J. Power Sources* **119** 121
[7] Braun P V, Cho J, Pikul J H, King W P and Zhang H 2012 *Curr. Opin. Solid St. M.* **16** 186
[8] Balaji S, Mutharasu D, Sankara N S and Ramanathan K 2009 *Ionics* **15** 765
[9] Higuchi M, Katayama K, Azuma Y, Yukawa M and Suhara M 2003 *J. Power Sources* **119** 258
[10] Beninati S, Damen L and Mastragostino M 2008 *J. Power Sources* **180** 875
[11] Song M S, Kang Y M, Kim J H, Kim D Y, Kwon H S and Lee J Y 2007 *J. Power Sources* **166** 260
[12] Zou H, Zhang G and Shen P K 2010 *Mater Res Bull* **45** 149
[13] Zaghib K, Guerfi A, Hovington P, Vijh A, Trudeau M, Mauger A, Goodenough J B, Julien C M 2013 *J. Power Sources* **232** 357
[14] Li H H, Jin J, Wei J P, Zhou Z, Yan J 2009 *Electrochem. Commun.* **11** 95
[15] Lim J, Kim J, Mathew V and Kim J 2010 *Phys Scripta* **2010** T139
[16] Aono S, Urita K, Yamada H and Moriguchi I 2012 *Solid State Ionics* **225** 556
[17] Gong Z and Yang Y 2011 *Energy Environ. Sci.* **4** 3223
[18] Murugan V A, Muraliganth T and Manthiram A 2008 *Electrochem. Commun.* **10** 903
[19] Dimesso L, Spanheimer C, Jacke S and Jaegermann W 2011 *Ionics* **17** 429
[20] Harrison K L, Bridges C A, Paranthaman M P, Segre C U, Katsoudas J, Maroni V A, Idrobo J C, Goodenough J B and Manthiram A 2013 *Chem. Mater.* **25** 768
[21] Devaraju K M and Honma I 2012 *Adv. Energy Mater.* **2** 284
[22] Ji H, Yang G, Ni H, Roy S, Pinto J and Jiang X 2011 *Electrochim. Acta* **56** 3093
[23] Bini M, Ferrari S, Capsoni D, Mustarelli P, Spina G, Giallo F, Lantieri M, Leonelli C, Rizzutie A and Massarotti V 2012 *RSC Advances* **2** 250
[24] Bao S J, Liang Y Y and Li H L 2005 *Mater. Lett.* **59** 3761
[25] Zhang Y,Dua P,Wang L,Zhang A, Song Y, Li X and Lv Y 2011 *Synthetic Met.* **161** 548
[26] Zhang Q, Jiang W, Zhou Z, Wang S, Guo X, Zhao S and Ma G 2012 *Solid State Ionics* **218** 31

Characterization of ceramics materials mixed with Co_3O_4

A F Guzmán[1], D A LandínezTéllez[2], J Roa-Rojas[2,*] and F Fajardo[1]

[1] Grupo de Estudios de Materiales – GEMA, Departamento de Física, Universidad Nacional de Colombia, AA 5997, Bogotá DC, Colombia

[2] Grupo de Física de Nuevos Materiales, Departamento de Física, Universidad Nacional de Colombia, AA 5997, Bogotá DC, Colombia

E-mail: jroar@unal.edu.co

Received 4 October 2013, in final form XX Novembrer 2013

Abstract. We have performed the preparation, structural, electrical and mechanical characterizations of ceramic materials composed of kaolinite $Al_2(Si_2O_5)(OH)_4$ and alumina (Al_2O_3) mixed with different concentrations of cobalt oxide (Co_3O_4). Ceramic samples were prepared from a base concentration of alumina 30% and kaolinite 70%, mixed with various concentrations of cobalt oxide in steps of 4% up to a value of 20%. The samples were sintered by the standard solid-state reaction method at a temperature of 1350 °C. In all samples with cobalt was found the presence of mullite. It was determined that alumina and cristobalite decreased when the cobalt concentration was increased due to the formation of the cobalt spinel. In order to determine the crystal structure of the samples, crystallographic analysis from X-ray diffraction experiments and also the semi-quantitative phase analysis were performed. Results were compared with theoretical parameters through the PowderCell 2.4 software. By increasing the concentration of cobalt oxide was found a significant increase in the resistance of materials to friction wear and a small decrease on the mean value of the dielectric constant. Through flexion measurements is observed the increases of the elasticity modulus by about 45% for the sample with 4% of cobalt oxide when compared with the samples without cobalt.

1. Introduction

Mullite ($3Al_2O_3.2SiO_2$) is a ceramic material of high resistance to thermal shock, which is characterized by very good physical and mechanical properties to be used in optical and electrical applications and as refractory, among others [1]. From the 80s the study of this material has been intensified in order to find an appropriate way to make the production process and sintering at a low cost and with high purity [2]. Different sintering processes have been used to achieve this goal, as for example, the methods of solid state reaction [3], sol-gel [4] and chemical vapor deposition [5]. These methods allow the sintering process but in most cases required high purity precursor oxides and the process is quite expensive, which is not viable for industrial production [6]. When it is done the sintering process of a low cost material such as kaolinite, primary mullite is obtained at specific temperatures and after the nucleation process the SiO_2 is released but different phases are present [7,8]. The interaction of cobalt oxide with zinc oxide on surface of kaolin is also interesting for applications as protective colour coatings [9]. Another important topic of study is to see how to modify the electrical, structural and morphological characteristics of ceramic materials when mixed with certain oxides such as Co_3O_4.

The aim of this work is to show the study of the reaction between the kaolinite and alumina at a sintering temperature of 1350 °C by the method of solid-state reaction and the effect of using cobalt oxide (Co_3O_4) as an additive in the mixture. The phase identification was performed from X-ray diffraction patters and the semi-quantitative analysis is carried out by considering the different concentrations of cobalt oxide. Electrical properties are measured by testing electric polarization. Mechanical properties of the material were investigated measuring the strength to flexion and hardness using the abrasive wear test.

2. Experimental

A fixed concentration of 70% kaolinite and 30% alumina was taken for the production of the samples in order to determine the effect of adding cobalt oxide in the sintering process of kaolinite and alumina. Then different percentages of Co_3O_4 were incorporated from 4 to 20% in increments of 4%. For all concentrations a total mass of *5 g* was used in the mixing process of the precursor oxides. Samples were pressed to form pellets of 2.5 cm diameter. For the flexion test some pellets with rectangular shape (0.3x2x8 cm) were prepared. Sintering of the samples was performed at 1350 °C during 4 hours, to reach this temperature a ramp with rise and fall times of 15 hours was employed in all cases. X-ray diffraction patters were obtained using the X-Pert PRO MPD Panalytical diffractometer, employing CuKα radiation (1.540598 Å) Crystallographic parameters were determined for each phase using the Rietveld refinement and compared with theoretical parameters through PowderCell software version 2.4. The weight concentration of each of the phases present in each sample depending on the concentration of Co_3O_4 was also determined. For measuring the bending of the sintered specimens an equipment SHIMADZU Universal Testing Machine was used, fitted with an AG-IS 5 kN load cell. For the abrasive wear test an experimental setup was used consisting of a motor with a disk on which is placed sandpaper having 180 aluminum oxide granules per square centimeter. When the disc is rotated at a determined angular velocity the sample is placed in contact with the abrasive surface for a period of 15 s. Polarization curves were measured through a Radiant Technologies Polarimeter applying electric fields from 10 to +10 kV/cm.

3. Results and discussion

Figure 1 shows the XRD pattern of the samples with different concentrations of Co_3O_4, the x-ray diffraction of the Co_3O_4 powder is also displayed. For the sample with 0% of cobalt oxide (70% kaolinite and 30% alumina) semi-quantitative analysis shows the presence of 64 wt% mullite which has an orthorhombic structure, 27 wt% alumina which has a rhombohedral structure and a 9 wt% of silicon oxide which correspond to a tetragonal crystal system. The characteristic peaks of mullite were found at the 2θ positions 16.44°, 26.33° and 35.27°. The characteristic peaks of alumina were found in the 2θ positions 25.58°, 35.27° and 43.34°. Peaks corresponding to cristobalite were identified at the 2θ positions 21.70°, 31.00° and 42.60°.

By comparing the diffraction patterns of samples with Co_3O_4 concentrations of 4, 8, 12, 16 and 20% as showed in figure 1, a reduction of the cristobalite peak located at the 21.70° was observed while the characteristic peak of mullite sited at 26.33° remains invariant. From the concentration of 8% of Co_3O_4 it was observed the appearance of three peaks at diffraction angles of 36.76°, 59.16° and 65.01° which are characteristics of the cobalt spinel. The results presented suggest that for Co_3O_4 concentrations higher than 4% the cobalt oxide in the mixture is highly reactive in the presence of alumina forming cobalt spinel which has a cubic crystal structure. Considering the peak of greatest intensity for each of the phases present in the samples, we obtained an average grain size of 54, 69 and 56 nm for the mullite, alumina and the cobalt spinel, respectively.

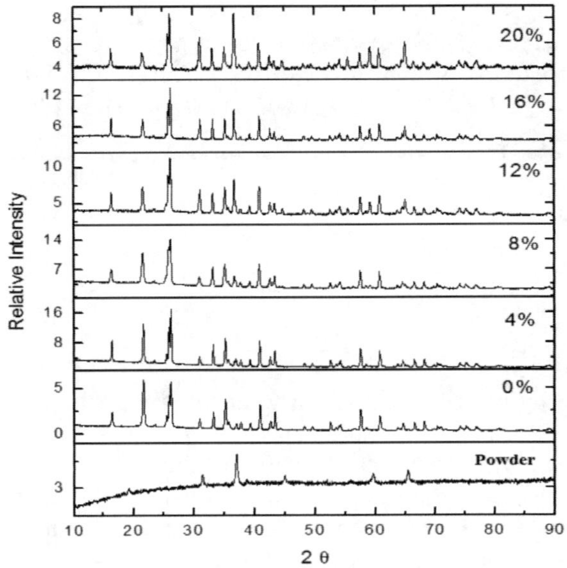

Figure 1. Comparison between the X-ray diffraction patterns of samples prepared at different Co_3O_4 percentages. The XRD for the Co_3O_4 powder is also displayed. The sample with 0% of Co_3O_4 corresponds to 70% of kaolinite and 30% of alumina.

From the semi-quantitative phase analysis is found that the weight percent of mullite after the nucleation process is approximately constant at all concentrations of cobalt oxide, as shown in figure 2. Note that the weight percentage of alumina and cristobalite phases decreases as cobalt oxide increases as a result of the formation of cobalt spinel.

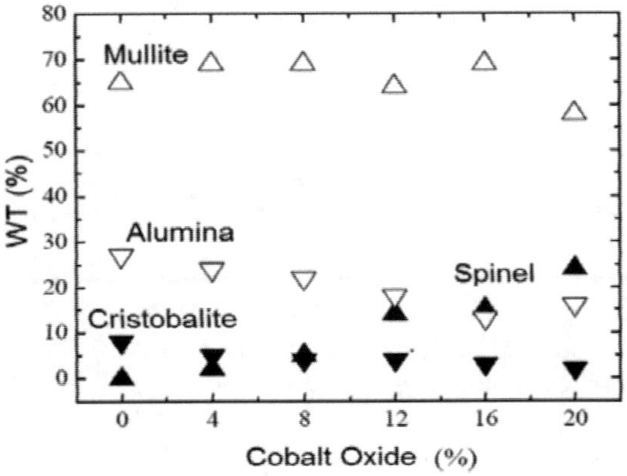

Figure 2. Quantitative analysis of the weight percentage of the phases present as a function of the concentration of cobalt oxide in samples sintered at 1350 °C.

Figure 3 shows the hysteresis behavior of polarization curves obtained for the sample with a concentration of 12% of cobalt oxide, showing a typical ferroelectric feature. A similar behavior was observed for all the percentages of cobalt oxide used in this work. Based on the applied voltage and the saturation polarization we obtained an average value of 26.50 for the relative dielectric constant. We found that with the increase of the Co_3O_4 in the samples is observed a small change in the dielectric constant, going from 27.82 to 26.30 for the samples with 0% and 20% of Co_3O_4, respectively.

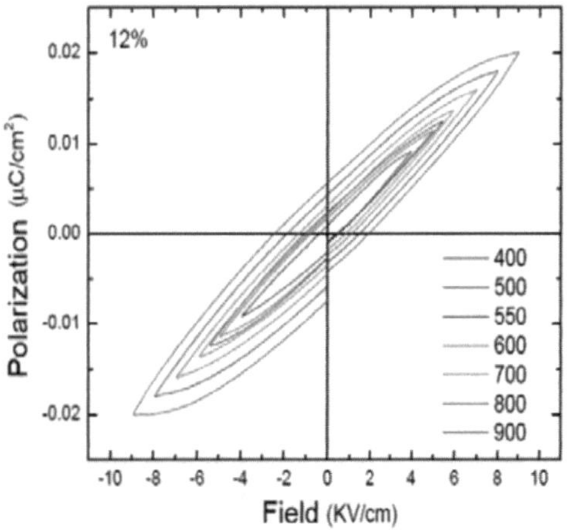

Figure 3. Hysteretic behavior of polarization as a function of applied field for a sample with 12% of cobalt oxide (Co3O4). The inset shows the applied voltage.

The analysis of the material hardness was done by means of a friction wear test for each of the samples. Figure 4 shows the wear of the ceramic pieces as a function of the percentage of cobalt oxide in the mixture. It was observed that there is a significant increase in the hardness of the material when the cobalt oxide is introduce in the mixtures as compared to the sample with 0% (70% kaolinite and 30% alumina). One can infer that the increased wear resistance occurs because the reaction of alumina and cristobalite with the Co_3O_4 during the sintering process produce the formation of the cobalt spinel.

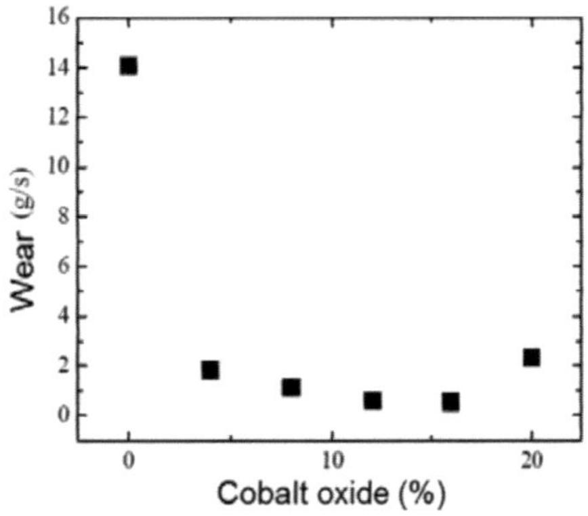

Figure 4. Frictional wear as a function of percentage of cobalt used to manufacture samples at a sintering temperature of 1350°C.

For the flexion test we used the three-point method. The distance between the supports was *40 mm*, the lower supports having a diameter of *5 mm* and the upper support has a diameter of *4 mm*. The pre-charge used for calibration of the equipment was 0.6 N. Bending measurement was performed for sintered samples with 0 and 4% of cobalt oxide.

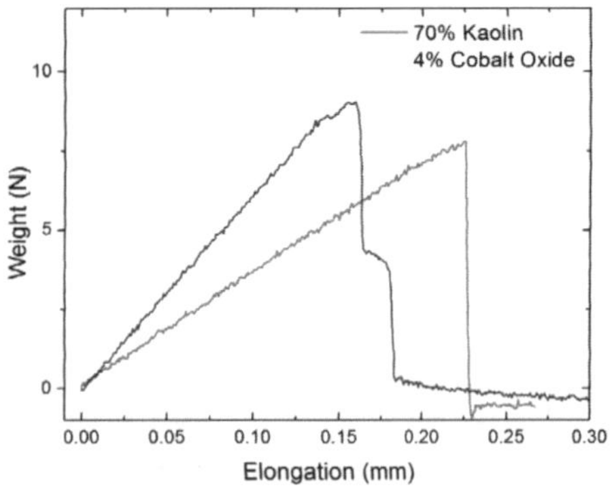

Figure 5. Applied load as a function of elongation for samples with O% (red) and 4% (blue) of Co_3O_4. The sample with 0% corresponds to 70% of kaolinite and 30% of alumina.

Figure 5 shows the applied load as a function of the elongation for two of the ceramic samples. Note that for the samples with 0 and 4% of Co_3O_4 in the mixture the maximum load that resists is 7.78 N and 11.28 N, respectively. This small increment on the maximum load is consistent with the increases of the material hardness when the percentage of Co_3O_4 is increased. Table 1 shows the maximum load applied, the maximum displacement and the value of the elasticity modulus for the samples shown in figure 5. We found that the sample with 4% of Co_3O_4 shows an increase of 45% in the elastic modulus as compared with the sample without cobalt oxide.

Table 1. Values of the maximum applied load, the maximum displacement and the elastic modulus for the characterized samples.

Co_3O_4 (%)	F_{Max} (N)	D_{Max} (mm)	Elas. Mod. (N/mm^2)
0	7.78	0.225	3774
4	11.28	0.205	5496

4. Conclusions

Was made a study of the effect of adding cobalt oxide on the structural, electrical and mechanical properties of samples prepared from a mullite matrix at a sintering temperature of 1350 °C. By XRD technique was determined the formation of cobalt spinel with a cubic structure when the Co_3O_4 cobalt oxide on percentages greater than 4% is incorporated on a ceramic matrix of 70% of kaolinite and 30% of alumina. The weight percent of mullite phase remained relatively constant when the percentage of $C0_3O_4$ varies from 4 to 20%. From the polarization curves we found a typical ferroelectric feature. Mechanical properties of the material, such as hardness and elastic modulus, are improved considerably when the cobalt oxide is incorporated on the ceramic.

5. Acknowledgments

This work was partially financed by DIB (Hermes code 13893) and “El Patrimonio Autónomo Fondo Nacional de Financiamiento para la Ciencia, la Tecnología y la Innovación Francisco José de Caldas” Contract RC No. 0850-2012”.

References

[1] L. Aksay, D. Dabbs, Mullite for Structural, Electronic, and Optical Applications. *J. Am. Ceram. Soc.***74**, 2344 (1991).
[2] R. Heimann, Classic and Advanced Ceramics, Wiley-VCH Verlag, Weinheim, Germany (2010).
[3] C. Chen, G. Lan, W. Tuan, Preparation of mullite by the reaction sintering of kaolinite and alumina. *J. Eur. Ceram. Soc.* **20**, 2519 (2000).
[4] D. Amutharani, F. Gnanam, Low temperature pressureless sintering of sol–gel derived mullite, *Mater. Sci. Eng.* A **264**, 254 (1999).
[5] Y. Hirata, I.A. Aksay, Processing of mullite powders processed by CVD, in Ceramic Transaction: Mullite and Mullite Matrix Composites, *Am. Ceramic Society*, Westerville, OH, **6** 287 (1990).
[6] H. Schneider, J. Schreuer, B. Hildman. Structure and properties of mullite. *J. Eur. Ceram. Soc.* **28**, 329 (2008).
[7] A. Rossíni, S. Arazi, Mullitización de Mezclas de Arcilla Caolinítica e Hidróxido de Aluminio. *Boletín de la Sociedad Española de Cerámica***5**, 579 (1970).
[8] B. Meng, J. Peng, J. Effects of in Situ Synthesized Mullite Whiskers on Flexural Strength and Fracture Toughness of Corundum-Mullite Refractory Materials, *Ceramics International***39**, 1525 (2013).
[9] Nivin M. Ahmed, Hesham T. Abdel-Fatah, Elham A. Youssef, Corrosion studies on tailored ZnCo aluminate/caolin core-shell pigments in alkyd based paints. *Progress in Organic Coatings***73** 76 (2012).

Journal of Physics: Conference Series **480** (2014) 012020 doi:10.1088/1742-6596/480/1/012020

Heat treatment influence on the structural and magnetic properties of the intermetallic $Fe_{56.25}Al_{43.75}$ alloy prepared by mechanical alloying and arc-melted

J. S. Trujillo Hernández*, J. A. Tabares and G. A. Pérez Alcázar

Departamento de Física, Universidad del Valle, Meléndez, A. A. 25360, Cali, Colombia.

*e-mail: trujillohernandezjuansebastian@gmail.com

Abstract. Alloys of the $Fe_{56.25}Al_{43.75}$ system were prepared by mechanical alloying (MA) using a high energy planetary ball mill, with milling times in the range from 12 up to 96 h named MA0 samples. The sample milled for 48 hours was heat treated at 700 °C for 9 days. Then this sample was milled for times of 1, 4, 8, 12, 24, and 48 h, named MA1 samples. Additionally, and for comparison, it was prepared a $Fe_{56.25}Al_{43.75}$ sample by arc-melting method. For all samples, the structural and magnetic study was conducted by X-rays diffraction (XRD) and Mössbauer spectrometry (MS). The XRD results show that the system is nanostructured and the MA0 samples present only the BCC disordered phase, whose lattice parameter remains relatively constant with milling time. For MA1 samples it was identify the FeAl, Fe_3Al, FeO and α-Fe phases. The Mössbauer spectra for all samples were fitted by using a hyperfine magnetic field distribution (HMFD), and a paramagnetic site for all the times used here. The ferromagnetism increases when milling time increases, and this is a consequence of the structural disorder induced by mechanical alloying.

1. Introduction

The Fe-Al binary alloy had been studied during many years due to its wide range of magnetic and structural properties and its potential industrial applications [1]. The magnetic properties of the FeAl alloys are strongly linked with their structural order [2]. It has been observed for alloys prepared by techniques such as mechanical alloying [3, 4], that the increase of their ferromagnetism is associated to the increase of their structural disorder. Most of the reported studies have been performed in ordered B2 (the simple cubic FeAl) alloys, and in particular on $Fe_{60}Al_{40}$ [5]. Recent studies have shown that when the FeAl alloy is thermally treated a change in the structure of the alloy and a paramagnetic site associated with the increase in thermally induced order were found [2]. The present work aims at the characterization of the structural and magnetic properties of the $Fe_{56.25}Al_{43.75}$ alloy samples produced by MA using different milling times, and by arc-melted. Also, we pretend to see how the structures and magnetic properties change with milling times and annealing temperature. In addition, we selected this composition because it corresponds to the B32 super lattice structure and their magnetic properties are easy simulated by the Monte Carlo method.

2. Experimental procedure

In this work we have prepared $Fe_{56.25}Al_{43.75}$ alloy by mechanical alloying, using high purity Fe and Al elemental powders (more than 99.7% from Alfa® AESAR), which were milled with times from 12 up to 96 h. The powders were mixed and then sealed in previously evacuated stainless steel jars in a high energy planetary mill Fritsch - Pulverisette 5 system equipped with balls of the same material. The ratio of the balls/powders weights was 20:1. To avoid overheating and oxidation, the process was performed evacuating the jails and then alternating 1 h milling followed by 1 h resting. The sample milled during 48 h was selected for thermal treatment at 700 °C during 9 days and then mechanically alloyed at 1, 4, 8, 12, 24 and 48 hours. Additionally, and for comparison, it was prepared a $Fe_{56.25}Al_{43.75}$ sample by arc-melted method. This sample was synthesized by a thermal treatment at 540°C during 10 hour using a vacuum of $6.0x10^{-2}$ mbar. All the prepared samples were characterized by MS and XRD measurements. Mössbauer measurements were performed at room temperature with transmission geometry using a $^{57}Co(Rh)$ source of 25 mCi. All the spectra were fitted with the MOSFIT program [6], taking an α-Fe foil as the calibration sample and using a HMFD and in some cases with the addition of a paramagnetic site. The X-ray measurements were obtained at room temperature for all samples using a PANalytical diffractometer equipped with a Cu target X-ray tube and a Ni filter to select the Cu-*Kα* radiation. X-ray profiles were collected from 20-90° in scattering angle using a 0.02° step and counting times of 2 s per point. The patterns were refined by the Rietveld method combined with Fourier analysis to describe the broadening of the lines using the Maud program [7].This yielded the average values of the lattice parameter, crystallite size and the structural phase.

3. Results and discussion

3.1 X-ray diffraction results

Fig. 1 (left) shows the evolution of the XRD patterns with milling time for all the MA0 samples. All the patterns shows a space group Fm3m which correspond to the BCC structural phase of FeAl with mean lattice parameter of 2.900 ± 0.001 Å, where It has also been observed by Zamora et al. [4], in particular for Al concentration up to 40 at.% [5]. For all the milling times the crystallite sizes are of nanometric order and exists a decreasing tendency in the mean crystallite size, this is a consequence induced by the increase of the milling time (Table 1). In Fig. 2 (left) are shown the XRD patterns for the synthesized MA1 samples. It can be noted in this figure that the angular position of the diffraction peaks allow us to identify the simple cubic (SC) FeAl, Fe_3Al, FeO and α-Fe phases. Also, some additional diffraction peaks are present that could tentatively associate to the ε phase reported in the equilibrium Fe-Al structural phase diagram [8], and this may be due to the stoichiometry used in this work and the heat treating employed. The Fig. 2 also shows that for 48 h only see α-Fe BCC phase, this is because the intensity of diffraction peaks vanish. It is attributed to the disorder induced by the increase of the milling time, very similar to the result obtained when sample was mechanically alloyed using elemental powders only (MA0 samples). Figure 3 (left) shows the XRD analysis for the melted and thermal treatment sample, and it allows us to identify the simple cubic (SC) FeAl ordered phase with space group Pm3m and lattice parameter of 2.900 ± 0.001 Å, the thermal treatment performed on the melted, showed an increase in the crystallite size of 30.3 nm.

3.2 Mössbauer results

Fig. 1 (right) shows the Mössbauer spectra of the MA0 alloyed samples and it is evident that they exhibit a ferromagnetic behavior with broad lines showing their disordered character, in accord with the report of Zamora et al. [4]. All the MS spectra were fitted with HMFDs, and for those of the 12 and 24 h samples it was necessary to add paramagnetic lines (a doublet and a singlet for 12 h and a doublet for 24 h) attributed to Fe sites rich in Al. For 48, 72, and 96 h the spectra are ferromagnetic due the structural disorder induced by MA which increase the ferro-ferro interactions and confirming that structural disorder promotes the ferromagnetism, as it was pointed above [3,4]. Table 1 shows the Mössbauer parameters obtained and it can be noted a decrease of the isomer shift and an increase of the mean hyperfine field (MHF) with the milling time, respectively, associated with the disorder generated by MA. These results are in accordance to Zamora et al. [9] and Kobeissi [10]. This magnetic disorder of the sample contributes to increasing in this way the mean number of Fe atoms around the Fe site and the microscopic consequence is the increase of the mean number of *s* electrons in the Fe nucleus [11, 12]. The increase of the ferromagnetic character is noted in the MHF values which change from 22 to 28 T. The presence of the paramagnetic sites for samples milled during 12 and 24 h was reported by Bohórquez et al. [13] in Fe-Al disordered samples obtained by melting. The doublet parameters, $\Delta = 0.14$ mm s^{-1} and $\Gamma_{exp} = 0.32$ mm s^{-1} are very similar to those reported by him. Fig. 2 (right) presents the Mössbauer spectra of the MA1 samples. For the 0 h sample a magnetic ordering grade is evident because the lines are narrower. After 1 h of milling it seems that some degree of magnetic disorder is induced because the lines are broad. The spectra fitting were performed with a doublet ($\Delta = 0.48$ mm s^{-1} and $\Gamma_{exp} = 0.50$ mm s^{-1}), and a HMFD. In Table 2 are shown the obtained Mössbauer parameters for the MA1 samples. It can be noted that the MHF increase with the milling time, and this is again associated to the disorder generated by the MA. It was necessary to add a paramagnetic site to fit the spectra for MA1 samples. This is a doublet with $\Delta = 0.48$ mm s^{-1} and $\Gamma_{exp} = 0.50$ mm s^{-1}, and for the spectra of 24 and 48 h the doublet presents $\Delta = 0.95$ mm s^{-1} and $\Gamma_{exp} = 0.30$ mm s^{-1} associated to the appearing of the FeO type oxide. The doublet obtained for samples milled up to 12 h is associated to the simple cubic (SC) FeAl ordered phase, detected by XRD. This phase is induced by the thermal treatment as was also discussed by Zeng et al. [7]. The doublet which appears at 24 and 48 h is associated to the FeO (wüstite) phase. Figure 3 (right) shows the Mössbauer spectra of the melted and thermal treated sample. For this sample an ordering is evident because the lines are narrower. The fit of the spectra were performed with a singlet, a sextet and a HMFD. The Mössbauer parameters of the singlet are $\Delta = 0.0$ mm s^{-1}, $\Gamma_{exp} = 0.30$ mm s^{-1} and the value of the MHF is 17 T.

Table 1. Structural parameter and Hyperfine parameters: isomer shift (IS), mean hyperfine field $<B_{hf}>$, quadrupole splitting (QS) of the MA0 samples.

Structural and Hyperfine Parameter						
Time [h]	a ± 0.001 [Å]	Size Crystallite [nm]	Component	$<B_{hf}>$ ± 0.2 [T]	IS ± 0.05 [mm/s]	QS ± 0.05 [mm/s]
12	2.910	16.1 ± 5	Singlet	---	0.01	
			Doublet	---	0.26	0.14
			Distribution	16.9	0.13	
24	2.898	16.4 ± 5	Doublet	---	0.27	0.57
			Distribution	20.2	0.12	
48	2.910	15.1 ± 5	Distribution	22.2	0.11	
72	2.903	15.7 ± 5	Distribution	26.7	0.08	
96	2.899	13.3 ± 5	Distribution	25.1	0.09	

Table 2. Hyperfine parameters: isomer shift (IS), mean hyperfine field $<B_{hf}>$, quadrupole splitting (QS) of the MA1 samples.

Hyperfine Parameter				
Time [h]	Component	$<B_{hf}>$ ± 0.2 [T]	IS ± 0.05 [mm/s]	QS ± 0.05 [mm/s]
0	Doublet	---	0.33	0.48
	Distribution	17.5	0.19	
1	Doublet	---	0.23	0.48
	Distribution	18.0	0.17	
4	Doublet	---	0.27	0.48
	Distribution	19.5	0.21	
8	Doublet	---	0.15	0.48
	Distribution	20.9	0.13	
12	Doublet	---	0.26	0.48
	Distribution	22.0	0.17	
24	Doublet	---	0.28	0.95
	Distribution	25.3	0.08	
48	Doublet	---	0.28	0.95
	Distribution	27.6	0.21	

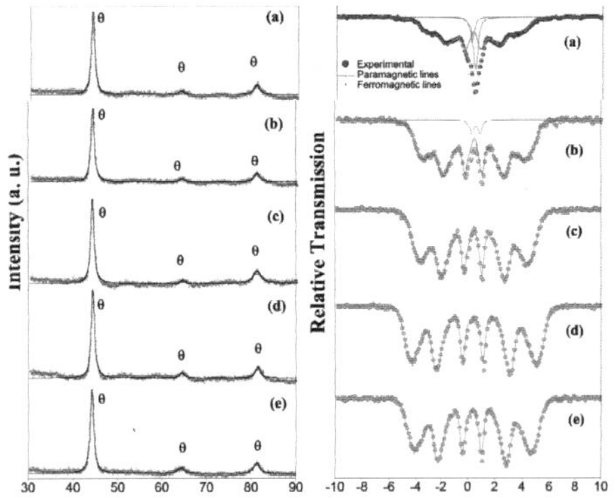

Figure 1. XRD patterns of MA0 samples, showing the BCC FeAl (θ) peaks (*left*). Mössbauer spectra of these samples (*right*).

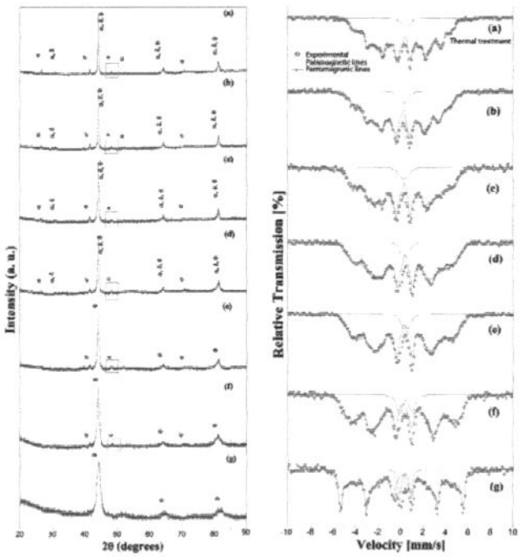

Figure 2. XRD patterns of MA1 samples, showing the Fe_3Al (**α**), α-Fe (**θ**), FeO (**σ**), FeAl (**δ**), ε Phase (**ε**), peaks (*left*). Mössbauer spectra of these samples (*right*).

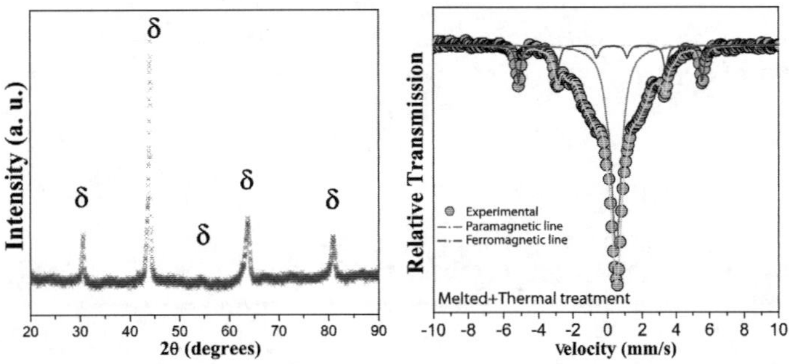

Figure 3. XRD pattern of melted sample, showing the simple cubic (SC) FeAl (**δ**) phase (*left*). Mössbauer spectra of the melted sample (*right*).

4. Conclusions

XRD results of all MA0 samples show a BCC disordered FeAl phase. Additionally, exists a decreasing tendency in crystallite size associated with the increase of the milling time. For the MA1 samples the results allow us to identify the simple cubic (SC) FeAl, Fe_3Al, FeO and α-Fe and the ε phases. For the sample corresponding to 48 h only see α-Fe BCC phase, this is because the intensity of some peaks vanish and this is attributed to the increase of the milling time. For the melted and thermally treated sample allows us to identify the simple cubic (SC) FeAl ordered phase. MS study of the MA0 samples showed a decrease of the isomer shift and an increase in the MHF with the milling time. For the MA1 and melted samples the results show a tendency to magnetic ordering with a doublet associated to the simple cubic (SC) FeAl phase obtained after the heat treatment, and the increase of the MHF with the increase of the milling time.

5. Acknowledgments

This work was supported by CENM (Center of Excellence for Novel Materials).

References

[1]. Martin Eggersmann, Helmut Mehrer 2000 *Philosophical Magazine A.* **80** 1219-1244.
[2]. Q. Zeng, I. Baker 2006 *Intermetallics* **14** 396-405.
[3]. S. S. Nayak, M. Wollgarten , J. Banhart, S. K. Pabi, B. S. Murty 2010 *Materials Science and Engineering* **A527** 2370-2378.
[4]. Ligia E. Zamora, G. A. Pérez Alcázar, G. Y. Vélez, J. D. Betancur, J. F. Marco, J. J. Romero, A. Martínez, F. J. Palomares, and J. M. González 2009 *Phys. Rev. B* **79** 1.
[5]. X. Amils, J. Nogués, S. Suriñach, J. S. Muñoz, M. D. Baró, A. Hernando, and J. P. Morniroli 2001 *Phys. Rev. B* **63** 052402.
[6]. F. Varret, J. Teillet. Unpublished MOSFIT Program.
[7]. L. Lutterotti, D. Chateigner, S. Ferrari and J. Ricote 2004 *Thin Solid Films* **450** 34-41.
[8]. U. R. Kattner.: Binary phase diagram of Al-Fe related literature [On-Line]. Illinois Institute of Tehcnology. http://tptc.iit.edu/index.php?option=com_content&view=article&id =222& Itemid=19. (2010). [cited 8 september 2010].
[9]. Ligia E. Zamora, G. A. Pérez Alcázar, J. A. Tabares, A. Bohórquez, J. F. Marco and J. M. González 2010 *J. Phys.: Condens. Matter* **12** 611-621.
[10]. Kobeissi M. A. 1991 *J. Phys.: Condens. Matter* **3** 4983.
[11]. G. Y. Vélez, G. A. Pérez Alcázar, Ligia E. Zamora, j. j. Romero, A. Martinez 2010 **195** 179-184.
[12]. V. V. Ovchinnikov 2006 *Mössbauer analysis of the atomic and magnetic structure of alloys* Cambridge International Science Publishing Ltda. **1** 12-13.
[13]. Bohórquez A, Tabares J. A, Pérez Alcázar G. A, and Gancedo J. R. 1994 *Hyperfine Interact.* **83** 311-14.

Journal of Physics: Conference Series **480** (2014) 012021 doi:10.1088/1742-6596/480/1/012021

Effect of argon ion bombardment on amorphous silicon carbonitride films

R G S Batocki[1], R P Mota[1], R Y Honda[1], D C R Santos[2]
[1] UNESP, Faculdade de Engenharia, Cep 12516-410, Guaratinguetá, SP, Brazil.
[2] CEETEPS, Faculdade de Tecnologia, Cep 12455-010, Pindamonhangaba, SP, Brazil.

E-mail: deborah.santos@fatec.sp.gov.br, rmota@feg.unesp.br.

Abstract. Amorphous silicon carbonitride (a-SiCN:H) films were synthesized by radiofrequency (RF) Plasma Enhanced Vapor Chemical Deposition (PECVD) using hexamethyldisilazane (HMDSN) as precursor compound. Then, the films were post-treated by Plasma Immersion Ion Implantation (PIII) in argon atmosphere from 15 to 60 min. The hardness of the film enhanced after ion implantation, and the sample treated at 45 min process showed hardness greater than sixfold that of the untreated sample. This result is explained by the crosslinking and densification of the structure. Films were exposed to oxygen plasma for determining of the etching rate. It decreased monotonically from 33 Å/min to 19 Å/min for the range of process time, confirming structural alterations. Hydrophobic character of the a-SiCN:H films were modified immediately after ion bombardment, due to incorporation of polar groups. However, the high wettability of the films acquired by the ion implantation was diminished after aging in air. Therefore, argon PIII made a-SiCN:H films mechanically more resistant and altered their hydrophobic character.

1. Introduction

Plasma-polymerized HMDSN films are widely employed in different applications due to attractive properties, such as transparency, biocompatibility, chemical inertness and electrical resistance, among others. However, for applications in sliding conditions, they cannot provide mechanical resistance to wear. In this sense, post-treatment such as plasma immersion ion implantation (PIII) is an advisable process, due to its operational and economical viability [1-3]. PIII is a known technique for surface treatment of metal, semiconductor and polymeric materials, but few works report its application to treatment of plasma-polymerized films. PIII process is performed in the same plasma system used to deposit plasma polymers, being the samples biased with high negative voltage pulses. Plasma ions attracted by target potential are driven to polymeric surface, and the interactions between ions and polymeric structure may be more intense than those induced on other kind of material [1-3]. Consequently, the surface properties of plasma-polymerized films may be completely altered. It knows that the degree of the modifications depends on the plasma parameters (power, pressure, ion specie, etc.) and pulse characteristics (shape, length, magnitude, frequency) [1-3]. In this work, it was investigated the effect of argon PIII on wettability, mechanical properties and oxidation resistance of the plasma-polymerized HMDSN. The surface properties alterations were correlated with the modifications in the chemical structure promoted by the ion implantation.

2. Materials and Methods

PECVD was performed in a cylindrical stainless chamber with two horizontal plate electrodes adjusted capacitively. The upper electrode is connected to RF power supply (13.56 MHz), and the lower one is ground and used as substrate holder. Power, pressure and time adjusted for the deposition of the plasma-polymerized HMDSN films were respectively of 80 W, 40 mTorr and 30 min. PIII was performed in the same reactor using a power of 50 W, argon pressure of 40 mTorr and sawtooth-like pulses of 30 kV-100 Hz. PIII treatment ranged from 15 to 60 min. After ion implantation, as-deposited and treated samples were exposure to oxygen discharges (50 W, 40 mTorr, and 30 min) to determination of the etching rate. Alpha-Step Tencor profilemeter was used to evaluate the thickness of the films after all the processes. Atomic chemical structure of the film surfaces was investigated by X-ray photoelectron spectroscopy (XPS), and molecular structure was estimated via deconvolution of XPS high resolution spectra. Hardness of the samples was evaluated by a Hysitron Triboindenter.

3. Results and discussions

Fig. 1 (a) shows the atomic concentration of chemical species identified in C 1s spectrum of the as-deposited and treated films. As can be seen, carbon concentration drops while oxygen concentration increases with process time. It means that ion bombardment induced C – C bond scission and C atom desorption. Besides, C – H bonds may be broken because hydrogen is light and weakly bound [4,5]. As result, free radicals or pendant bonds left in the polymeric structure are passivated by reactions with oxygen from environment. However at longer treatment time both carbon and oxygen atomic concentration trend to saturation. Moreover, silicon concentration was not altered during the bombardment, which indicates that silicon atoms are strongly bound in the polymeric chain. Nitrogen concentration is lower than others and diminishes after the implantation. It is likely that volatile nitrogen compounds were formed and etched from the surface during the bombardment [6].

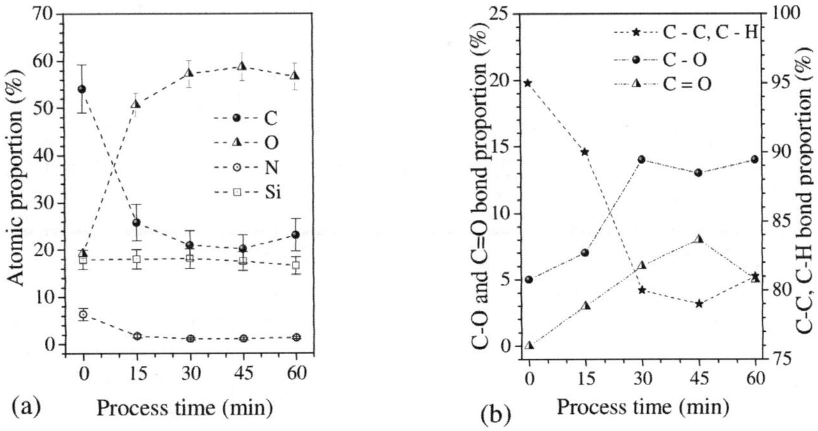

Fig. 1. C, O, N, Si atomic concentrations (a) and chemical bond concentration (b) of the plasma polymerized HMDSN films. As-deposited sample is shown in $t = 0$.

Recombination processes are confirmed in Fig. 1 (b), in which are depicted the chemical bond concentration. As can be observed, C – C or C – H bonds decreased with treatment time, as discussed above. Saturation behaviour is perceived after 30 min of process, indicating the stabilization of the chemical structure. During the implantation, interactions between ions and solid promote electronic and nuclear collisions. The former is responsible for crosslink and the later causes bond breakage or polymeric degradation [4]. In the beginning of implantation, free radicals are spaced from each other and the scissions are predominant. Increasing the time of process, free radical recombinations become more intense, leading to crosslink of the chains. At longer treatment, however, scissions restarts and the polymer degradation is inevitable [5]. Thus, it is believed that treatments longer than 15 min favour the structure strength, while treatment longer than 60 min may not contribute to enhancement of any surface property. Oxygen-containing groups, C – O and C = O, increases with treatment time. C – O species are sp^3 hybrid states and are related to crosslink of the chain [6]. As can be seen, such species also trend to saturation after 30 min of process. C = O bonds, on the other hand, decrease in 60 min of process. Although the bombardment favours unsaturated processes, the decrease of C = O bonds may be an indication about the beginning of degradation of plasma-polymer films. Thus, it is expected the improvement of the structure rigidity at treatment time shorter than 60 min.

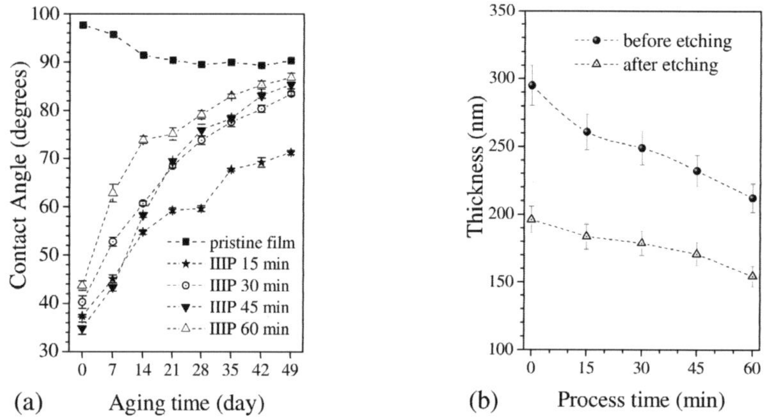

Fig. 2. Temporal evolution of contact angles (a) and thickness of the pristine ($t=0$) and PIII-treated plasma-polymerized HMDSN films (b).

Fig. 2 (a) shows the contact angle measurements as function of the aging time in atmosphere air. Plasma-polymerized HMDSN films are typically hydrophobics [7], which is evidenced by contact angle of 98°. Upon bombardment, this measure decreases drastically with treatment time, which is explained by incorporation of oxygen-containing groups, as informed by chemical analysis. After aging, however, the sample treated for 15 min shows the character more hydrophilic than others, presenting a contact angle around 70°, while the reminiscent samples present a contact angle larger than 80°. This behavior is interpreted as a reorganization of the structure after ion implantation. The incorporation of oxygen species in the sample surface generates a gradient of polar group concentration between the surface and the bulk of the polymer film, altering its enthalpy and entropy [8-10]. As result, the mobile polymeric chains tend to bury the polar groups of the surface into the bulk, which diminishes the surface wettability. In this sense, it is believed that hydrophobic recovery depends on the degree of crosslink, which causes the anchoring of the chains and may avoid their movement. But, it does not justify the lower recovery rate for the sample treated at shorter time. This is probably due to chemical and morphological alterations [9,10]. In the Fig. 2(b) is depicted the thickness of the films before and after exposure to oxygen plasma. As can be seen, the thickness of the samples tends to decrease monotonically after ion implantation, which is explained by densification of the samples due to increase of the chain crosslink. Moreover, argon ions usually induce ablation, which also contribute to decrease of the thickness [10]. As can be observed in the same figure, the thickness of the oxygen plasma-exposed films diminished at lower rate, which means that ion implantation enhanced the chemical resistance of the plasma-polymerized films. Fig. 3(a) shows the etching rate of the samples, indicating the improvement of the chemical resistance up to 57%. This is explained by the rigidity enhancement of the polymeric structure.

Hardness of plasma-polymerized HMDSN films as function of treatment time is depicted in Figure 3 (b). An improvement of sixfold (3.2 GPa) is observed in sample treated within 45 min of processing. It is interpreted as consequence of a highly connected network formed by polymeric chain crosslinking, improving dimensional stability of the structure, as well as mechanical and tribological properties [4-6,9,10]. The crosslink increase was detected by XPS analysis, where C – O bond concentration increased almost threefold within 15 min of implantation, but unsaturated bonds, such C = O, also contributes to stability of the chains. However, it is believed that longer treatment time does not favor the improvement of crosslink due to beginning of polymeric degradation[6,10].

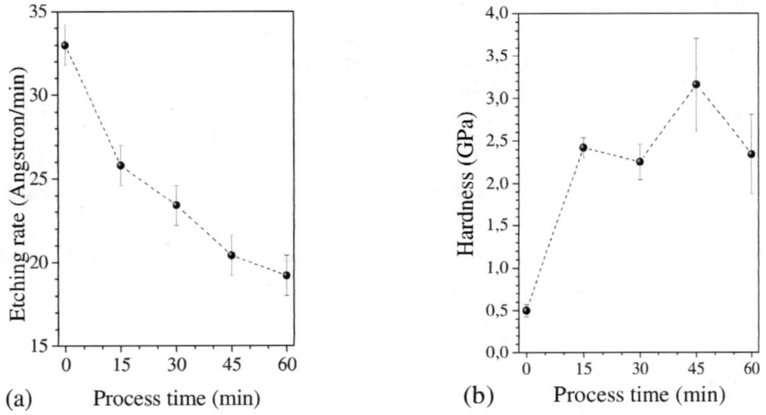

Fig. 3. Etching rate (a) and hardness (b) of the PIII-treated plasma-polymerized HMDSN film

4. Conclusions

Argon PIII is an efficient method to modify the molecular chemical structure of plasma-polymerized HMDSN films and their surface properties. Chemical analysis shows that ion implantation promoted the increase of sp^3 hybrid states and, as consequence, improved the crosslinking of polymer chains. Crosslinking and unsaturated bonds enhanced hardness and oxidation resistance, but they were not enough to avoid the hydrophobic recovery. All of the samples acquired a hydrophilic character upon implantation, but this hydrophilicity diminished after aging in air. In general, the modification induced by argon ion implantation on plasma-polymerized HMDSN films depends on treatment time.

Acknowledgements

The authors would like to thank the INPE (National Institute for Space Research - Brazil) for the thickness measurements and Professor Elidiane Cipriano Rangel, from UNESP/Sorocaba - Brazil, for the nanoindentation measurements.

References

[1] Nastasi M, Mayer J W and Hirvonen J K 1996 *Ion-Solid Interactions, Fundamentals and Applications* (Cambridge University Press)

[2] Burakowski T and Wierzchon T 1999 *Surface Engineering of Metals – principles, equipment, technologies* (Florida, CRC Press LLC)

[3] Rimini E 1995 *Ion Implantation: basics to device fabrication* (Massachusetts, Kluwer Academic Publishers)

[4] Dong H and Bell T 1999 State-of-the-art overview: ion beam surface modification of polymers towards improving tribological properties *Surf. Coat. Technol.* **111** 29

[5] Lee E H, Rao G R and Mansur L K 1996 Super-hard-surfaced polymers by high-energy ion-beam irradiation. *Elsevier Science, Trends in Polymers* **4-7** 229

[6] Santos D C R, Rangel R C C, Mota R P, Cruz N C, Schreiner W H and Rangel E C 2004 Modification of plasma polymer films by ion implantation *Mater. Res.* **7-3** 493

[7] Bankovic P, Demarquette N R and Silva M L P 2004 Obtention of selective membranes for water and hydrophobic liquids by plasma enhanced chemical vapor deposition on porous substrates *Mater. Sci. Eng.* B **112** 165

[8] Siow K S, Brietcher L, Kumar S and Griesser H J 2006 Plasma Methods for the Generation of Chemically Reactive Surfaces for Bimolecular Immobilization and Cell Colonization - A Review *Plasma Process. Polym.* **3** 392

[9] Rangel E C, Santos N M, Bortoleto J R, Durrant S F, Schreiner W H, Honda R Y, Rangel R C C and Cruz N C 2011 Treatment of PVC using an alternative low energy ion bombardment procedure *Appl. Surf. Sci.* **258** 1854

[10] Santos D C R, Mota R P, Honda R Y, Cruz N C and Rangel E C 2013 Plasma-polymerized acetylene nanofilms modified by nitrogen ion implantation *Appl. Surf. Sci.* **275** 88

Crystallographic and Electronic Structure of the $Sr_3Sb_2CoO_9$ Triple Perovskite

W. González[1,2], R. Cardona[2], D. A. Landínez Téllez[1], J. Roa-Rojas[1],*

[1] Grupo de Física de Nuevos Materiales, Departamento de Física, Universidad Nacional de Colombia, AA 5997, Bogotá DC, Colombia

[2] Grupo de Estudios de Materiales, Departamento de Física, Universidad Nacional de Colombia, AA 5997, Bogotá DC, Colombia

E-mail: jroar@unal.edu.co

Abstract. Compounds perovskites are materials with physical and chemical characteristics that make them optimal for application in the technological and scientist. When the ideal formula of perovskite ABO_3 is modified by introducing a special structural arrangement can get to get triple perovskites, which correspond to the formula $A_3B_2B'O_9$. In this work we report the synthesis process and the study of electronic structure and crystal $Sr_3Sb_2CoO_9$ new triple perovskite. From the experiments of X-ray Diffraction and the application of the Rietveld refinement method was revealed that the system crystallizes in a perovskite structure with a characteristic triple given by the space group Immm (#71) and lattice parameters a=9.791(9) Å, b=5.656(7) Å and c=16.957(8) Å. Ab initio calculations of density of states (DOS) and electronic structure were carried out for this perovskite-like system by using the Quantum EXPRESSO code. The exchange-correlation potential was treated using the Generalized Gradient Approximation (GGA). All calculations were carried-out using spin polarization.

Keywords: Ceramic materials; crystal structure; electronic properties.

1. Introduction

Ceramic materials represent a great percentage of systems which are actually investigated by the physics and chemistry of solids. Particularly, the perovskite family has concentrated important attention in last decades. From the point of view of chemical composition, perovskites have characterized by the ideal formula ABX_3, where A generally is an alkaline earth element, B represents a transition metal or rare earth element and X, more of times, is the oxygen [1]. Modifications of atomic radii of A and B, introduces structural distortions and new crystalline phases. Inclusions of rare earth elements instead the alkaline earth in the A site, and magnetic elements in the B site, give the possibility to produce materials with exotic electric and magnetic properties [2-3]. Partial substitutions of the A and B cations give rise to complex materials as the triple perovskites $A_3B_2B'O_9$ [4]. It chemical configuration supplies multiple chances to combine different elements, generating the possibility to synthesize new materials, which involve a more large gamma of physical properties. Depending on magnetic and electric characteristics of B and B' it is relatively easy to create new perovskite systems with half-metallic properties [5], magnetoelectric response [6] or magnetic ordering [7], which offer promissory perspectives in the new spintronics technology [8].

In order to analyze the possibility to create new magnetic perovskite materials, in this work we report the synthesis, the structural characterization and the electronic properties of the new triple perovskite $Sr_3Sb_2CoO_9$, which was idealized as the introduction of the alkaline earth Sr in the A' site, the transition metalloid Sb in the B location and the magnetic element Co in the B' site of the $A_3B_2B'O_9$ formula, to construct a triple complex perovskite.

2. Experimental

The samples were synthesized by the solid state reaction recipe. The precursor powders $SrCO_3$, Co_3O_4 and Sb_2O_3 (Aldrich 99.9%) were mixed in stoichiometric proportions according to the chemical formula $Sr_3Sb_2CoO_9$. Mixture was ground to form a pellet and annealed at 900 °C for 24 hours. The samples were then regrinded, repelletized and reannealed at 1000 °C for 24 hours. At last, the sample was sintered at 1100 °C for 24. X-Ray diffraction (XRD) experiment was performed by means a PW1710 diffractometer with $\lambda_{CuK\alpha}$=1.5406 Å. Rietveld refinement of diffraction pattern was carried out by the GSAS code [9].

3. Theoretical calculations

The calculations have been performed using the Quantum ESPRESSO code [10], which is based on the density functional theory (DFT) [11]. The electronic exchange-correlation potential was calculated using the generalized gradient approximation (GGA) based on the Perdew–Burke–Ernzerhof expression [12]. The electron-ion interaction was treated by using Vanderbilt ultrasoft ab-initio pseudopotentials [13], with the following valence electron configuration, Sr($5s_2$), Sb($5s_2$ $5p_3$), Co($3d_7$ $4s_2$) and O($2s_2$ $2p_4$). The irreducible Brillouin zone was sampled using the Monkhorst–Pack scheme [14]. The atomic-projected DOS were calculated by the Lowdin populations [15].

4. Results and Discussion

The analysis of XRD pattern showed in figure 1 reveals the presence of characteristic peaks of complex perovskite systems. In figure 1, crosses represent the experimental data and line corresponds to simulated pattern by means of GSAS code. Base line is the difference between theoretical and experimental results. Rietveld refinement permitted to establish that this material crystallizes in a monoclinic triple perovskite with space group $P2_1/c$ (#14) and lattice constants are a=9.791(9) Å, b=5.656(7) Å, c=16.957(8) Å and tilt angle β=125.339°. These results show deviation of no more than 3% when compared with the theoretical values obtained from the Structure Prediction Diagnostic Software SPuDS [16], which predicts that lattice constants a=9.608(3) Å, b=5.478(4) Å, c=16.808(0) Å and tilt angle β=125.645° for the $Sr_3Sb_2CoO_9$ material. Parameters of refinement are: R_F^2=0.1407; x^2=1.145 and R_{WP}=0.

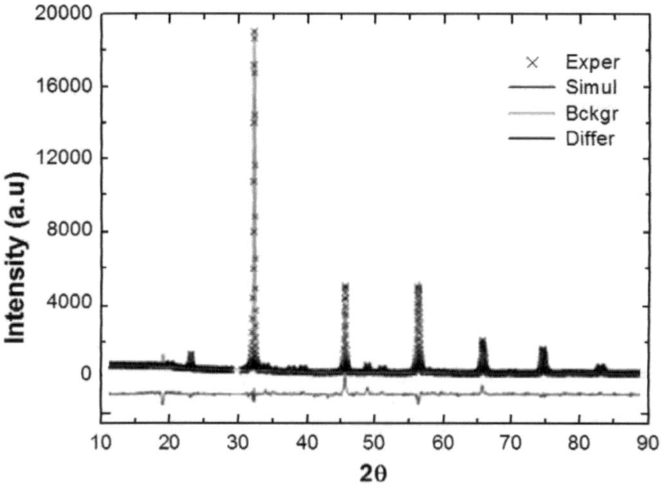

Figure 1. XRD pattern for the $Sr_3Sb_2CoO_9$ triple perovskite. Symbols represent the experimental diffraction data and base line corresponds to the difference between experimental and simulated patterns (continuous line).

The numeric results of the Rietveld analysis are shown in the table 1. Figure 2 shows the schematically representation of the structure of $Sr_3Sb_2CoO_9$ which was drawn from the lattice parameters, tilt angle, atomic positions and bond distances obtained from the Rietveld refinement of the experimental XRD data.

TABLE 1. Atomic positions for the Sr3Sb2CoO9 complex perovskite.

Ion	*Site*	*x*	*y*	*z*
Sr	4e	0.2500	0.5014	0.0777
Sr	4e	0.7500	0.0014	0.0777
Sr	4e	0.2500	0.0014	0.2444
Co	2a	0.0000	0.0000	0.0000
Co	2d	0.5000	0.0000	0.5000
Sb	4e	0.5047	0.5000	0.0000
Sb	4e	0.0048	0.0000	0.3381
O	4e	0.9400	0.6900	0.2500
O	4e	0.5600	0.8100	0.2900
O	4e	0.2500	0.6200	0.2500
O	4e	0.9400	0.6774	0.9125
O	4e	0.4400	0.2974	0.8725
O	4e	0.5600	0.8226	0.9525
O	4e	0.0600	0.2026	0.9125
O	4e	0.7374	0.1200	0.9125
O	4e	0.2374	0.6200	0.9125

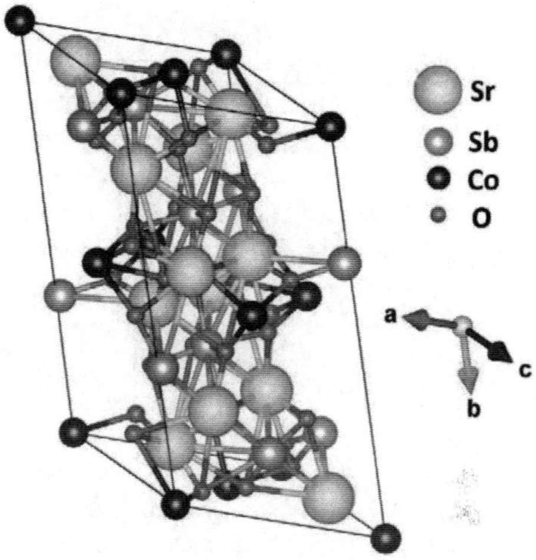

Figure 2. Structure of the monoclinic $Sr_3Sb_2CoO_9$ triple perovskite obtained from the Rietveld refinement.

Figure 3 shows the total DOS calculated from the charge density for the $Sr_3Sb_2CoO_9$ material close to the Fermi level which is the reference for energies. It is observed on the picture that the $Sr_3Sb_2CoO_9$ perovskite behaves as a conductor material for both up and down spin orientations. Due to the difference between the up and down spin contributions in the vicinity of the Fermi level it can be suggested that this material is magnetic. In order to examine in detail the contributions of each of the constituent atoms of the $Sr_3Sb_2CoO_9$material has been developed a picture of partial DOS which is presented in Figure 4.

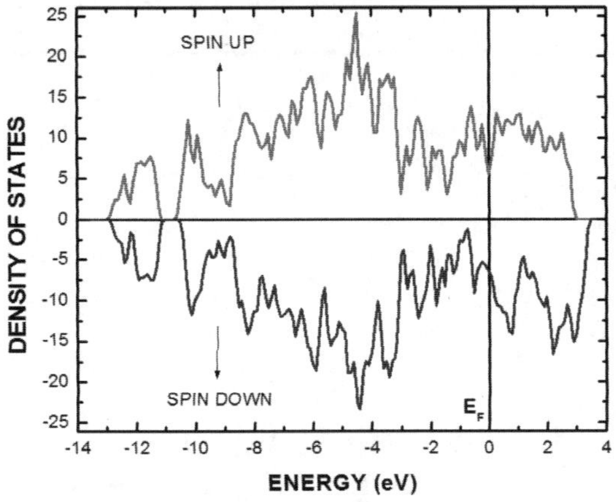

Figure 3. Total DOS with spin polarization for the $Sr_3Sb_2CoO_9$ tri-perovskite.

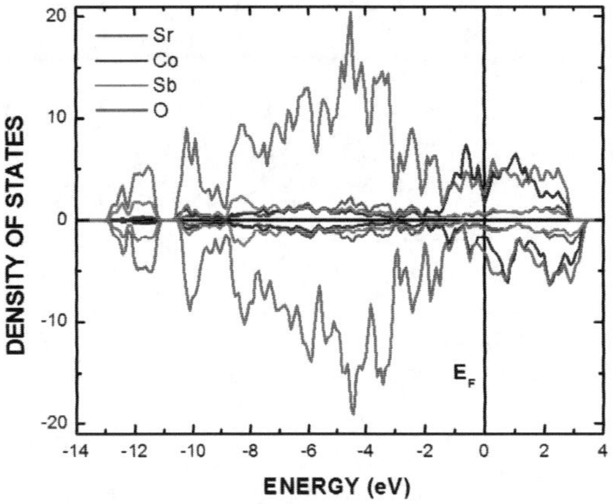

Figure 4. Partial DOS due to contributions of Sr, Co, Sb and O orbitals including spin polarization for the $Sr_3Sb_2CoO_9$ material.

From the analysis of our results we can to affirm that de main contributions to the DOS, in the valence and conduction regimes, close to the Fermi level are due to 3d-Co and 2p-O orbitals. This may be because in this type of structures the distortions that occur in the Co-O octahedra have significant effects on all physical properties of the material. Important contributions due to 5s-Sr and 5p-Sb orbitals throughout the region near the Fermi level were not observed.

5. Conclusions

The synthesis and structural characterization of the new $Sr_3Sb_2CoO_9$ perovskite-like material was performed. The Rietveld analyses reveal that this material crystallizes in an orthorhombic complex perovskite which corresponds to the $P2_1/c$ (#14) space group. Calculations of density of states by the Quantum EXPRESSO code reveal the conductor character of this material, mainly due to 3d-Co and 2p-O orbitals. Difference between the DOS of spin-up and spin-down orientations close to the Fermi level suggests the magnetic character of the $Sr_3Sb_2CoO_9$ tri-perovskite.

Acknowledgements

This work was partially financed by DIB (National University of Colombia), Hermes Code 13893, and "El Patrimonio Autónomo Fondo Nacional de Financiamiento para la Ciencia, la Tecnología y la Innovación Francisco José de Caldas" Contract RC No. 0850-2012".

References

[1] R.M. Hazen, Scientific American 258 (1988) 54.
[2] P.A. Sharma, J.S. Ahn, N. Hur, S. Park, S.B. Kim, S. Lee, J.G. Park, S. Guha, S.W. Cheong, Phys. Rev. Lett. 93 (2004) 177202.
[3] X. Lu, J. Xie, H. Shu, J. Liu, C. Yin, J. Lin, Mater. Sci. Engineer. B 138 (2007) 289.
[4] J.C. Albornoz, D.A. Landínez Téllez, J. Roa-Rojas, J.A. Munévar, E. Baggio-Saitovich, J. Supercond. Nov. Magn. 26 (2013) 2313.
[5] M. Bonilla, D.A. Landínez Téllez, J. Arbey Rodríguez, J. Albino Aguiar, J. Roa-Rojas, J. Magn. Magn. Mater. 320 (2008) e397.
[6] V.R. Palkar, S.K. Malik, Solid Stat. Commun. 134 (2005) 783.
[7] S. E. Lofland, T. Scabarozi, Y. Moritomo, Sh. Xu, J. Magn. Magn. Mater. 260 (2003) 181.
[8] A. Di Trolio, R. Larciprete, A.M. Testa, D. Fiorani, P. Imperatori, S. Turchini, N. Zema, J. Appl. Phys. 100 (2006) 13907.
[9] A. C. Larson, R. B. Von Dreele, General Structure Analysis System (GSAS), Los Alamos National Laboratory Report LAUR, 2000, pp. 86-748.
[10] S. Baroni, A.D. Corso, S. de Gironcoli, P. Giannozzi, C. Cavazzoni, G. Ballabio, S. Scandolo, G. Chiarotti, P. Focher, A. Pasquarello, K. Laasonen, A. Trave, R. Car, N. Marzari, A. Kokalj, http://www.pwscf.org/i.
[11] W. Kohn, L.J. Sham, Phys. Rev. 140 (1965) A1133.
[12] J.P. Perdew, K. Burke, M. Ernzerhof, Phys. Rev. Lett. 77 (1996) 3865.
[13] D. Vanderbilt, Phys. Rev. B 41 (1990) 7892.
[14] H.J. Monkhorst, J.D. Pack, Phys. Rev. B 13 (1976) 5188.
[15] D. Sanchez-Portal, E. Artacho, J.M. Soler, Solid State Commun. 95 (1995) 685.
[16] M.W. Lufaso, P.M. Woodward, "Prediction of the crystal structures of perovskites using the software program SPuDS", Acta Cryst. B 57 (2001) 725.

Journal of Physics: Conference Series **480** (2014) 012023 doi:10.1088/1742-6596/480/1/012023

Influence of Y substitution in $Gd_{2-x}Y_xFe_{17}$ compounds

J. C. Tróchez, C. Solano, and E. Ortiz Muñoz

Facultad de Ciencias Básicas, Universidad del Atlántico, Km 7 vía a Puerto Colombia, A.A 1890 Barranquilla, Colombia

J. C. Tróchez, juliotrochez@mail.uniatlantico.edu.co

Abstract. $Gd_{2-x}Y_xFe_{17}$ (x = 0; 0.6; 1.0) compounds were prepared by standard arc melting method. Thereafter, samples were subjected to annealing for two days at T=1273 K, finishing with quenching to 273 K. The samples were cut into small portions to perform structural, electrical, and thermal characterization. Heat flow and Curie temperature (T_c) arise from measurements of thermal properties through Differential Scanning Calorimetry (DSC), structural properties were tested via X-ray diffraction (XRD) at T=273 K, electrical characterization was carried out through the Impedance Spectrometry (IS) method, upon zero applied magnetic field and temperature range between 300 K<T<500 K. Results show that T_c decreases when Y concentration increases. T_c is obtained as the maximum temperature variation of heat flux. The Gd_2Fe_{17} compound shows two endothermic heat flux peaks; this fact can be attributed to not identifying an impurity phase from X-ray diffraction measurements. From Rietveld refined data, the crystal structure for all compounds is established, resulting in a hexagonal Th_2Ni_{17}-type structure. Measurements of impedance spectrometry as function of frequency were analyzed in terms of equivalent circuit method, finding that for temperatures below T_c they fit according to the model composed of one RL and two RC parallel circuits, in contrast to temperatures above T_c given by the model that consists of only two RC parallel circuits.

1. Introduction

In recent decades, intermetallic compounds have received special attention due to several applications of magnetic, thermal, and electrical properties [1]. Among these materials, compounds with high concentrations of Fe, like R_2Fe_{17} (with R rare earth) have emerged as more recognizable, given that excess Fe will provide more contribution to magnetic moment. Research on these materials is of interest for the manufacture of electric engines and, more recently, in magnetic refrigeration [2]. Features of R_2Fe_{17} compounds are related to their crystalline structure and magnetic behavior, given that the former depends on the rare earth element. For instance, heavy rare earth has a hexagonal $P6_3/mmc$ space group and Th_2Zn_{17}-type crystal structure in which the rare earth magnetic moment is antiparallel to the magnetic moment of the Fe sublattice. On the other hand, light rare earth crystallizes as Th_2Ni_{17}-type rhombohedral structure belonging to space group $R\bar{3}m$; the rare earth magnetic moment is parallel to the magnetic moment of the Fe sublattice. Compounds with Y, Gd, and Tb can be crystallized in either Th_2Ni_{17} or Th_2Zn_{17}, depending on the annealing process [3],[4]. Magnetic ordering temperature in these compounds is mainly due to $J_{Fe\text{-}Fe}$ interactions because of high concentrations of Fe ions that form a sublattice with several sites for Fe, while R in Th_2Zn_{17}-type structure has one site and two in the Th_2Ni_{17}-type structure. The contribution of $J_{Fe\text{-}R}$ interactions to

Journal of Physics: Conference Series **480** (2014) 012023 doi:10.1088/1742-6596/480/1/012023

magnetic moment and, thus, magnetic ordering temperature could be tested when rare earth substitution is done. Then, increased T_c is observed for these compounds from Y to Gd. Magnetic interactions on the Gd_2Fe_{17} compound has been studied by *ab-initio* calculation, revealing that in rhombohedral structure the magnetic moment is shorter than the hexagonal structure, given that it is expected for magnetic ordering temperature to have the same behavior. Some experimental works have reported T_c=475 K and T_c=466 K for hexagonal and rhombohedral structures, respectively. From theoretical calculation for both structures, T_c shows differences between hexagonal and rombohedral cases, but values are not exact to experimental values [5]. On the other hand, Y_2Fe_{17} compounds have T_c=370 K; therefore, it is expected that substitution of Gd by Y in this compound decreases T_c. Substitution is reflected in changes in local magnetic moment, reducing total magnetic moment and magnetic ordering temperature.

R_2Fe_{17} compounds present relatively low magnetic ordering temperature and anisotropy; two features that must be taken into account to evaluate if they are suitable for use as material for permanent magnets. Therefore, two ways to improve these features exist: substitution of Fe sites by ions like Ga and Mn or by N or C insets in interstitial sites at the Fe sublattice. In both cases, modification of composition results in strengthening ferromagnetic interactions against antiferromagnetic ones; due to enlarged separation between Fe ions that changes interactions between Fe-Fe ions. Also, exchange interactions between rare earth (R) and Fe ions ($J_{R\text{-}Fe}$) contribute to magnetic moment and T_c in these materials. Indeed, substitution of Y in Gd_2Fe_{17} will be an appropriate way to modify magnetic transition temperature. The rare earth element contributes to magnetic anisotropy because its s and f electrons provide a stronger effect of crystalline electric field. In $Gd_{2-x}Y_xFe_{17}$ it was found that hyperfine field of Fe sites arise when x increases; however, in this case the structure corresponds to Th_2Ni_{17}-type compounds [6]. Recently, potential technological uses for these materials have been released in magnetocaloric-based devices, given that profound knowledge of thermal properties is required to decide their suitability for this purpose. In this sense, experimental data is available on thermal properties of Y_2Fe_{17}, but not for $Gd_{2-x}Y_xFe_{17}$ compounds, thus, this fact is a motivation for this work. In order to evaluate the viability for this practical application, two conditions must be considered for successful magnetocaloric effect (MCE) materials; firstly, ΔS must be high and secondly, the ΔT range taking place at the phase transition will be wide. Y_2Fe_{17} presents a large value for ΔT, thereby, it is suitable for MCE uses.

2. Experimental procedure

Samples were prepared in an arc melting furnace, where five melting processes for all samples were conducted; turning samples over after each melting to attain homogenized samples. Then, samples were evacuated in sealed quartz tube with 10^{-2} torr pressure and annealed to 1273 K for two days, ending with rapid quenching to 273 K. Thereafter, the samples were fractionated into small portions for characterization. XRD at room temperature was performed by using a PANanalytical diffractometer, differential scanning calorimeter (DSC) from T=203 K to T=473 K, at zero magnetic field to determine T_c and enthalpy. Impedance measurements were conducted in the same temperature range using Solartron frequency response analyzer (FRA) model 1260 under continuum frequency variation from 1 to 10^5 Hz. XRD experimental data was analyzed through Rietveld method by using Fullprof© software [8]. For a correct analysis of impedance measurements, we made sure the range of frequencies was within the region where no dispersive behavior takes place.

3. Results and Discussion

From XRD patterns, we can identify samples of Gd_2Fe_{17}, $Gd_{1.4}Y_{0.6}Fe_{17}$, and $GdYFe_{17}$ as hexagonal structures with space group $P6_3/mmc$, where eight of every nine peaks were identified belonging to this structure. Figure 1 shows XRD data for $GdYFe_{17}$. Determination of T_c for all samples was done by taking the maximum variation of heat flux as a function of temperature, as displayed in Figure 2. From this figure, it can be observed that T_c decreases as a function of x. For Gd_2Fe_{17}, T_c was found at 437 K. The endothermic peak from heat flux allows determining T_c; nevertheless, in the Gd_2Fe_{17}

compound a broader endothermic peak is present, apparently due to contribution of an impurity phase, as shown in Figure 2. This fact, as mentioned before, was detected from XRD given that we identified eight of nine peaks from XRD patterns. For $Gd_{1.4}Y_{0.6}Fe_{17}$ and $GdYFe_{17}$, T_c is 406 and 387 K, respectively. From thermal analysis results, it can be observed that T_c decreases as a function of x and this behavior is related to the weakness of $J_{R\text{-}Fe}$ interactions. Indeed, influence of rare earth is very important near to room temperature because T_c for Y_2Fe_{17} is 324 K and 466 K for Gd_2Fe, as reported from Mössbauer spectroscopy measurements [6].
Electric transport measurements in samples with two flat surfaces were carried out to study the physical properties of these compounds. We conducted impedance measurements as a function of temperature in the range of T=303 to 523 K. Impedance spectroscopy analysis was done through the equivalent circuit method, taking into account that $Z = Z' + Z''$, where Z' is the real part and $Z'' = X(t,w)$ is the imaginary part of impedance.

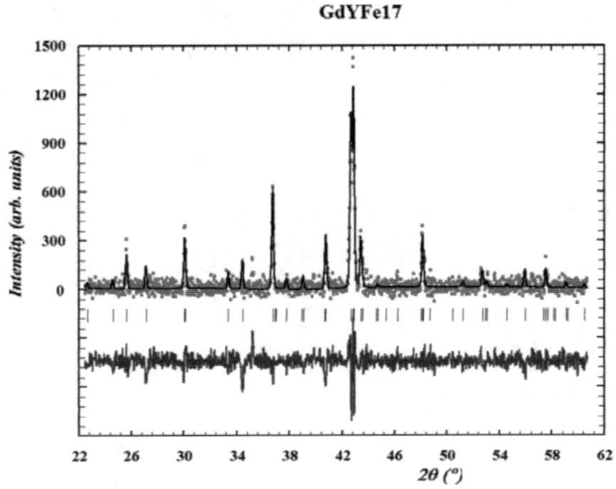

Figure 1. XRD patterns for $GdYFe_{17}$ showing refined experimental data. For on-line version experimental data, calculated spectra, and difference between experimental and calculated it is red, black, and blue, respectively.

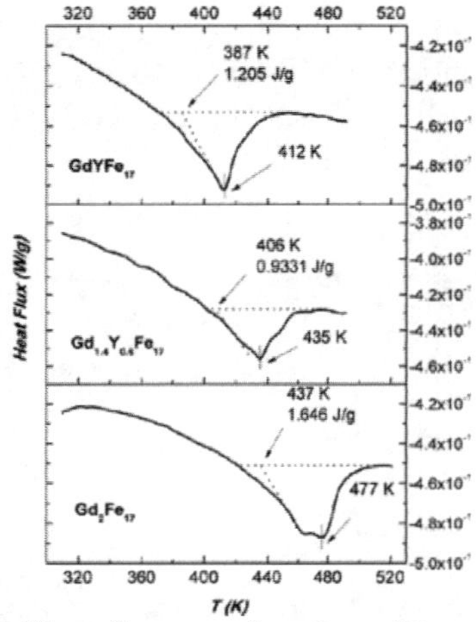

Figure 2. Heat flux as a function of temperature for $Gd_{2-x}Y_xFe_{17}$ compounds. T_c is a maximum variation of heat flux and enthalpy arises from the calculated area from a region limited by dotted lines.

In our analysis we used combinations of RC and RL parallel circuits as equivalent circuits, where each has a characteristic relaxation time that determines the electrical behavior of samples when an AC signal is applied to the samples. Impedance measurements data can be fitted by using the analytical expression obtained from a combination of RL and RC parallel circuits. For this propose, we used ZView® software [9]. We can see that all measurements were fitted to achieve small values for chi-squared, assuring a good fit. For our measurements in the Gd_2Fe_{17} compound, we found that at temperatures below T_c all measurements of Z'' vs. Z' were fitted by using a model that consists of a combination of two parallel RC circuits and one parallel RL circuit. We also tried analyses with other models like three RC and two RL parallel circuits, but without good fit. In fact, whether all samples can, thus, be analyzed is a consequence of their magnetic behavior, where at temperatures below T_c all have an inductive part in their impedance. Although for temperatures above T_c, we analyzed via one model that included two parallel RC circuits because the inductive part is nonexistent. Again, we tried to analyze with other models, but the best results were achieved when using two parallel RC circuits. We indicate that, in this case, the R and C values obtained from fitting for RC parallel circuits for either below and above T_c are in the same range. In this case, we found that for temperatures below T_c the best fit was for the model with a combination of parallel RL and RC circuits, while for temperatures above T_c the best fit was obtained by using two parallel RC circuits as a model. Figure 3 shows the behavior of Z'' vs Z' for several temperatures for $GdYFe_{17}$. It is observed that an inductive

component of impedance, related to the negative component of Z'' is present for temperatures below T_c (Figures 3 and 4). For high temperature values, resistivity decreases and reactance shifts from negative values to positive values in the Z'' axis. The inset shows measurements for high temperatures. The same behavior is observed for $Gd_{1.4}Y_{0.6}Fe_{17}$, and Gd_2Fe_{17}, as shown for the first in Figure 4. For insight into the behavior of frequency response analysis it is important to take into account the nature of the materials. These samples are polycrystalline, therefore, with grains that will explain the observed capacitive behavior. Additional contribution to capacitive behavior comes from microstructure defects like cracks and surface oxidation processes. Moreover, the contribution of two RC circuits is not associated to electrode effect because the R and C values observed are very low according to the nature of the materials. In fact, the results are consistent with magnetic behavior because for temperatures above T_c, Z'' shifts from inductive to capacitive behavior.
In summary, we have found that substitution of Y by Gd in Gd_2Fe_{17} compounds decreases T_c by weakening $J_{Fe\text{-}R}$ interactions. Heat flux measurements show a broader behavior close to T_c. These features imply magnetocaloric effect applications where there is a wide range for application near to Curie temperature. Evidence of magnetic behavior through impedance spectroscopy measurements is related to fitting data by using the equivalent circuits method, where the $T<T_c$ model consists of one RL and two parallel RC circuits; instead, the $T>T_c$ model contains only two parallel RC circuits.

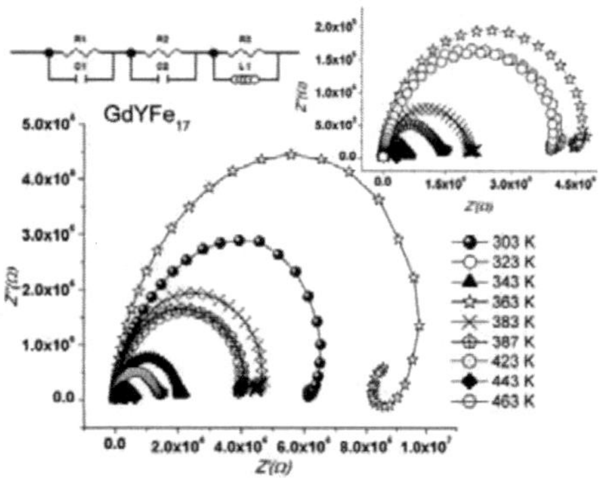

Figure 3. Z'vs. Z'' for a $GdYFe_{17}$ compound for several temperatures close to T_c,. Inset shows behavior at high temperatures.

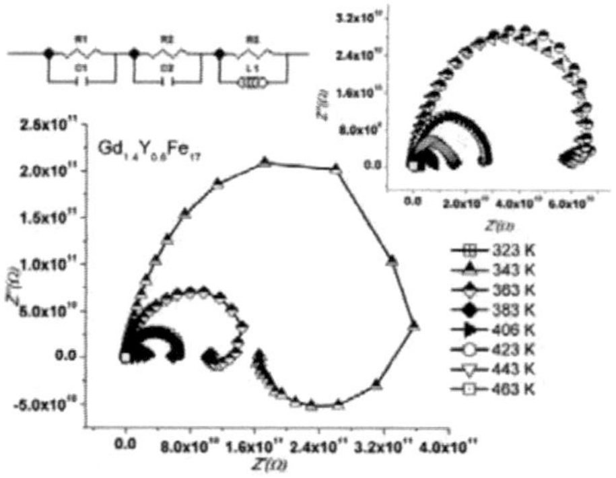

Figure 4. Z'vs. Z'' for a $Gd_{1.4}Y_{0.6}Fe_{17}$ compound for several temperatures close to T_c. Inset shows behavior at high temperatures.

4. Acknowledgments

This work was supported in part by Universidad del Atlántico. We also express our gratitude to Dr German Perez for his support in preparing samples and Dr. David Landinez for XRD measurements.

References

[1] A V Lukoyanov, E E Kokorina, M V Medvedev and I A Nekrasov 2009 *Phys Rev. B.* **80** 104409-1
[2] A G Kuchin, W Iwasieczko 2010 *Solid State Comm.* **150** 1580
[3] K H J Buschow, R Coehoorn, D B de Mooij, K de Waard, T H Jacobs 1990 *J. Magn. Magn. Mater.* **92** L35
[4] K H J Buschow 1996 *J Less Commun. Met.* **11** 204
[5] Yu V Knyazev, A V Lukoyanov, Yu I Kuz`min, A G Kuchin, and I A Nekrasov 2006 *Phys. Rev. B.* **73** 094410
[6] F Richomme, J M Le Breton, M S Ben Kraiem, A Cheikhrouhou 2010 *J. A. and Comp.* **494** 5
[7] W E Wallace 1985 *Prog. Solid State Chem.* **16** 127
[8] Rodríguez-Carvajal J 1992 *Physica B* Vol. **192** 55
[9] ZView®, impedance analysis software of http://www.scribner.com/zview-for-windows.html

Journal of Physics: Conference Series **480** (2014) 012024 doi:10.1088/1742-6596/480/1/012024

Optical and Structural Study of CuSe and CuSe/In Thin Films

J.I. Montes-Monsalve[1,2], R. Bernal Correa[1,2] and A. Pulzara Mora[1]
[1] Laboratorio de Nanoestructuras Semiconductoras, Universidad Nacional de Colombia, sede Manizales. A.A. 127.
[2] Escuela de Materiales, Facultad de Minas, Universidad Nacional de Colombia, sede Medellín.
e-mail: jimontesmo@unal.edu.co, aopulzaram@unal.edu.co

Abstract. The preparation of (Cu,Se)(In,N) thin films by r.f magnetron sputtering technique opens an interesting research area with potential application in solar cells. In this work, we presented a systematic study of growth and characterization of CuSe thin films and layers of CuSe/In deposited on glass substrates by varying the power r.f source and substrate temperature. The effects of experimental parameters on morphology and the optical properties of the films were investigated by XRD, UV-Visible and SEM. The results show that the samples are polycrystalline with a hexagonal structure for the CuSe films, and tetragonal structure for the CuSe/In layers. SEM images took on the surface of the films, showed a change in their morphology, evidencing a strong dependence of the experimental parameters. From UV-Visible measurements determine the bandgap and the optical behaviour of the films.

1. Introduction

The ability to deposit a wide variety of thin film alloys, in a homogeneous and uniform way, has enable the development semiconductor alloys applied in fields such as optoelectronics, photonics, fiber optics, data storage devices and energy conversion [1]. CuSe alloys has been one of the most studied in recent years, with stoichiometric compositions (α-Cu_2Se, Cu_3Se_2, CuSe and Cu_2Se) and non-stoichiometric ($Cu_{2-x}Se$) reporting a continuous change of physical properties. Moreover, different crystalline phases have been reported with orthorhombic structure [2], cubic, hexagonal, [3] and tetragonal [4], depending on the stoichiometry and the growth methods. These features make the electrical and optical properties are of interest for applications in solar cells [5], super ionic conductor [6], optical filters [7] and lasers. [8] The CuSe semiconductor could be a direct gap of 2.2 eV or 1.4 eV indirect [9]. On the other hand, it is worth to note that the semiconductor alloys $CuInSe_2$ (CIS) thin film, are one of the most material promising because of their high energy conversion efficiencies, high absorption coefficient, and its low toxicity (without post-selenization) and optical and electrical properties, which make it a good candidate for solar cells [10].
We report the preparation of CuSe films and In/CuSe layers, by systematic variation of deposition parameters such as rf source power. It was found that the crystallographic orientation and morphology show a strong dependence with the power growth. Furthermore, from optical absorption determined the band gap energy in UV/Vis range.

[1] aopulzaram@unal.edu.co

2. Experimental procedure

Films CuSe and CuSe/In were deposited on glass substrates by magnetron sputtering system rf (13.56 MHz). The substrates were cleaned with acetone in an ultrasound chamber. CuSe and In targets of high purity with diameters of 2 inches and 1 inch were used in the process. Vacuum chamber base pressure before deposition ≈ 1x 10^{-6} Torr. CuSe layers were deposited in Argon atmosphere at a working pressure of 5 x 10^{-3} Torr for 120 minutes, with a 150 °C substrate temperature and a power of 15 Watts (sample M1) and 35 Watts (sample M2). The target-substrate distance was set at 5 cm, tilted 35 ° with respect to the normal of the substrate.
CuSe/In bi-layers were deposited on glass substrates in order to study the influence of the In layer on their structural properties. Three samples were deposited at In target power source (P_{In}) of 5 Watt (sample M3), 15 Watt (sample M4), and 30 Watt (sample M5). The deposition time of each CuSe and In layer was of 20 min and 2 min, respectively. The residual pressure for all layers is holding in $1.2x10^{-3}$ Torr. The experimental conditions are show in the table 1.

Table 1. Experimental conditions for CuSe/In layers

Layer	r.f Power (Watts)		Ts (°C)	
	In	CuSe	In	CuSe
M3	5	20	300	150
M4	15	20	300	150
M5	30	20	300	150

The crystal structure was determined by X-ray diffraction (System XRD Cu Ka radiation, Rigaku = 1.5405 A). The surface morphology was analyzed by in-plane view scanning electron microscopy (SEM). Optical transmission spectra were measured with a UV-Visible spectrophotometer (LAMBDA PERKINELMER) in a wavelength range 400 - 900nm.

3. Results and discussion

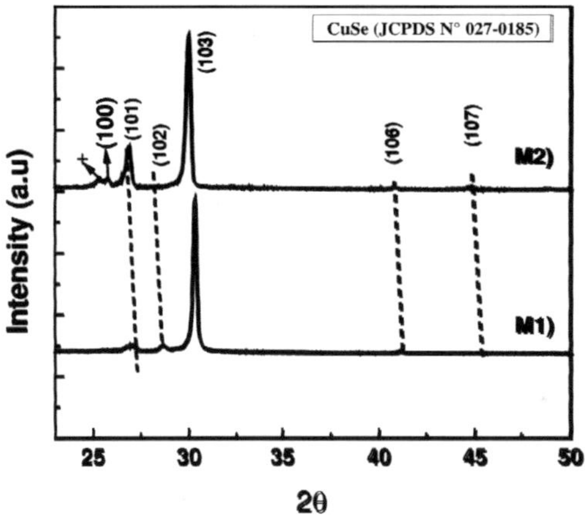

Figure 1. XRD Difractograms for CuSe samples. M1) 15 Watt M2) 35 Watt

Figure 1 shows the X-diffraction patterns of CuSe films. The samples were found to be polycrystalline and all orientations of the planes correspond to a hexagonal crystal structure, according to the database (JCPDS N0 027-0185) for CuSe. XRD revealed the presence of three predominant planes corresponding to directions (101), (102), (103) and (106) at 2θ of 26.91 °, 28.66 °, 30.32 ° and 41.14 ° respectively, for the film prepared at a power of 15 Watts (sample M1). For sample deposited at 35 watts (M2), are observed six planes, which two of them present low intensities and correspond to the crystallographic directions (101) and (107) at 25.81 ° and 44.76 °. One possible explanation for this behavior may refer to the presence of an impurity phase which could be increased with the sputtering power [11].

The shift observed in the position of the all plane can be due to the micro-tensions that gave line defects such as dislocations edge as a consequence of the difference of the lattice parameters. The plane labeled by (+) in Fig. 1, was not possible to identify. From these spectra the grain sizes were estimated using the Debye-Scherrer equation taking into account the instrumental width, obtaining values of 35 nm for M1 and 30 nm for M2, where can see a decrease of the crystallite size due to the increase of the sputtering power.

SEM images taking in-plane view of the M1 (P_{In} = 15 Watt) and M2 (P_{In} = 35 Watt) CuSe samples are shown in the Fig. 2, a) 50 kX, b) 100kX and c) 200kX magnification for sample M1, and Fig. 2, d) 10kX, e) 15kX and f) 30kX magnifications for sample M2, respectively. The images show crystals with different shapes and geometries, being visible laminar pentagons and hexagons. Hexagons have sizes up to 400 nanometers to larger crystals (Fig.2c), distributed over the surface. For M2 sample, the images reveal the formation of hexagonal crystals with average sizes between 0.5 μm and 1 μm, showing a clear increase in the size of the crystals with the sputtering power. It is worth to note, that SEM images confirm XRD results regarding the shape of the hexagonal crystal structure.

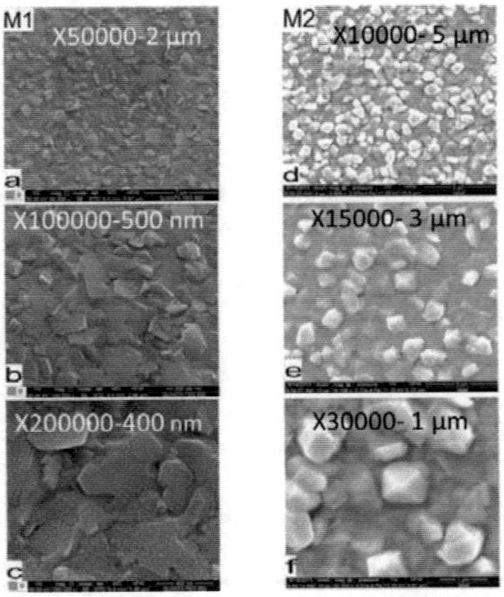

Figure 2. SEM - Micrograhps for M1 and M2 CuSe samples at different magnify-cations.

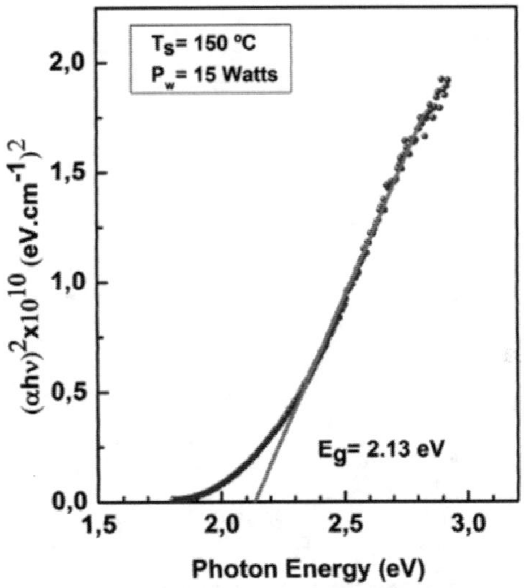

Figure 3. Plot of $(\alpha h\nu)^2$ vs Photon Energy for M1 CuSe sample.

In order to analyze the optical behavior of the thin films, the energy band gap was determined from the absorption coefficient vs photon energy, as shown in Figure 3. For high absorbing region where α obey the equation $\alpha h\nu = cte(h\nu\text{-}Eg)^{1/2}$, by plotting $(\alpha h\nu)^2$ as a function of photon energy hν, as shown in the Figure 3, and extrapolating the linear region of this curve to $(\alpha h\nu)^2 = 0$ we obtain the value of E_g, = 2,13 eV, in good agreement with the value of data reported [9,11].

Figure 4 shows the XRD results of the M3, M4 and M5 layers, where we can see a quasi-crystal growth. XRD spectra revealed the presence of two planes corresponding to (112), and (103) orientations at 2θ = 25.93 ° and 25.87 ° for M1-M3 samples, respectively. The crystallographic orientations of the planes correspond to tetragonal structure, according to the database for tetragonal $CuInSe_2$ (JCPDS No. 089-5646). It is clear that all samples present a preferred orientation in the [112] direction. The crystallite size estimated from the Debye-Scherrer equation for the samples M3, M4 and

M5 was of 28, 37 and 29 nm. The x-ray results have also showed the incorporation of In atoms in the CuSe matrix forming the InCuSe alloy. The CuInSe semiconductor alloy likely is formed by indium (In) diffusion into CuSe layers. As far as we know does not exist report of the quasi-crystalline growth of CuInSe by Magnetron sputtering.

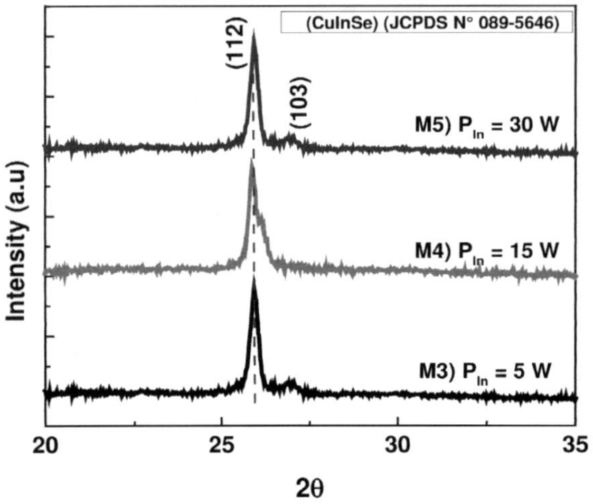

Figure 4. XRD spectra for CuSe/In layers at power source of 5 (M3), 15 (M4) and 30 (M5) Watt.

4. Conclusions

CuSe films and CuSe/In layers were deposited by rf magnetron sputtering technique. CuSe films have a high-crystalline quality, a hexagonal structure and crystallite sizes between 35 and 30 nm. The absorption measurements show that CuSe thin films present a direct band gap transition at 2.13 eV. From CuSe/In bilayers we obtain the CuInSe alloy, which is formed by indium (In) diffusion into CuSe layers.

Acknowledgments

J.I. Montes-Monsalve and R. Bernal-Correa thank to Colciencias by financial support of the scholarship - Francisco Jose de Caldas – 2011. We also appreciate the technical support of the Labs. Materiales Nanoestructurados y Funcionales and Microscopía - UIS.

5. References

[1] U. Mishra and J. Singh 2008 *Springer*
[2] Y. Zhang, Z.-P. Qiao, and X.M. Chen 2002 *J. Mater. Chem* **12**, 2747–2748
[3] V. Milman 2011 *Acta Cryst. B* **58**. 437
[4] L.B. O. Madelung, M. Schulz, H. Weiss, Landolt-Börnstein 2008 *Springer-Verlag*, Berlin.
[5] S. Lakshmikumar and A. Rastogi 1994 *Sol. Energ. Mat. Sol. Cells*, **32**, 7
[6] M. Fadel, S. A. Fayek, M. O. Abou-Helal, M. M. Ibrahim and A. M. Shakra 2009 *J. Alloys Compd.* **485** 604.
[7] V. S. Gurin, a. a. Alexeenko, S. a. Zolotovskaya, and K. V. Yumashev 2006 *Mater. Sci. Eng. C* **26** 952
[8] C. Heske, U. Winkler, H. Neureiter, M. Sokolowski, R. Fink, E. Umbach, C. Jung, and P. R. Bressler 1997 *Appl. Phy Lett*, **70** 1022
[9] A. M. Hermann and L. Fabick 1983 *J. Cryst. Growth* **61** 664
[10] L. E. Oikkonen, M. G. Ganchenkova, a. P. Seitsonen, and R. M. Nieminen 2012 *Phys. Rev. B*, **86** 165115
[11] Y. Z. Li, X. D. Gao, C. Yang, and F. Q. Huang 2010 *J. Alloys Compd.***505** 623

Determination of the Optical Properties of ZnSe Thin Films Using the Transfer Matrix Method

R Castiblanco, J Vargas, J Morales, J Torres and A. Pardo
Departamento de Física, Universidad Nacional de Colombia, Sede Bogota
E-mail: recastiblancos@unal.edu.co, jhvargasb@unal.edu.co

Abstract. The transfer matrix method (TMM) is presented as a highly effective alternative in the treatment of problems of electromagnetic wave propagation through stratified media. This method allows to determine the transmission and reflection coefficients of the system. In this paper the TMM is applied to simulate the experimental transmittance spectra of ZnSe thin films prepared with high vacuum evaporation. From the simulation the spectral variation of the refractive index, the absorption coefficient and material gap of the ZnSe were obtained. The TMM was applied to a series of samples prepared at substrate temperatures between (200 and 350)^oC. Absorption coefficients values in the order of 10^{-3} cm^{-1} were found in the high absorption region while the refractive index was around 2.7 at $\lambda = 600$ nm. The material gap found for all the samples was about 2.65 eV.

1. Introduction

In II-VI compound semiconducting material, ZnSe is used as a window layer for the fabrication of thin film solar cells[1, 2]. It is mainly used as protective and antireflection coating for infrared-operating electrochromic thermal-control surfaces[3]. ZnSe is an important promising material for optoelectronic devices such as light-emitting diodes (it emits blue light)[4], and dielectric mirrors[5, 6]. ZnSe doped with chromium (ZnSe:Cr) has been used as an infrared laser gain medium emitting at about 2.4 μm[7]. Characterization of ZnSe thin films have been widely studied with variations such as: deposition onto transparent substrates[8], transparent substrates for specific regions of interest[9], doping with other elements, different techniques of deposition, growth and deposition temperature[10]. The variants described in the optical characterization take great importance due to the behavior of ZnSe in the regions of visible and infra-red, behaving as a dielectric with refractive index decreasing when the wavelength increases under the influence of multi-photon absorption[10].

In the characterization of thin films, some authors who analyze the transmittance or reflectance spectra have developed elaborate numerical and analytical methods[11, 12], including both the absorption of different areas as well as different angles of incidence on the sample. In the same way, the transfer matrix method (TMM) can be used for the analysis of wave propagation of quantum particles, such as electrons[13], electromagnetic[14], acoustic and elastic waves. This method is particularly effective in the analysis of transmission and reflection phenomena in layered systems, mostly in one dimension[15]. This paper proposes a model based on TMM to calculate the spectral variation of the refractive index, extinction coefficient, thickness of the samples and the gap of the material.

2. Experimental details

The samples were deposited on glass substrates, under high vacuum at a pressure of 2.5×10^{-5} $mbar$, with substrate temperatures $(100 - 350)^oC$. Evaporator temperature was held constant at 850^oC during growth of the material. ZnSe thin films were deposited by high vacuum evaporation. The samples were deposited in common glass substrates at temperatures between $(200 - 350)^oC$. The growth of the ZnSe thin film was performed using an evaporation chamber, which is formed by a coaxial oven made of graphite and a cell *Knudsen* also constructed in graphite. The material is located in the interior of the Knudsen cell, the latter one is placed inside the oven. The material is heated by radiation, passing a electrical current of about $120A$ through the oven . The material is sublimated and then released into the substrates, where the film is formed at a specific substrate temperature. During the deposition process a temperature controller mantains the substrate temperature at the desired set point with a precision of $\pm 1^oC$. Before the deposition process, the precursor material is heated at 300^oC during one hour in order to remove impurities acquired during storage. In this process, the shutter prevents the impurities released from reaching the substrates. During this process, the substrates are also heated to 100^oC and then raised to the desired temperature. After cleaning the system, the temperature is raised in the evaporator and when it reaches 850^oC the shutter is opened and the film begins to form. The deposition time were for 5 minutes in all samples. The transmittance measurements were performed by a Spectrometer Varian Cary 5000 in the spectral range of $(330 - 2500)nm$, at normal incidence to the sample surface.

3. Transfer matrix method TMM

From the requirements of the continuity of the tangential components of the electric $\mathbf{E}$ and the magnetic $\mathbf{H}$ fields, the transfer matrix M_{1-2} is derived from a single interface between two media. Its elements determine the transmission and reflection amplitudes for both the electric and the magnetic fields. The boundary conditions on the tangent components of the field vectors $\vec{E}$ and $\vec{H}$ hold that they are the same on both sides of the interface. Thus the relation between the electric fields on opposite sides of the interface can be expressed as:

$$\begin{pmatrix} E_2^+ \\ E_2^- \end{pmatrix} = M_{(1-2)} \begin{pmatrix} E_1^+ \\ E_1^- \end{pmatrix}; \qquad M_{(1-2)} = \frac{1}{2} \begin{pmatrix} 1 + \frac{\mu_2}{\mu_1}\frac{k_{1z}}{k_{2z}} & 1 - \frac{\mu_2}{\mu_1}\frac{k_{1z}}{k_{2z}} \\ 1 - \frac{\mu_2}{\mu_1}\frac{k_{1z}}{k_{2z}} & 1 + \frac{\mu_2}{\mu_1}\frac{k_{1z}}{k_{2z}} \end{pmatrix} \tag{1}$$

$M_{(1-2)}$ corresponds to the ***transfer matrix*** for a single interface, where for normal incidence $k_{(iz)} = (\omega \hat{N}/c)\cos(\theta)$ will be $k_{(i)} = (\omega/c)(n_i + \imath\kappa_i)$ (for the general case). The transfer matrix $\mathbf{M}_{(i-j)}$ is defined for the general case having the electromagnetic waves coming from the left and the right of the interface. The transmission $t_{(1-2)}$ and the reflection $r_{(1-2)}$amplitudes are defined from the matrix transfer as:

$$t_{(1-2)} = \frac{detM_{(1-2)}}{M_{22\,(1-2)}} = \frac{2\mu_2 k_{1z}}{\mu_1 k_{2z} + \mu_2 k_{1z}}; \quad r_{(1-2)} = -\frac{M_{21\,(1-2)}}{M_{22\,(1-2)}} = \frac{\mu_2 k_{1z} - \mu_1 k_{1z}}{\mu_1 k_{2z} + \mu_2 k_{1z}} \tag{2}$$

where 2 are the familiar *Fresnel* relations for a single interface. For multilayer systems, it is possible to determine a simple method of analysis. First of all is important recognize that the last matrix transfer found in the previous section corresponds to the matrix transfer for a single interface $\mathbf{M}_{(1-2)}$, and it is always useful that the electromagnetic wave changes the media. Thus, for two slab $Fig.1$ we will need three of these matrixes $\mathbf{M}_{(1-2)}$ from the media 1 to 2, $\mathbf{M}_{(2-3)}$ from the media 2 to 3 and $\mathbf{M}_{(3-4)}$ so on. However it is necessary to introduce a new matrix called the *propagation matrix* $\mathbf{M}^{(P)}$ which describes the wave propagation within the thin film (ϵ_2, μ_2) and the substrate (ϵ_3, μ_3). This matrix can be deduced following the last treatment.

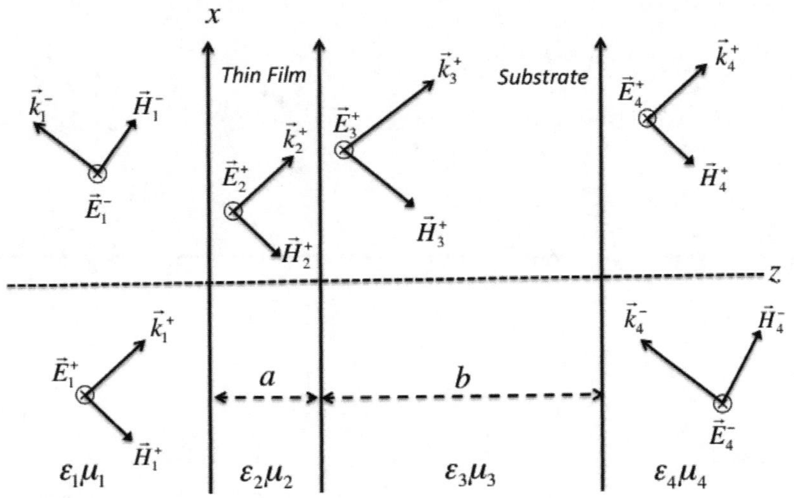

Figure 1. Transmission of an electromagnetic wave through a thin film deposited on a substrate with width a and b respectively.

Since the matrix transfer for the entire structure will be obtained from the multiplication of five matrixes, three for the interfaces and two for the propagation in the media.

$$\begin{pmatrix} E_4^+ \\ E_4^- \end{pmatrix} = M_{(12)} \cdot M^{(P_2)} \cdot M_{(23)} \cdot M^{(P_3)} \cdot M_{(34)} \begin{pmatrix} E_1^+ \\ E_1^- \end{pmatrix} \tag{3}$$

where the matrices $M_{(ij)}$ are given by the equations 1, and the diagonal matrix $\mathbf{M}^{(P_j)}$

$$\begin{pmatrix} e^{ik_{jz}l} & 0 \\ 0 & e^{-ik_{jz}l} \end{pmatrix} \tag{4}$$

Correspond to the matrix propagation for a homogeneous media (ϵ_j, μ_j). Having solved this matrix system, the *matrix transfer* is therefore obtained.

$$\begin{pmatrix} E_4^+ \\ E_4^- \end{pmatrix} = M^{System} \begin{pmatrix} E_1^+ \\ E_1^- \end{pmatrix} \tag{5}$$

In order to find the transmission and reflection amplitudes, these are calculated in the same way than for a single interface with the equations 2 respectively.

$$t_s = \frac{detM^s}{M_{22}^s} \quad and \quad r_s = -\frac{M_{21}^s}{M_{22}^s} \tag{6}$$

and for the transmission coefficient we have

$$T_s = \frac{\mu_1}{\mu_4} \frac{Rek_{4z}}{Rek_{1z}} |t_s|^2 \tag{7}$$

Finally for the system embedded in a homogeneous medium $(k_4 = k_1)$, $(\mu_4 = \mu_1 \approx 1)$ and for a thin film that presents absorption $\left(\alpha = \frac{4\pi}{\lambda}\kappa\right)$.

$$T_s = \frac{1}{A + B + C + D + E + F + G + H} \tag{8}$$

where

$$A = \frac{\cos^2(\gamma)}{2}\left[\cos(\alpha)+\cosh(\beta)\right] \qquad C = \frac{\cos(\gamma)\sin(\gamma)}{2}\left[o\sinh(\beta)-p\sin(\alpha)\right]$$
$$B = \frac{\cos^2(\gamma)}{2}\left[h\sinh(\beta)-m\sin(\alpha)\right] \qquad D = \frac{\cosh(\beta)-\cos(\alpha)}{8}\left[q\sin^2(\gamma)+r\cos^2(\gamma)\right]$$
$$G = \frac{k_{13}\sin^2(\gamma)}{4}\left[p\sinh(\beta)+o\sin(\alpha)\right] \qquad E = \frac{k_{13}}{4}\cos(\gamma)\sin(\gamma)\left[h\sin(\alpha)+m\sinh(\beta)\right]$$
$$H = \frac{k_{13}^2}{8}\sin^2(\gamma)\left[\cosh(\beta)+\cos(\alpha)\right] \qquad F = \frac{\cos(\gamma)\sin(\gamma)}{2}\left[\cosh(\beta)-\cos(\alpha)\right]u$$

whit the constants

$$\alpha = \frac{2n_2\omega a}{c} \qquad k_{13} = \left(\frac{k_1}{k_3}+\frac{k_3}{k_1}\right) \qquad m = \frac{n_2(\kappa^2+n_2^2+n_1^2)}{n_1(n_2^2+\kappa^2)}$$
$$\beta = \frac{2\kappa\omega a}{c} \qquad u = \frac{\kappa(n_2\kappa(n_3^2-n_1^2)}{n_1n_3(n_2^2+\kappa^2)} \qquad o = \frac{\kappa(\kappa^2+n_2^2-n_3^2)}{n_3(n_2^2+\kappa^2)}$$
$$\gamma = \frac{k_3b}{c} \qquad h = \frac{\kappa(\kappa^2+n_2^2-n_1^2)}{n_1(n_2^2+\kappa^2)} \qquad p = \frac{n_2(\kappa^2+n_2^2+n_3^2)}{n_3(n_2^2+\kappa^2)}$$

$$q = \frac{n_3^4+2n_3^2(n_2^2-\kappa^2)+(n_2^2+\kappa^2)^2}{n_3(n_2^2+\kappa^2)} \qquad r = \frac{n_1^4+2n_1^2(n_2^2-\kappa^2)+(n_2^2+\kappa^2)^2}{n_1(n_2^2+\kappa^2)}$$

4. Results and discussion

The Figure 2 shows a typical transmission spectrum of ZnSe thin film (solid black line) deposited by evaporation, which was obtained by using a spectrophotometer PERKIN-ELMER Lambda 9 and theoretical fitting by the TMM (dashed red line). Transmittance experimental spectra were simulated theoretically by equation 8 to determine the optical properties of the material, the spectral variation of the real part of the refractive index n_ω and absorption coefficient α. Using the model proposed by *Wemple and DiDomenico*[16], which consists of fitting the n_2 data, obtaining the following expression for the medium and weak regions of absorption.

$$n^2 - 1 = \frac{(E_0E_d)}{E_0^2-(\hbar\omega)^2} \tag{9}$$

such expression was deduced using the model of a single-effective-oscillator; where $\hbar\omega$ is the photon energy, E_0 is the single oscillator energy and E_d is the dispersion energy.

Figure 3 shows the theoretical results by TMM for the refractive index of a ZnSe sample. The full line corresponds to the fitting according to the W-D model. TMM method allowed us to obtain the values for the refractive index n_2, absorption coefficient κ and thickness a reported in (Table I), by adjusting the spectrum in all of range of wavelength according to the experimental data. Figure 4 shows the variation of the refractive index as a function of the photon energy for substrate temperatures corresponding to $(200, 250, 300\, and\, 350)^oC$ with values of $(2,54)$, $(2,71)$, $(2,79)$ and $(2,86)$ respectively at $600nm$ of wavelength. Similar results have been reported by Lakshmikumar and Rastogi [18] and [17]. The figure 4 shows similarly, as well how the refractive index increases with the temperature of the substrate, which is attributed to an improvement in the crystallinity of the deposited films of ZnSe[8, 17].

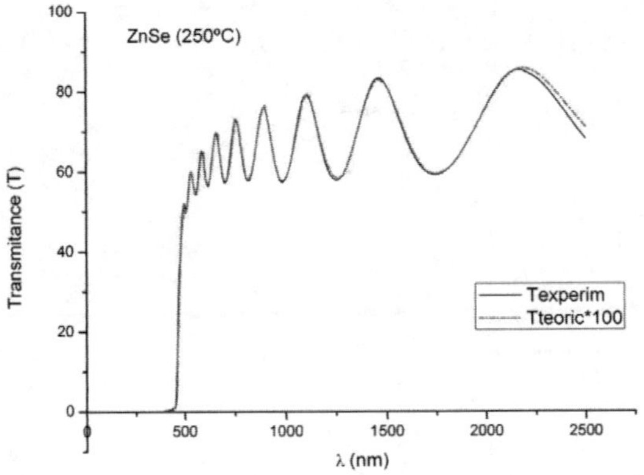

Figure 2. Typical transmission spectrum of a ZnSe thin Film (solid line) deposited by evaporation (substratum temperature 250°C) and theoretic curve fit (dashed line)

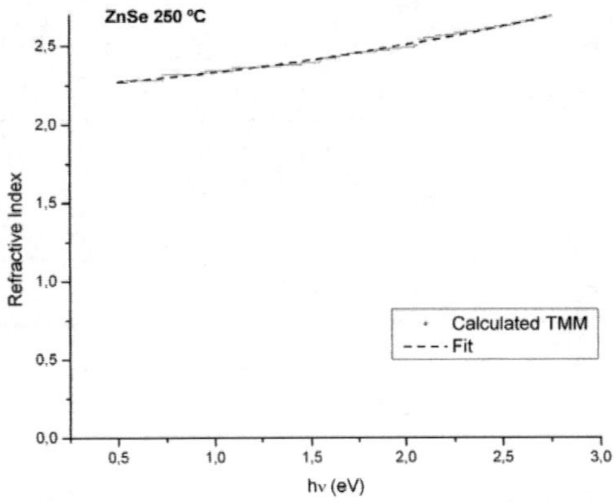

Figure 3. Variation of the refractive index of a ZnSe sample, calculated using TMM.

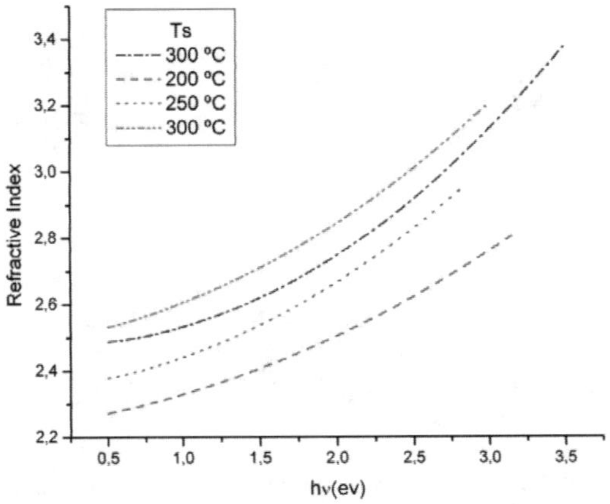

Figure 4. Variation of the refractive index n as a function of the photon energy for ZnSe thin films with four different substratum temperatures.

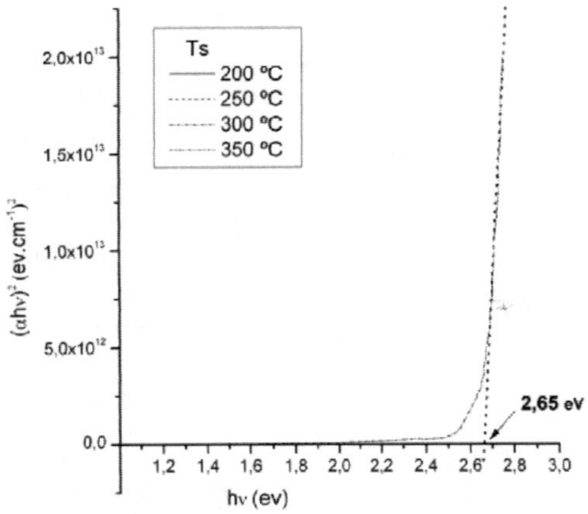

Figure 5. Curves of $(\alpha h\nu)^2$ Vs hν corresponding to ZnSe thin films with four different subtratum temperatures.

Figure 5 shows the curves of $(h\nu\alpha)^2$ vs $h\nu$ corresponding to the four samples of (ZnSe) before mentioned. The optical gaps E_g of the four samples were determined from the intercepts of the $(h\nu)$ axis with the straight lines which fit the $(h\nu\alpha)^2$ and $h\nu$ curves in the region of strong absorption, considering here that these semiconductors are direct band gap type. The opticals gaps obtained for the four samples were $2,65\,eV$.
Since the results presented in this work are in agreement with those reported in the literature for ZnSe thin films [8, 12, 17], it can be concluded that the practical method proposed in this article permits the determination with good reliability of the optical constants (n, α) thickness and the optical gap of ZnSe thin film prepared by evaporation.

Journal of Physics: Conference Series **480** (2014) 012025 doi:10.1088/1742-6596/480/1/012025

Table 1. Values parameters refraction index n_ω and thickness a for ZnSe thin films, deposited at different substrate temperatures.

Temperature ($°C$)	Refraction $n_{(\omega)}$	Thickness a (nm)
200	2,54	900
250	2,71	1000
300	2,79	650
350	2,86	450

5. Conclusions

it was used the transfer matrix method to theoretically simulate the transmittance spectra of ZnSe thin films. From the simulation the spectral variation of the refractive index and the extinction coefficient were extracted and the thickness of the samples and the material gap was found as well. The optical values obtained were comparable with those found in the literature [8, 17].

References

[1] Chin-Hau Chia, Wen-ChungFan, Yen-ChenLin, Wu-ChingChou, Radiative recombination of indirect exciton in type-II ZnSeTe/ZnSe multiple quantum wells. Journal of Luminescence 131 (2011) 956-959

[2] V. S. Nagarethinam, M. G. Syed Basheer Ahamed, Structural, optical and electrical properties of $ZnTe_{1-x} Se_x$ thin films, J Matter Sci: Matter Electron (2011) 22:607?613.

[3] L. Yan, J. A. Woollam, E. Franke, J. Vac Sci Technol A 20 (3) (2002) 693

[4] Cardona M, Yu P.Y, Comprehensive Semiconductor Science and Technology, Volumen 2, (2011), p. 125-195.

[5] T. Shi rakawa, Mater. Sci Emg. B 91-92 (2002) 470.

[6] M. Godlewiski, E Guziewiez, K. Kooalko, E. Lusakowska, E. Dynowska, 102-103 (2003) 455

[7] G. Grebe, G. Roussos and H.-J. Schulz, J. Phys. C: Solid State Phys. vol. 9 pp. 4511-4516 (1976)

[8] S. Venkatachalam, D. Mangalaraj, Sa. K. Nayandass, K. Kim, Physica B, 358 (2005) 27-35

[9] G. Riveros, H. Gomez, R. Henriquez, R. Schrebler, Solar Energy Material and Solar Cells 70 (2001) 255-268

[10] G. Hawkins, Roger Hunneman, Infrared Physics and Technology 45 (2004) 69-79

[11] A. Penzkofer, E. Drotleff, W Holzer, Optics Communications 158 (1998) 221-230

[12] J. Torres, J. I. Cisneros, G Gordillo, F. Alvarez, Thin Solid Films 289 (1996) 238-241

[13] P. A. Mello, N. Kumar, Quantum transport in mesoscopic systems. Oxford University Press (2004)

[14] P. Yeh, Opticals Waves in Layered Media, 2nd ed. Wiley Series in Pure and Applied Optics (2005)

[15] P. Markos, C. M. Soukoulis, Wave Propagation Princeton University Press (2008)

[16] S. W. Wemple, M. Didomenico, J. Physics Rev. B., 3 (4) (1971) 1338

[17] Characterization Optics, Electric and Morphologic of ZnSe thin Films at Differents Temperatures of Deposition

[18] S. T. Lakshmikumar, A. C. Rastogi, Thin Solid film 259 (1995) 150

Titanium dioxide thin films: a study of film morphology and structural changes upon photocatalytic degradation of methylene blue

A Naranjo[1], G Zambrano[1], W Torres[2] and M E Gómez[1]

[1]Physics Department, Thin Films Group,Universidad del Valle, Cali, Calle 13 No 100 - 00, Colombia
[2]Chemistry Department, Electrochemistry Group,Universidaddel Valle, Cali, Calle13 No 100 - 00, Colombia

E-mail: andrea.naranjo@correounivalle.edu.co

TiO_2 thin films were deposited by reactive DC-magnetron sputtering at different Ar/O_2 ratios in the gas mixture. The photocatalytic activity (under UV light irradiation) of these films was evaluated upon degradation of aqueous solutions of methylene blue (MB), a model organic pollutant. Upon photocatalysis, the TiO_2 films showed structural, morphological and compositional changes that were characterized by X-ray diffraction (XRD), Atomic Force Microscopy (AFM), Scanning Electron Microscopy (SEM) and energy dispersive X-ray (EDX).

1. Introduction

Research on TiO_2 and TiO_2-based materials as photocatalysts for decontamination of water and air with UV and visible light has been active since the 1970's. Most of work on the field has focused on transport processes at the semiconductor interface, kinetics of the contaminant removal reactions, and fabrication of reactors for commercial applications [1]. Catalyst lifetime is an important characteristic [2]. However, TiO_2 deactivation is seldom studied presumably because this catalyst is perceived as a low cost material. Although deactivation is often observed in photo-assisted processes at the TiO_2/gas phase interface [3], catalyst deactivation and reactivation have hardly been studied in the TiO_2/aqueous interface probably because it is believed that the competitive adsorption of water and its ability to dissolve degradation intermediates help to keep the TiO_2 surface clean. However, chemical contamination of the TiO_2 surface may prevent photon absorption or inhibit interfacial charge transfer [4, 5].Therefore, one question that has to be addressed is the modification of the photocatalyst surface and changes in the kinetics of the degradation reactions as a function of the structure of common chemical contaminants. Our aim here is to study the structural, morphological and compositional changes of TiO_2 thin films, obtained by reactive DC-magnetron sputtering, after the photocatalytic degradation of a methylene blue (MB) a model organic pollutant.

2. Experimental procedure

TiO_2 thin films were deposited by reactive DC-magnetron sputtering, on quartz (100) substrates at 400 °C, with a high purity titanium target (99.9%), at 20 W/cm^2, setting the base pressure to $2x10^{-5}$ mbar and the work pressure in the deposition vacuum chamber to $2.0x10^{-2}$ mbar for 60 minutes. The Ar/O_2 gas mixture ratio was varied (from 80/20 through 50/50) to study the influence of O_2 partial pressure on the crystallographic structure and catalytic activity of the as deposited TiO_2 films which were used as photocatalysts for the degradation of methylene blue, a model pollutant, under UV radiation.

For photocatalytic experiments, a quartz cell containing the MB test solutions and a TiO_2 film was irradiated with UV light and an incandescent lamp (4 W each) located 4 cm from the cell. Air was bubbled in the solution for 10 min before irradiation. Upon irradiation, the MB concentration was followed by measuring the solution absorbance at 292 nm in a Shimadzu UV-vis 1601 spectrophotometer. Chemical composition of the deposited films was studied in a Philips XL30 SEM with an EDX microprobe for chemical analysis (12 kV), equipped with a window for the detection of light elements. Morphological characterization was made by using an Asylum research Atomic Force Microscopy (AFM) in non-contact mode. AFM image data was processed with a fractal algorithm and the WSxM freeware developed by NanotecElectrónica, Spain [6], to obtain surface parameters like grain size, saturation roughness (**σ**), correlation length (**ε**) and roughness exponent (**α**).The crystalline structure of the TiO_2 films was determined by grazing angle X-ray diffraction (GAXRD) at 2° incidence angle, over the range [20 – 80]° using the $K\alpha_1$ line of Cu (λ = 1.5406 Å, 45 kV and 40 mA) in a X'pert-Pro Panalytical X ray diffractometer.

3. Results and discussion

3.1 Photocatalytic activity of TiO_2 films

The electronic spectrum of MB in water exhibits two characteristic absorption maxima at 292 and 668 nm, respectively. Upon photocatalytic treatment with UV light, in the presence of O_2, MB is known to decompose into a series of fragments before complete mineralization to produce inorganic acids (HCl, H_2SO_4, and HNO_3), CO_2 and water.

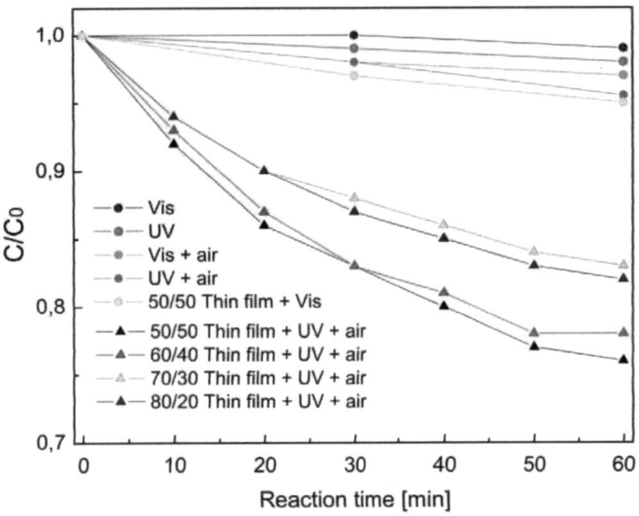

Figure 1. Relative concentration curves of MB aqueous solutions (initial concentration: $2.5x10^{-5}$ M) upon different experimental conditions.

Figure 1 shows the relative concentration, C/C_0, (at 292 nm) of $2.5x10^{-5}$ M MB aerated test solutions subject to UV irradiation in the presence of TiO_2 films prepared in Ar/O_2 mixtures and several control experiments [7]. After 1 h of UV irradiation, in aerated solutions, the best reaction rates are achieved with the 50/50 TiO_2 films, indicated by the 24% decrease in absorbance (black triangles), which, in addition, means that complete mineralization of MB does not happen under our experimental conditions. For the 60/40 films (red triangles), the absorbance change is slightly lower (22%). Apparent reaction rates are lower with the 70/30 and 80/20 films (blue and green triangles, respectively, absorbance decrease of ca. 18%, each). A control experiment, performed by irradiating the test solution with visible light in the presence of a 50/50 TiO_2 film (light blue circles), showed a total absorbance decrease of 5 %. Other control experiments, including irradiation of the aerated and deaerated test solution with UV and visible light in the absence of TiO_2 showed absorbance changes lower than 5 %. Henceforth, we will use the 50/50 samples to compare the stuctural, compositional and morphological changes before and after photocatalytic activity in MB solutions.

3.2Changes of TiO_2 films upon photocatalytic activity

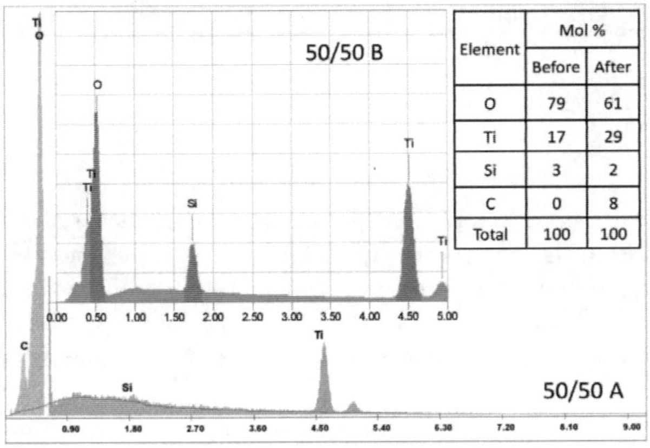

Figure 2.EDX spectra of a 50/50 TiO_2 thin film before and after photocatalytic treatment of MB.

Figure 2 shows the Energy Dispersive X-ray (EDX) analysis of a 50/50 TiO_2 thin film before and after photocatalytic treatment of MB solutions. After reaction, the major spectral change is the appearance of a carbon peak in the 50/50 TiO_2 film, indicating adsorption of MB decomposition intermediates on the film.

The XRD patterns for a 50/50 TiO_2 thin film, before and after photocatalysis, are shown in Figure 3.

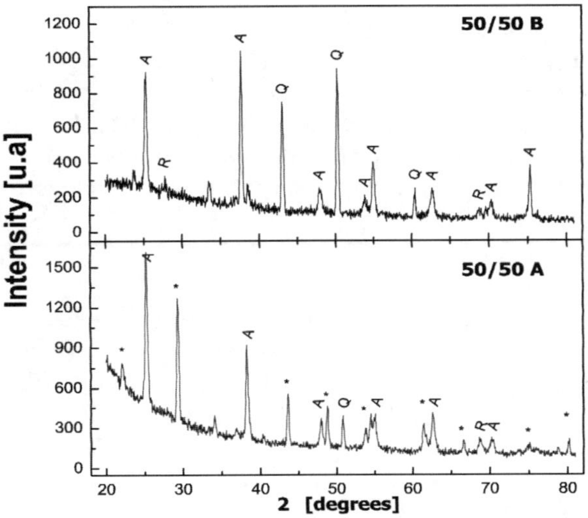

Figure 3. X-ray diffraction patterns of a 50/50 TiO_2thin film, before and after photocatalytic treatment of MB.A=Anatase, R=Rutile Q=Quartz and* additional phases.

As grown 50/50 TiO_2 thin films exhibit 90% anatase and 9% rutile domains. The diffraction patterns show that, upon photocatalytic activity, the peaks of the quartz substrate (Q) and the rutile (R) phase practically disappear and new diffraction peaks (* at 22.09, 29.35, 43.62, 53.7, 62.49, 75.07, and 80.01 degrees) of other phases emerge, presumably associated to the chemical intermediates of the degradation of MB. In addition, the higher background in the diffraction pattern of the TiO_2 sample after photocatalysis is consistent with the appearance of some amorphous phases in the sample.

Figure 4 shows the changes in roughness and grain size of the TiO_2 films, before and after photocatalysis, studied by atomic force microscopy in noncontact mode.

Figure 4.AFM images of a 50/50 TiO_2film before and after photocatalytic oxidation of methylene blue. Window size: 500 nm.

Figure 4 indicates that particle agglomeration occurs upon photocatalysis. Control experiments show that this phenomenon does not result from exposure to UV or visible light alone. Table 1 summarizes the morphological parameter values obtained from the fit of the size dependent roughness as a function of window size for the 50/50 TiO_2 thin film. The TiO_2 grain size is twice as large, while the saturation roughness decreases, suggesting abrasion of the surface as if the density of the film decreased along the vertical axis.

Table 1. Summarized morphological AFM parameters.

L=500nm	50/50 B	50/50 A
Grain size [±1nm]	39	81
σ [±0.1nm]	3.1	2.9
ε [±0.01nm]	0.3	0.51
α [±0.02nm]	0.35	0.43

5. Conclusions

Relative concentration curves of MB aqueous solutions in the presence of TiO_2 films deposited at different Ar/O_2, show that the 50/50 TiO_2 thin films, composed of 90% anatase and 9% rutile phases (as determinate from XRD analysis), exhibit the highest photocatalytic activity toward the degradation of MB. In addition, data processing of AFM images indicates that the TiO_2 film grain size doubles upon one single photocatalytic experiment, while the saturation roughness decreases. Further studies on whether rutile to anatase phase conversion and formation of new Ti-C and/or Ti-S phases can occur under the photocatalytic treatment are under way.

6. Acknowledgements

Funding for this work has been provided by "El patrimonio Autónomo Fondo Nacional de Financiamiento para la Ciencia, la Tecnología y la Innovación Francisco José de Caldas" (under contract RC 275-2011 with The Center of Excellence for Novel Materials, CENM) and Universidad del Valle through grant CI7899-2012.

7. References

[1] Kumar S G and Devi L G 2011 *J. Phys. Chem.* A.**115** 13211–13241
[2] Mills A and Lee S-K 2002*J.Photochem.Photobiol.*A**152** 233-247
[3] González L A *et al.*2004*Applied Catal.* B.**52** 69-77
[4] Chen S-Z*et al.* 2006*Applied Surf. Sci.***252** 7532-7538
[5] Uyguner S and Bekbolet C 2009*M. J. Adv. Oxidation Tech.***12(1)** 87-92
[6] Horcas I, Fernández R, Gómez-Rodríguez J M, Colchero J, Gómez-Herrero J, and Baro A M2007 *Rev. Sci. Instrum.***78** 013705
[7] Franco L M, Zambrano G A and Gómez M E 2012 *Superficies y Vacío* **25(3)** 161-165

Journal of Physics: Conference Series **480** (2014) 012027 doi:10.1088/1742-6596/480/1/012027

Applied electric field and hydrostatic pressure effects on quasistationary states in single and coupled GaAs–(Ga,Al)As quantum wells

A M Schönhöbel, J A Girón and N Porras-Montenegro

Group of Solid-State Theoretical Physics, Departamento de Fisica, Universidad del Valle, A.A 25360, Cali,Colombia

E-mail: jaime.giron@correounivalle.edu.co

Abstract. We have studied the effects of hydrostatic pressure and an uniform electric field on the electron energy levels GaAs–(Ga,Al)As single quantum wells (QWs) and coupled double quantum wells (DQWs) by using the Enderlein's method to solve exactly the Schrödringer equation. Numerical results were obtained using the density of states (DOS) as a function of the applied electric field, hydrostatic pressure, Al concentration, and the geometry as well. We found that the quasistationary ground and excited states energy diminish with and the applied electric field, increase with the confinement potential and the width of central barrier in the DQW. In the latter structure we observed the anti-crossing between the first and second quasistationary energy levels. We found that the applied electric field and the hydrostatic pressure modify the period of Pulsations in QWs.

1. Introduction

The physics of low dimensional semiconductor systems has been widely studied since the last century due to its potential application on the optoelectronic devices construction. Most of the studies on semiconductor systems have been carried out on GaAs–(Ga,Al)As heterostructure, especially single and multiple quantum wells, where GaAlAs is used as the barrier material and therefore the electron mobility is confined within the GaAs.

There have been several works devoted to the understanding the role of the applied electric field without considering the exactly solution of the wave function in the barrier regions [1–3] and others who have taken into account the influence of hydrostatic pressure on the electron states in these structures [4–9]. D. E. Aspnes [4], S. Adachi [5], B. Welber *et al.* [6] and R. C. Miller *et al.* [7] determined the carrier's masses (electrons and holes), the height barriers, the bandgap and other properties of heterostructures experimentally as a function of hydrostatic pressure, besides adjusting the respective constants. In this paper, we used the exact solution proposed by R. Enderlein *et al.* [10, 11] to calculate the effects of hydrostatic pressure and uniform electric fields on the electron energy levels in GaAs–(Ga,Al)As in QWs and DQWs.

2. Theoretical Framework

We performed the study of the effects of an electric field F applied on the grown direction $-z$, as well as those of an applied hydrostatic pressure in QWs and DQWs. The DQW heterostructure

Journal of Physics: Conference Series **480** (2014) 012027 doi:10.1088/1742-6596/480/1/012027

is compound by two GaAs wells, a central $Ga_{1-y}Al_yAs$ barrier, and by semi-infinite $Ga_{1-x}Al_xAs$ barriers in its extremes. The Schrödinger equation is given by

$$\left(-\frac{\hbar^2}{2m}\frac{d}{dz}\frac{1}{m^*(z)}\frac{d}{dz} - eFz + V(z)\right)\varphi_E(z) = E\varphi_E(z), \tag{1}$$

where $V(z)$ is the confinement potential for a QW or DQW, and the action of hydrostatic pressure is included in the electron effective-mass m^*. We will focus on theory corresponding to DQWs, considering the length of the DQW heterostructure is L, the electron wave function can be written as

$$\varphi_E(z) = \begin{cases} A_i\left(-\frac{z}{l_{bF}} + \frac{V_0-E}{\hbar w_{bF}}\right), & z < -L/2, \\ \phi_E(z), & -L/2 \leq z \leq L/2, \\ a_3 A_i\left(-\frac{z}{l_{bF}} + \frac{V_0-E}{\hbar w_{bF}}\right) + b_3 B_i\left(-\frac{z}{l_{bF}} + \frac{V_0-E}{\hbar w_{bF}}\right), & z > L/2, \end{cases} \tag{2}$$

where ϕ_E is the solution to Schrödinger's equation in $-L/2 < z < L/2$, $w_{b,wF} = \left(\frac{e^2F^2}{2\hbar m^*_{b,w}}\right)^{1/3}$ and $l_{b,wF} = \left(\frac{\hbar^2}{2eFm^*_{b,w}}\right)^{1/3}$, the subscripts b and w are for barrier and well, respectively. ϕ_E can be written within each well or barrier, as a lineal combination of Airy functions of first and second type. The interfaces of the heterostructure can be connected through the transfer matrix $S(E)$ [10, 11] and the corresponding coefficients can be found using the continuity conditions of ϕ_E and ϕ'_E/m^*. From above, we can determine the DOS for a DQW under the action of an electric field [10, 11].

$$\rho(E) = \frac{1}{\hbar w_{bF} l_{bF}}\frac{1}{|S_{11}|^2 + |S_{21}|^2}, \tag{3}$$

where S_{11}, S_{21} are related with structural and energy parameters of the heterostructure. The energies are calculated finding the energy values which the denominator of Eq. 3 has a local minima.

The application of hydrostatic pressure modifies the lattice constants, barrier height, effective masses and dielectric constants. $m^*(P,T)$ is the pressure- and temperature-dependent electron effective-mass given by [5, 12]. We use for the the pressure - and temperature - dependent energy gap for the GaAs the expression given by [9], also the gap diferences between both semiconductors, and it dependence on P and x are given in reference [4, 6]. Finally the hydrostatic pressure-dependent QW height is obtained from the fractional change in the volume of the zinc-blende structure as given in reference [13].

3. Results and Discussion

For a GaAs-(Ga,Al)As QW, the applied electric field makes the energy of the ground and excited states to decrease, being more notorious for ground and first excited state, even for weak electric fields, as it is shown in Fig. 2(a), the lines E_a, $E_{a'}$, E_b and $E_{b'}$ are related with the points marked in the energy profile of the heterostructure as displayed in Fig. 1. The energy of ground, first, and second electron excited states as a function of the electric field for three values of the applied hydrostatic pressure is presented in Fig. 2(b). As expected, the energy of the ground and excited states diminishes with the well width and the applied electric field for a given value of x, as it is presented in Fig. 2(c). On the other hand, Fig. 2(d) presents the competition between the electric eld which makes the energy of the electron states to diminish and the confinement potential, which makes it to increase with the barrier width. As it is observed, both effects together make the energy of the ground state to stabilize, for x values greater than a particular one, depending on the applied electric field. The behavior of the ground and first excited

Journal of Physics: Conference Series **480** (2014) 012027 doi:10.1088/1742-6596/480/1/012027

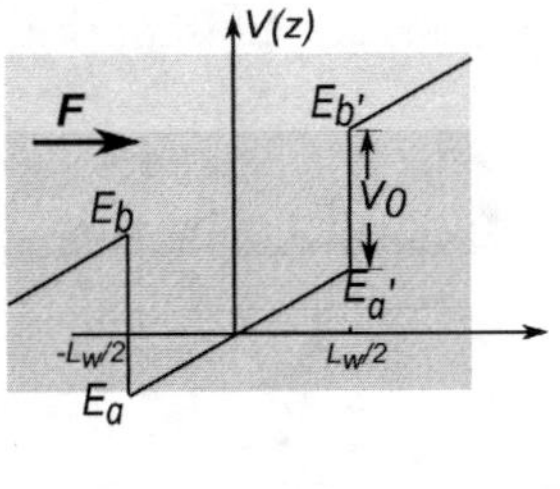

Figure 1. QW Energy Profile

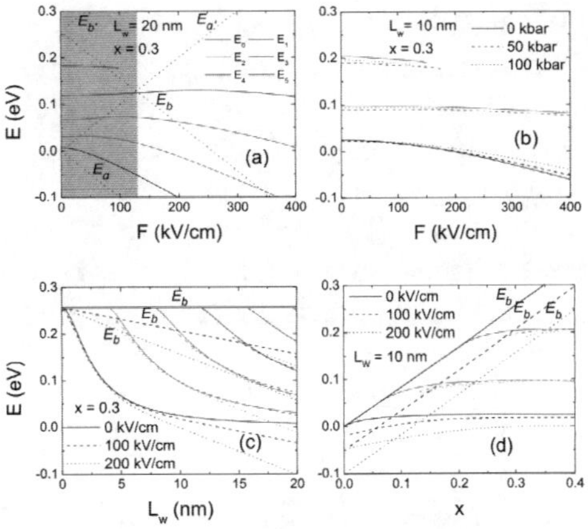

Figure 2. Quasistationary ground and first excited states in a GaAs–$Ga_{1-x}Al_xAs$ QW as a function of (a) applied electric field, F, (b) applied electric field for different values of applied hydrostatic pressure, P (c) well width, L_W, for different values of electric field (d) Al concentration, x, for different values of electric field.

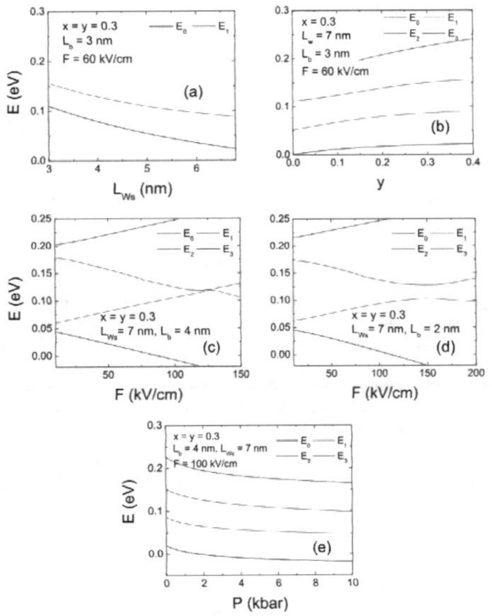

Figure 3. Ground and first excited states in GaAs–(Ga,Al)As DQW as a function of (a) well widths,L_{Ws} (b) Al concentration of the central barrier, y, (c) applied hydrostatic pressure, P,(d) electric field, F, the anti-crossing phenomenon is observed at F_c=130.2 kV/cm, (e) electric field, the anti-crossing phenomenon is observed at F_c=149.3 kV/cm

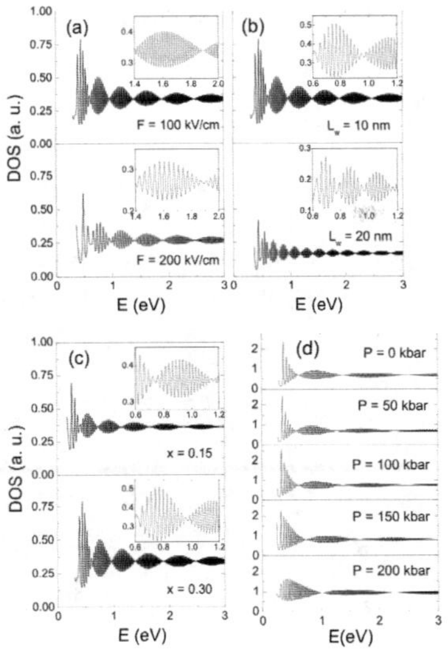

Figure 4. DOS in a QW (pulsation region) as a function of energy for different (a) applied electric fields with $x = 0.3$, $L_W = 10nm$ and $P = 0kbar$ (b) well widths with $x = 0.3$, $F = 100kV/cm$ and $P = 0kbar$ (c) Al concentrations with $F = 100kV/cm$, $L_W = 10nm$ and $P = 0kbar$, and (d) applied hydrostatic pressure, P with $x = 0.3$, $F = 100kV/cm$ and $L_W = 10nm$.

state energy in a DQW as a function of the well width for $x = 0.3$ and $y = 0.3$, is presented in Fig. 3(a). Note as expected, that for both states diminish with the well width and for both states the energy is lower than in the QW due to the lower confinement the electron feels in the DQW, despite the presence of the central barrier. On the other hand, the height of the central barrier, which increases with y, modifies the energy of the ground and excited states augmenting its value, as it is shown in Fig. 3(b), due to the increment of the confinement potential. The applied hydrostatic pressure makes the energy of the ground and excited states to decrease, as it is shown in Fig. 3(e). The anti-crossing between the first and second excited electron states is observed in Figs. 3(c) and 3(d), as a function of the applied electric field for two values of central barrier width, $L_b = 4nm$ and $L_b = 2nm$, respectively. Note that the anti-crossing strengthens for the case of smaller barrier width because the ground and excited states behave like they were QW states.

In Fig. 4 we present the DOS for a QW as a function of energy in the region of pulsations ($E > E_b$, as displayed in Fig. 1). In panel (a) it is observed that the period of the fast oscillations increases with the applied electric field for a given well width. The number of nodes in a given energy region is proportional to the well width as it is shown in panel (b). However, panel (c) shows that while the number of nodes does not vary with x, the DOS of the pulsations increases with it. On the other hand in panel (d) it is noted a tendency in increasing the number of nodes with pressure up to 50 kbar, but to diminish for pressures greater than this value. This result is related with the confinement potential with the applied hydrostatic pressure, as established in the work by Reyes *et al*[3]. As commented for QW the period of the fast oscillations increases with the applied electric field. These results can be used to interpret transitions between rotational and vibrational molecular states, which are in the energy range of the fast oscillations.

4. Conclusions

In summary, in this work by using the Enderlein's method we have studied the effects of hydrostatic pressure and an uniform electric field on the electron energy levels in GaAs–(Ga,Al)As QWs and DQWs. We found that the applied electric field makes the energy of the quasistationary ground and excited states to decrease, being more notorious for the ground and first excited state, even for weak electric fields. The ground and excited states energy diminish with the well width and the applied electric field, increase with the confiniment potential and width of the centarl barrier. In the latter structure we have found the anti-crossing phenomenon between the first and second quasistationary energy levels for a specific electric field, being more apparent for smaller central barrier width. We found that the period of Franz-Keldysh oscillation type in QWs increases with the applied electric field and that the number of nodes of pulsations augment with the well width.

References

[1] Bastard G, Mendez E E, Chang L L and Esaki L 1983 *Phys. Rev. B* **28** 3241
[2] López-Gondar J, d'Albuquerque e Castro J and Oliveira L E 1990 *Phys. Rev. B* **42** 7069
[3] Raigoza N, Duque C A, Reyes-Gómez E and Oliveira L E 2005 *Phys. B* **367** 267
[4] Aspnes D E 1976 *Phys. Rev. B* **14** 5331
[5] Adachi S 1985 *J. Appl. Phys.* **58** R1
[6] Welber B, Cardona M, Kim C K and Rodriguez S 1975 *Phys. Rev. B* **12** 5729
[7] Miller R C, Kleinman D A, Tsang W T and Gossard A C 1981 *Phys. Rev. B* **24** 1134
[8] Kopf R F, Herman M H, Schnoes M, Perley A P, Livescue G and Ohring M 1992 *J. Appl. Phys.* **71** 5004
[9] Ehrenreic H J 1961 *J. Appl. Phys.* **32** 2155
[10] Enderlein R, Holz T and Gondar J L 1989 *Phys. Stat. Sol. B* **156** 259
[11] Enderlein R 1990 *Phys. Rev. B* **42** 4708
[12] Lopez S Y, Porras-Montenegro N and Duque C A 2005 *Phys. B* **362** 41
[13] Elabsy A M 1994 *j. Phys. Condens. Matter* **6** 10025

Journal of Physics: Conference Series **480** (2014) 012028 doi:10.1088/1742-6596/480/1/012028

Conductance fluctuations in high temperature critical superconducting system based on $Bi_2Sr_{1.4}(Pb)_{0.6}Ca_{2-x}Y_xCu_3\,O_{10+y}$. thick films

W Saldarriaga[1], W Castro[2] and E Baca[3]

[1] Laboratorio de Materiales Cerámicos y Vítreos, Universidad Nacional de Colombia, Sede Medellín, A.A. 568, Medellín, Colombia.

[2] Proyecto Curricular de Ingeniería Topográfica. Universidad Distrital Francisco José de Caldas. A. A. 8668. Bogotá D. C.

[3] Grupo de Ingeniería de Nuevos Materiales, Universidad del Valle, A.A. 25360 Cali, Colombia.

E-mail adress: wjsaldarriagaa@unal.edu.co.
Received 4 October 2013, in final form XX Novembrer 2013

Abstract. $Bi_2(Sr_{1.4})Pb_{0.6}Ca_{2-x}Y_xCu_3O_{10+y}$ superconducting thick films were prepared and deposited on MgO (100) substrates using a Melt-Quench-Annealing method (MQA).The superconducting transition was found around 89 K. Electrical conductance fluctuations and the thermodynamic fluctuations, close to Tc, were analyzed from results of R vs T measurements on the samples. Ytrium concentration range was in the range $0 \leq x \leq 0.8$. Additionally, considering the Aslamazov-Larkin law, with Tc value fixed according to d^2R/dT^2 criterion, the conductance trend in the samples was revealed. So, when $x \leq 0.075$ a 3D behavior was found, on the other hand when the concentration increases to $x = 0.8$ a 2D behavior is stabilized. From these characteristics a correlation to the structural properties of the sample can be inferred.

1. Introduction

One characteristic and relevant effect of the high-Tc superconductors (HTSCs) concerns the thermal fluctuations near the normal-superconducting transition. In the HTSCs, the combination of short coherence length, quasi-two-dimensional (2D) structures and high operating temperatures associated with large thermal energy $\sim K_BT$ causes a large increase in the effect of thermal fluctuations. Different models were proposed to explain the high magnitude of fluctuation effects in these high-Tc superconductors. Most of the features of fluctuation phenomena in the conductivity of HTSCs are explained on the basis of Aslamazov-Larkin (AL)[1] and Maki-Thompson (MT) theories [2]. The widely undertaken analysis of excess electrical conductivity, $\Delta\sigma$, HTSCs from different [1-3] fluctuation theories has concentrated mostly on investigating a crossover [4-6] from a 3D fluctuation (exponent λ, approaching 0.5) at low temperatures to a 2D one (λ approaching 1.0) on warming across a crossover temperature, T_0, and on the inter-layer coupling constant J [6-9]. The electrical measurement may provide relevant information on the dynamic behavior of vortices. The study of the temperature dependence on the fluctuation-enhanced conductivity above Tc is a way to probe the dimensionality of the order-parameter fluctuations.

Journal of Physics: Conference Series **480** (2014) 012028 doi:10.1088/1742-6596/480/1/012028

The layered structure of these systems with complex crystal chemistry and the inherent granularity lead to strong structural disorder at the microscopic and the mesoscopic levels, respectively. A large number of studies have been devoted to investigate the interplay between the superconducting order parameter fluctuation (SCOPF) and the inhomogeneities in HTSCs. The inhomogeneities occur at widely varying length scales as in single crystals, grain aligned films and melt-textured and sintered samples. mesoscopic inhomogeneities like grain boundaries, cracks, voids etc, have a much larger scale of length than that of coherence length ξ and, being temperature independent, do not influence the SCOPF region [10]. These inhomogeneities strongly influence the transition region below Tc. The microscopic inhomogeneities such as structural defects (twin boundaries, stacking faults) and chemical imperfections (i.e oxygen deficiencies) occur inside the grains. The scale of length of these inhomogeneities is smaller than that of the mesoscopic inhomogeneities, but still larger than ξ. These inhomogeneities therefore are expected to have negligible effect on the SCOPF region [10]. However, many studies have shown that both the mesoscopic and the microscopic inhomogeneities crucially influence the critical and the mean field region of SCOPF [11–14].

For a highly anisotropic layered material like $Bi_2Sr_2CaCu_2O_{8+y}$ (BSCCO) where the anisotropy ratio γ =0.02, the conductivity fluctuation is well-described [15,16] by the 2D AL theory in the wide temperature range $-4 \leq \ln(\varepsilon) \leq -2$, where $\varepsilon = (T - T_C)/T_C$ is the reduced temperature. However, fluctuation phenomena have been explained [17-19] in $YBa_2Cu_3O_{7-x}$ (YBCO) and in similar systems taking the MT theory into account as well. For very anisotropic materials, the MT contributions, which increase the apparent width of the transition, have been ignored because they are probably absent in HTSC as a consequence of the pair-breaking effect of strong inelastic electron scattering [20] and these are treated as a very clean system because ξ_0/l (where ξ_0 and l is the zero-temperature in plane coherence length and electronic mean free path, respectively) is very small.

The aim of this work is to analyze the contribution of the thermal fluctuations on the electric conductivity at zero magnetic field in the mixed state near Tc of the $Bi_2(Sr_{1.4})Pb_{0.6}Ca_{2-x}Y_xCu_3O_{10+y}$ (Bi -2223) thick film.

2. Experimental procedure

Superconductor precursor material was prepared by solid state reaction. Stoichiometric ratios of the precursor powders Bi_2O_3, $SrCO_3$, PbO_2, $CaCO_3$, Y_2O_3 and CuO (purity 99.99%) were finely ground and thoroughly mixed to finally obtain $Bi_2Sr_{1.4}(Pb)_{0.6}Ca_{2-x}Y_xCu_3O_{10+y}$.

A melt-quench and annealing process was applied for growing *c*-axis textured Bi-2223/MgO substrate with a dimension of 10 × 10 × 1 mm^3. In the melt-quench step, the Bi-2223/MgO precursor was pushed into a furnace which was preheated to a melting temperature of 1050 °C. After keeping the sample at the melting temperature for about 10 minutes, the sample was rapidly quenched to room temperature. The samples consisted mainly of the Bi-2223 phase. In the annealing step, the melt-quenched sample was annealed at temperatures of 865 °C in air to be converted to *c*-axis textured Bi-2223/MgO thick film.

The electrical resistance of the samples was obtained using the standard four-terminal configuration (Van Der Pauw technique) in a He closed cycle cryostat within the range of 14 K-300 K, a 2400 SourceMeter (Keithley) and a temperature controller 9700 (Leybold) (0.1 K resolution) were used.

3. Results and discussion

We have analyzed our resistance vs. temperature experimental data by adopting the simplest fluctuation conductivity approach [21] (see Fig.1). The fluctuation induced enhancement in conductivity, termed as conductance excess or paraconductivity, $\Delta G = G - G_R$, where G is the measured conductance, i.e., $G=1/R$, and $G_R=1/R_R$ is the regular term extrapolated from the high temperature behavior. According to the Aslamazov-Larkin (AL) proposal [1], the fluctuation conductivity diverges as a power law of the type $\Delta G = A\varepsilon^{-\lambda}$, where A is a constant and λ is the critical exponent.

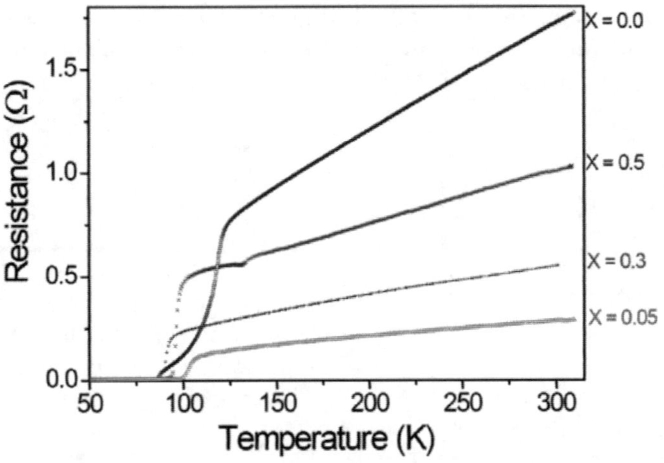

Figure 1. Temperature dependence of electrical resistance of $Bi_2Sr_{1.4}(Pb)_{0.6}Ca_{2-x}Y_xCu_3O_{10+y}$. thick films for different Y contents.

The main problem in the SCOPF analysis is related to the difficulty of assessing the mean-field critical temperature *Tc*. The *Tc* in each theoretical model is never directly accessible due to the presence of fluctuation effects, which round the critical behavior of any observable near *Tc*. In the attempt to directly determine *Tc* from the experimental data, Han *et al* [22] extrapolated the excess conductivity raised to the inverse of critical exponent λ, where *Tc* is determined from the intercept on the temperature axis from the mean field region. Direct determination of *Tc* has also been attempted by analyzing the temperature dependence of the inverse of the logarithmic temperature derivative of fluctuation conductivity [21]. Both methods, however, do not rule out error in *Tc* determination since the exponent in the SCOPF region varies over a large range. We have used the derivative of the resistance data as a function of temperature (*dP*/*dT*), with *Tc* defined as the peak position of this derivative (Fig. 2) as has been done by many [21,23,24]. The effect of uncertainty in the determination of the *Tc* results in a temperature range corresponding to the critical region, where the excess conductivity depends sensitively on the reduced temperature [25,26]. Therefore, we have selected a temperature window around the mean field region and restricted our data analysis mainly to this region.

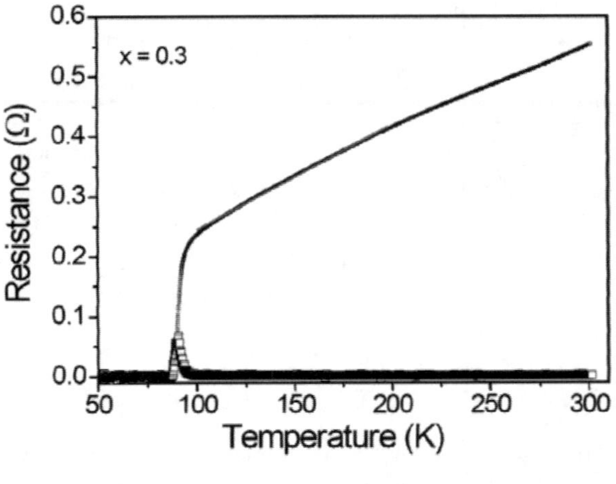

Figure 2. Representative plot of the resistance and temperature derivative of resistance near Tc as a function of temperature in $Bi_2Sr_{1.4}(Pb)_{0.6}Ca_{2-x}Y_xCu_3O_{10+y}$ thick films.

The AL relation is based on the Ginzburg–Landau (GL) mean field theory and is valid only in a mean field temperature range between ~1.01Tc and 1.1Tc (see Fig. 3). We can summarize the results in Table 1, for values of $\varepsilon < 0.00001$ the paraconductivity does not vary significantly, indicating that the coherence length begins to be significant to these temperatures. λ values as obtained from the adjustment can be seen in the Table 1 are of the order of unity for x> 0.075, which corresponds to a

2D conduction process as predicted by the equation $\lambda = 2 - D/2$ (AL model), where D represents the dimensionality for the fluctuation spectrum. Log-Log plots of excess conductivity as a function of reduced temperature for $Bi_2Sr_{1.4}(Pb)_{0.6}Ca_{2-x}Y_xCu_3O_{10+y}$ on MgO substrates for different Ytrium contents can be seen in fig 4.

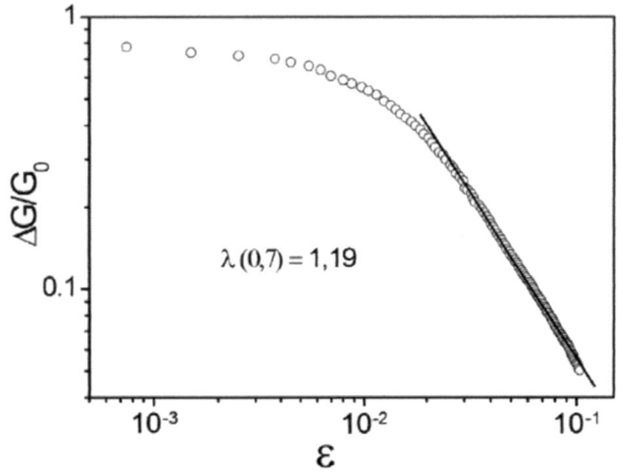

Figure 3. Effect of different Tc values on excess conductance. The sensitivity on the transition temperature is more pronounced in the critical region close to Tc.

Table 1. λ values obtained from the AL relation.

Ytrium Concentration	λ	T_C (dR/dT)	Ytrium Concentration	λ	T_C (dR/dT)
X = 0.000	0.255	106.78	X = 0.200	0.876	93.01
X = 0.025	0.472	121.24	X = 0.300	0.904	89.70
X = 0.050	0.431	103.42	X = 0.500	1.190	95.82
X= 0.075	0.329	92.26	X = 0.700	1.190	94.58
X = 0.100	1.060	89.00	X = 0.800	1.190	94.83

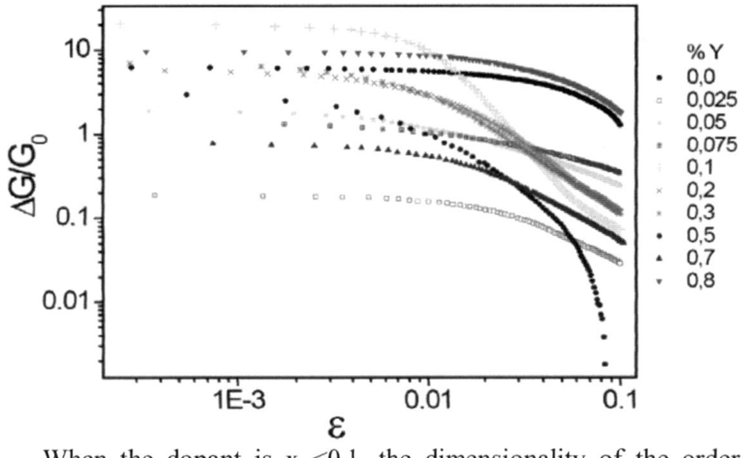

Figure 4. Log-Log plots of excess conductivity as a function of reduced temperature for $Bi_2Sr_{1.4}(Pb)_{0.6}Ca_{2-x}Y_xCu_3O_{10+y}$ thick films for different Y contents.

When the dopant is x <0.1, the dimensionality of the order parameter tends to 3D, while the concentration x > 0.1 tends to 2D. (see Table 1). According to results reported to YBCO, the fluctuations of dimensionality are 2D or 3D, depending on the nature of the samples while for BSCCO are quasi 2D for polycrystalline samples. Two-dimensional behavior in the Bi system exhibits a decoupling of the CuO planes and is similar to the YBCO system, showing 3D behavior. This analysis shows systematic changes on the paraconductivity in the critical exponent for the yttrium-doped

Journal of Physics: Conference Series **480** (2014) 012028 doi:10.1088/1742-6596/480/1/012028

BSCCO system; in our case, λ tends to increase with the concentration of yttrium, having a value of close to half (3D behavior) in 0 <x <0,1 tends to saturate between 1 and 3/2 (possibly 2D) 0.1 <x <0.8 for higher levels of dopant. Other changes of the exponent have already been reported [23]. Where it has been found that the increase of the concentration of Ni in the system (Bi, Pb)SrCaCuO λ increases, it is consistent with our results.

In summary, the conduction behavior, even with low dimensionality of the order parameter observed in our samples, is due to yttrium doping which induces the percolative transport. This interpretation suggests that the yttrium ions substitute Cu not only in the pyramid but also in the arrangement of square planar Cu-O centers

References

[1] G L Aslamazov, A I Larkin, *Phys. Lett.* **26** (1968) 238.

[2] K Maki, R S Thompson, *Phys. Rev. B* **39** (1988) 2767.

[3] W E Lawrence, S Doniach, in: E. Kanada (Ed.), *Proceedings of the 12th International conference on Low temperature Physics*, Kyoto, Japan, Keigaku, Tokyo, (1980) 361.

[4] J Y Juang, M C Hsieh, C W Luo, T M Uen, K H Wu,Y S Gou, *Physica C* **329** (2000) 45.

[5] [5] S K Bandyopadhyay, P Barat, S K Kar, Udayan De, P Mandal, B Ghosh, C K Majumdar, *Solid State Commun.* **82** (1992) 397.

[6] S K Bandyopadhyay, P Sen, P Barat, Udayan De, K Mandal, S K Kar, C K Majumdar, *Physica C* **267** (1996) 303.

[7] D Sanyal, K Mandal, Udayan De, *CPTA-98*, Saha Institute of Nuclear Physics, Calcutta, Feb, 1998.

[8] Udayan De, K Mandal, D Sanyal, D Banerjee, C K Majumdar, *Regional Meeting on Radiation Physics*, Bose Institute, Calcutta, November 1995.

[9] Udayan De, A Sarkar, D Sanyal, D Banerjee, R Kumar, *Proceedings of the International Workshop on High-Tc Superconductors*, Rajshahi University, Bangladesh, 2-6 November (1998) 483.

[10] Maza J and Vidal F *Phys. Rev.* B **43** (1991) 10560.

[11] Cukauskas E J and Allen L H. *J. Appl. Phys.* **84** (1998) 6187.

[12] Aswal D K, Singh A, Sen S, Kaur M, Viswandham C S, Goswami G L and Gupta S K *J. Phys. Chem. Solids* **63** (2002) 1797.

[13] Sato T, Hakane H, Yamazaki S and Mori N *Physica* C **372** (2002) 1208.

[14] Sidorenko A S, Zdravkov V I, Ryazanov V V, Klemm M, Horn S, Tideeks R, Muller C and Wixforth A *J. Supercond.* **17** (2004) 211

[15] D H Kim, A M Goldman, J H Kang, K E Gray and R T Kampwirth. *Phys. Rev. B* **39** (1989) 12275.

[16] V Calzona, M R Cimberle, C Ferdeghini, D V Livanov, D Marre, M Putti, A S Siri, G Balestrino and E Milani, *Solid State Commun.* **87** (1993) 397.

[17] Y Matsuda, T Hirai, S Komiyana, T Terashima, Y Bando, K Iijima, K Yamamoto, and K Hirata, *Phys Rev. B* **40** (1989) 5176.

[18] U Welp, S Flesher, W K Kwok, R A Kñemm, V M Vinokur, J Downy, and G W Crabtree, *Phys. Rev. Lett.* **67** (1991) 3180.

[19] K H Kim, K E Gray, and M D Trochet, *Phys. Rev. B* **45** (1992) 10801.

[20] Thomas Lemberger and Liam Coffexy, *Phys. Rev. B* **38** (1988) 7058 .

[21] P Pureur, R Menegotto Costa, P Rodríguez, Jr., J Schaf and J V Kunzler, *Phys. Rev. B* **47** (1993) 11420.

[22] S H Han and M Anderson, *Physica B* **194** 196 (1994) 1513.

[23] Vidal F, Torron C, Vina J and Mosqueira J *Physica* C **332** (2000)166.

[24] Demirel A I *Supercond. Sci. Technol.* **14** (2001) 978.

[25] N Mori, J A Wilson, H Ozaki, *Phys Rev B*, **45** (1992) 10633.

[26] Q Y Ying and H S Kwok, *Phys. Rev. B.* **42** (1990) 2242.

Journal of Physics: Conference Series **480** (2014) 012029 doi:10.1088/1742-6596/480/1/012029

Study of exchange bias effect in bilayers based on $La_{2/3}Ca_{1/3}MnO_3$ and $BiFeO_3$

C Domínguez, J E Ordoñez, M E Gómez, W Lopera
Thin Film Group, Department of Physics, Universidad del Valle, A.A. 25360 Cali, Colombia.
E-mail: claribeldominguezo@gmail.com

Abstract.
$BiFeO_3$ (BFO) is an interesting material due to the coexistence of magnetic and ferroelectric order at room temperature. Moreover, mixed manganese perovskite oxides like ferromagnetic $La_{2/3}Ca_{1/3}MnO_3$ (LCMO-F) present remarkable properties, like colossal magnetoresistance and non-volatile memory. Bilayers of BFO/LCMO-F and LCMO-F/BFO were grown on (100) $SrTiO_3$ (STO) substrates. LCMO-F films were deposited by DC sputtering and BFO by RF magnetron sputtering technique at high-oxygen pressures. Samples were structurally characterized by x-ray diffractometry. Electrical properties were evaluated by resistivity measurements as function of temperature, showing how the transition temperature of manganite is affected by the presence of the BFO layer. Magnetic characterizations were carried out by magnetization measurements as function of temperature and magnetic field. BFO/LCMO-F/STO bilayer showed a higher Curie temperature than LCMO-F layers grown by similar parameters. Magnetic hysteresis curves for LCMO-F/BFO/STO bilayers gave us indication of the existence of exchange bias effect with a Hex of 200 Oe after field cooling at 1 T; we can also conclude that this bilayer configuration offers posibilities to obtain bilayers exhibiting this polarization effect.

1. Introduction

Multiferroics materials are those which exhibit multiple functional properties in which two or all three of the properties ferroelectricity, ferromagnetism and ferroelasticity occur simultaneously in the same phase [1]. $BiFeO_3$ (BFO) is a multiferroic material which structure and magnetic ordering has been study by having two types of long range order: antiferromagnetic ordering below $T_N = 370°C$ and ferroelectric ordering, with a high Curie temperature about $820 - 850°C$. $BiFe0_3$ is known to have a rhombohedrally distorted perovskite structure owing to R_{3C} symmetry [2]. Moreover, mixed manganese perovskite oxides present remarkable properties, like colossal magnetoresistance and non-volatile memory. $La_{2/3}Ca_{1/3}MnO_3$ (LCMO-F) is widely studied due to its magnetic and transport properties; has a ferromagnetic phase and perfect cubic perovskite structure and exhibits paramagnetic-ferromagnetic transition simultaneously with a metal-insulator transition around 270K for thin films systems [3].

Previously, we have grown multiferroic BFO [4] and ferromagnetic LCMO-F thin films with high quality [5], for that reason we can explore, in this work, the behavior of ferromagnetic LCMO and antiferromagnetic BFO bilayers to seek Exchange Bias (EB). This phenomenon manifests itself by a shift in the hysteresis loop of a ferromagnetic in contact with an antiferromagnetic and arises from the exchange coupling at the FM/AF interface [6]. In this report, we shall give an indication of the possible existence of Exchange Bias effect in LCMO-F/BFO/STO bilayers.

2. Experimental Details

We have grown BFO/LCMO-F and LCMO-F/BFO bilayers on (100) $SrTiO_3$ (STO) substrates by sputtering technique at oxygen atmosphere (O_2 99.999%). The LCMO-F layer was deposited using DC sputtering with 830°C susbtrate temperature, oxygen pressure at 3.5 mBar, and a estimated deposition rate of 80 nm/h, whereas BFO layer was grown using RF magnetron sputtering with 600°C as subtrate temperature, oxygen pressure at $4.0x10^{-1}$ mBar and 72 nm/h deposition rate [4][7]. Samples were structurally characterized by x-ray diffraction (XRD), electrical properties were evaluated by resistivity measurements as function of temperature by DC four points resistance, Van der Pauw technique, in a physical properties measurements sytems (PPMS) quantum design. Magnetic characterizations were carried out by magnetization measurements as function of temperature (10 to 300K) and Field colling (FC) isothermal M(H) loops for temperatures of 10 and 60K using the vibrating sample magnetometer (VSM) probe of the PPMS systemm and SQUID magnetometer (Superconducting Quantum Interference Device) with a magnetic field applied in the plane of the film. We used LCMO-F/BFO/STO and BFO/LCMO-F/STO notation for the two possible configurations of our bilayers.

3. Results

Figure 1, shows the X-ray diffraction pattern for both configurations of our bilayers. LCMO-F/BFO/STO (a) shows characterisitic peaks like [104], [110] of the BFO layer, whereas Bragg peaks from LCMO-F layer are not visible. However, this sample show magnetic properties as you can see later. BFO/LCMO-F/STO bilayer (b), presents a highly oriented growth of the ferromagnetic layer of manganite around (200) reflection of the substrate, respect to BFO layer peak we observed [012] anchored to the [100] peak of the substrate. This sample show us a textured growth regarding to the (100) substrate direction. Bragg peaks noticed with asterisk corresponds to other phases of bismuth oxide present on BFO films.

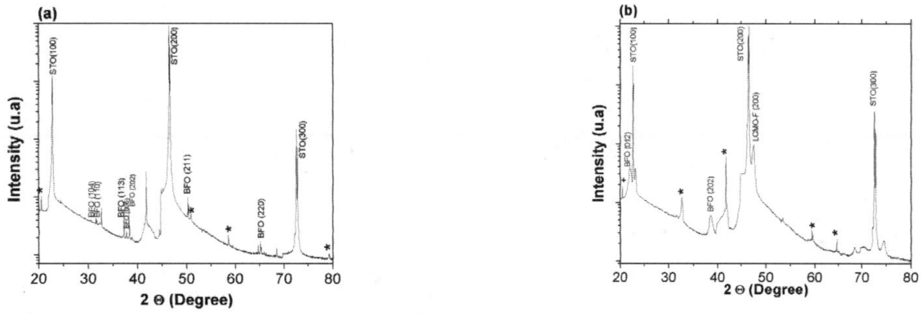

Figure 1. XRD patterns for LCMO-F/BFO/STO (a) and BFO/LCMO-F/STO (b) configurations.

Figure 2, displays the temperature dependence of the resistance curves, in the range of 10 - 300 K for 100 nm LCMO-F film (a), LCMO-F/BFO/STO (b) and BFO/LCMO-F/STO (c). LCMO-F/BFO/STO (b) shows typical ferromagnetic metal isolator transition characteristic of the LCMO system, with a transition temperature, T_{IM} of 225±25K; this transition temperature is lower than that for a 100 nm LCMO-F film (a) grown under identical parameters. Moreover, BFO/LCMO-F/STO sample (c) contact electrodes are on the BFO insulator layer and the resistance curve exhibits predominant behavior of insulating phase at low temperatures, even then it is possible to observe a decrease of the resistivity for temperatures below 150 K arising from the metal behavior of the LCMO-F layer. Our resistance measurements are carried out in the plane of the layers. Thus, we have contribution to the resistance from the two layers and also from the interface, but we cannot discriminate each.

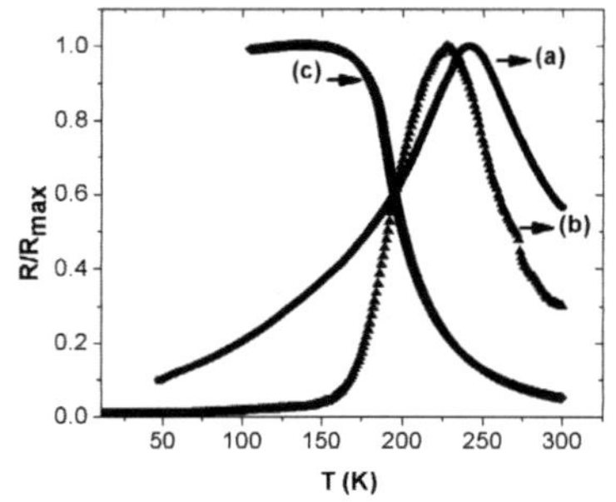

Figure 2. Resistance (normalized) as function of temperature forLCMO-F (a), LCMO-F/BFO/STO (b) and BFO/LCMO-F/STO (c).

Figure 3, shows the zero filed cooling (ZFC) thermal demagnetization behavior for LCMO-F/BFO/STO sample (b), and BFO/LCMO-F/STO sample (c). The behavior are dramatically different. BFO/LCMO-F/STO bilayer, where the LCMO-F layer grew directly on substrate, shows the typical demagnetization behavior of the LCMO-F system (a), whereas the LCMO-F/BFO/STO bilayer an atypical ferromagnetic response, with a maximun in M a very low temperature (50 K) and for temperatures below the magnetization drecreases dramatically; these results indicate that the magnetic property of the LCMO-F film is strongly dependent of its crystalline structure and morphology. LCMO-F grew very well on STO substrates; but after growing the BFO layer on top of LCMO-F, its magnetic properties decrease. We could determine the Curie temperature of 247.1 ± 0.1 K for the BFO/LCMO-F/STO bilayer using a least squares fit with a gaussian distribution of ferromagnetic regions [7].

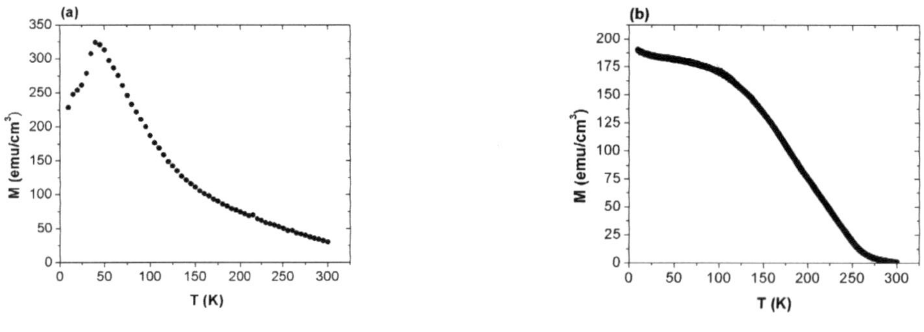

Figure 3. Magnetization as function of temperature, ZFC, for LCMO-F/BFO/STO sample (a), and BFO/LCMO-F/STO bilayer (b).

Figure 4, shows isothermal magnetic loops at 10 K (blue) and 60 K (red) for LCMO-F/BFO/STO sample (a) and BFO/LCMO-F/STO bilayers (b). Both samples are ferromagnetic at the measured temperatures. However, BFO/LCMO-F/STO bilayer is magnetically very poor, and we can not measured a shift in the hysteresis loop after field cooling, as we can see in the inset. On the other hand, LCMO-F/BFO/STO bilayer show clearly ferromagnetic behavior exhibiting a weak exchange bias effect, see inset. Coercive field, width of the hysteresis curve, exchange bias magnitud increase with the decreasing of the temperature as expected for the LCMO system.

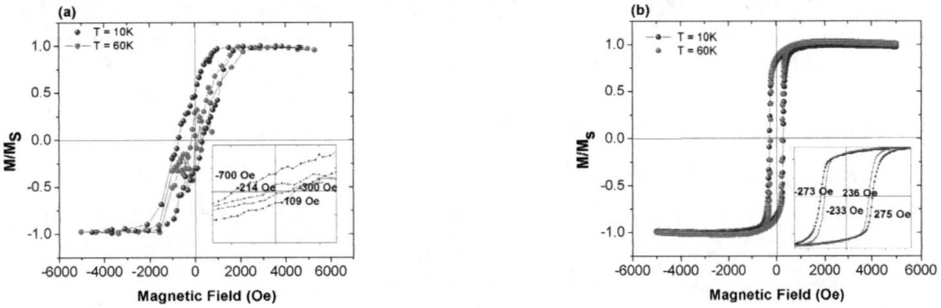

Figure 4. Magnetic hysteresis loops at 10 and 60K for LCMO-F/BFO/STO (a) and BFO/LCMO-F/STO (b).

The magnitude of the shift, denoted as exchange field (H_{EX}) was measured over hysteresis isotherms according to the average between the two coercive fields and we found the most significant result for the LCMO-F/BFO/STO bilayer at 10K obtaining Hex = 200 Oe.

4. Conclusions

We have suscefully grown BFO/LCMO-F and LCMO-F/BFO bilayers on (100) $SrTiO_3$ substrates, where LCMO-F/BFO/STO system are exhibiting exchange bias effect with a Hex of 200 Oe after field cooling at 1 T. We can also conclude that from magnetic properties point of view, this bilayer configuration offers posibilities to obtain bilayers exhibiting this polarization effect.

Acknowledgments

This work was supported by COLCIENCIAS with the projects 1106-56-933104, 1106-48-925531, Universidad del Valle with the project CI 7917, Center of Excellence for Novel Materials (CENM) and U.S. Department of Energy, under Grant No. DE FG03-87ER-45332. C. D. thanks to S. Guénon and Ivan K. Schuller for isothermal magnetic loops at the University of California, San Diego (UCSD).

References

[1] H Schmid 1994 *Ferroelectrics.* **162** 317
[2] I Sosnowskat, T Peterlin-Neumaier and E Steichele 1982 *J. Phys. C: Solid State Phys.* **15** 4835-4846
[3] P Vivas, E Baca, G Campillo, B Maiorov, N Haberkorn and J Guimpel 2002 *Revista Colombiana de Física.* **34** 625-629
[4] J Zapata, J M Caicedo, W Lopera, M E Gomez, F Cuellar and P Prieto 2008 *Colombian Journal of Physics.* **40** 77-80
[5] M E Gómez, G Campillo, J G Ramírez, P Prieto, A Hoffmann, J Guimpel, N Haberkorn, A Condó and F Lovey 2006 *IEEE Transactions on Magnetics.* **42** 10
[6] J Nogués and I K Schuller 1999 *J. Magn. Magn. Mater.* **192** 203-232
[7] G Campillo, A Berger, J Osorio, J E Pearson, S D Bader, E Baca and P Prieto 2001 *Journal of Magnetism and Magnetic Materials.* **237** 6168

Journal of Physics: Conference Series **480** (2014) 012031 doi:10.1088/1742-6596/480/1/012031

Enhanced electric and magnetic properties in LCMO/BTO bilayer deposited by Pulsed Laser Deposition

J E Ordoñez[1], M E Gomez[1], W Lopera[1] and P Prieto[1,2]

[1]Department of Physics, Universidad del Valle, A. A. 25360 Cali, Colombia

[2]Centro de Excelencia en Nuevos Materiales - CENM, Universidad del Valle, A. A. 25360 Cali, Colombia
E-mail: john.ordonez@correounivalle.edu.co

Abstract. We have deposited bilayers of the FM $La_{2/3}Ca_{1/3}MnO_3$ and FE $BaTiO_3$ as a route to design systems with artificial magnetoelectric coupling. We have optimized the growth parameters for depositing LCMO/BTO/(001)Nb:STO by PLD at pure oxygen atmosphere and a substrate temperature of 820°C, keeping fixed the magnetic layer thickness (t_{LCMO}~50 nm) and varying the thickness of the ferroelectric layer (t_{BTO}=17.8, 44.4, 98.4 nm). Magnetization and transport measurements indicate possible multiferroic behaviour in the bilayer. We found an increase in the Curie and metal-insulator transition temperature in the bilayer in comparison with those for LCMO (48nm)/STO. Hysteresis loops on bilayers show ferromagnetic behaviour.

1. Introduction

Multiferroic materials exhibit, simultaneously, two or more "ferroic" properties, *i.e.*, ferroelectricity (FE), ferromagnetism (FM), ferroelasticity, or ferrotoroidicity [1]. However, multiferroics are limited in nature: Perovskite $BiFeO_3$ (BFO) has attracted much attention given the coexistence of FE and magnetic orders at room temperature. Another compound is $BiMnO_3$ (BMO), this material exhibits unusual orbital ordering [2] leads to FM below 105 K approximately, also presents FE properties [3], makes this compounds potentially interesting for technological applications and for the study of magnetoelectric interactions. However, this property in both compounds is present below room temperature, limiting the practical applications.

Ferroelectricity and magnetic ordering require, to some extent, antagonistic conditions. Therefore heterostructures combining ferromagnets and ferroelectric properties well defined have emerged as a new route to design systems with artificial magnetoelectric coupling. Among the most attractive materials for practical implementation of multiferroic heterostructures and fundamental studies of magnetoelectric coupling are the ferromagnetic phase of the Calcium doped Lanthanum Manganite, $La_{2/3}Ca_{1/3}MnO_3$ (LCMO) and the ferroelectric $BaTiO_3$ (BTO) bilayers [4], grown on single-crystalline oxide substrates via PLD [5] [6]. In this work, we investigated the effect of the presence of BTO with different thicknesses (t_{BTO}) on the magnetotransport properties in LCMO/ BTO bilayers grown on (001) Nb:$SrTiO_3$ substrates. Hysteresis magnetic loop and thermal demagnetization measurements carried out on these bilayers show a dependence of both FM behavior and FM-PM transition temperature with BTO thickness.

[1] To whom any correspondence should be addressed.

Journal of Physics: Conference Series **480** (2014) 012031 doi:10.1088/1742-6596/480/1/012031

2. Experimental

LCMO/ BTO bilayers were deposited on (001) single-crystal Nb:STO via PLD at pure O_2 atmosphere at pressure on the order of 10^{-3} Torr and a substrate temperature of 820 °C . An excimer laser KrF operating at 248 nm, 10 Hz and 20-50 mJ was used. In our studies, t_{LCMO} was chosen around 50 nm and t_{BTO} varied systematically in the range from 20 to 100 nm. We choose three different values for the BTO thickness: 17.8 nm, 44.4 nm y 98.4 nm, that is, t_{BTO}/t_{LCMO} is around ½, 1/1 and 2/1, respectively. Magnetotransport measurements, thermal demagnetization and magnetic loops, were carried out in a Physical Property Measurement System (Quantum Design -PPMS–6600) in the range of 10 to 300 K. Resistance as a function of temperature was measured by the standard four-probe technique with current-in-plane (CIP) geometry. Magnetoresistance (MR) was examined with an applied magnetic field (H) of up to 20 kOe, perpendicular to the current direction and perpendicular to the plane. In all samples, resistance decreases with the increase of the applied magnetic field, at the same temperature. In order to compare the temperature dependence of the resistance curves, we normalized the resistance (for H = 0) at the transition temperature. For MR, we use the expression MR = R(H)/R(H=0), where R(H=0) is the resistance value at H=0 at the transition temperature.

3. Results

3.1. Transport properties

Temperature dependence of the resistance (normalized) measured in CIP geometry on LCMO/BTO bilayers grown on (001) Nb:STO substrates are shown in Figure 1, for different values of applied magnetic field (H): 0 kOe (open circles), 5 kOe (squares), 10.0 kOe (triangles), and 20.0 kOe (pentagons). We maintained t_{LCMO}, around 50 nm. We varied t_{BTO}, from 17.8 nm (Figure 1-. a), 44.4 nm (Figure 1-.b) to 98.4 nm (Figure 1-.c). For all samples, the curves display the metal insulator transition typical for the ferromagnetic phase of the LCMO system. They also show the dependency of the transition temperature with the applied magnetic field and also with t_{BTO} layer.

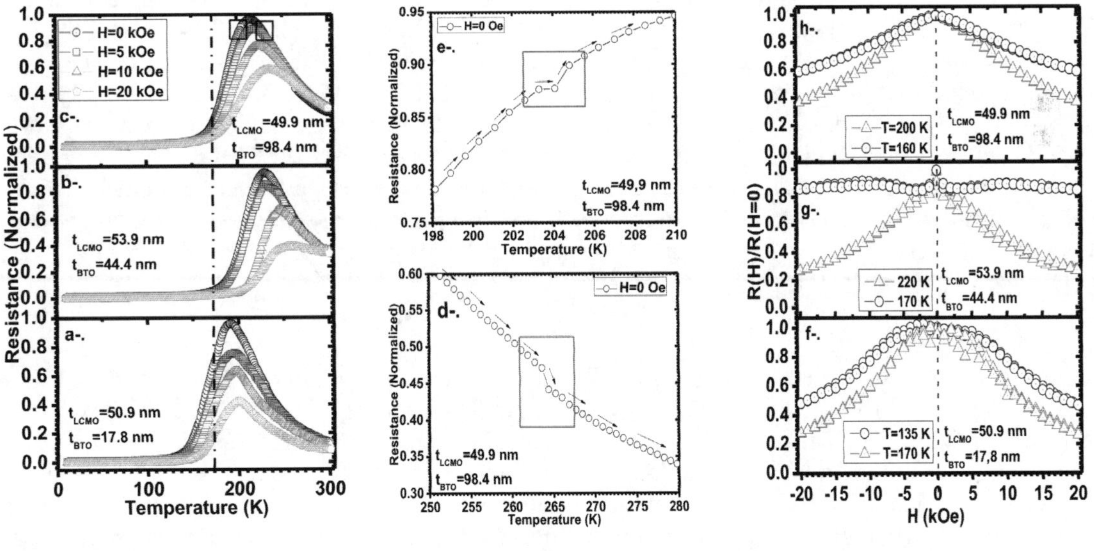

Figure. 1. Electric properties for LCMO (50 nm)/BTO/Nb(001)STO bilayer. Temperature dependence of the resistance (normalized) at applied magnetic field of zero (open circles), 5 kOe (squares), 10 kOe (triangles), 20 kOe (pentagons) for t_{BTO}=17.8 nm (a), 44.4 nm (b) and. 98.4 nm (c) Dotted line indicates the T_{MI} for LCMO(48 nm)/STO film. Zoom at resistance (normalized) versus temperature curve for LCMO (49.9 nm)/BTO(98.4 nm) at H=0 on insulator region (d) and on metal region (e). Isothermal MR as a function of magnetic field for sample with t_{BTO} = 17.8 nm (f), 44.4 nm (g), and 98.4 nm (h).

We observe the following remarkable features in these results: transition temperature, $T_{M\text{-}I}$, is higher than that for a LCMO(48 nm)/(001)STO film, shown in the plots by the dotted vertical line; $T_{M\text{-}I}$

increases with the increase of the applied magnetic field H, as expected for the optimally doped manganite [4], but the shift toward higher values of T_{M-I} is more dramatic for the sample with thickness ratio (t_{LCMO}/t_{BTO}) about 1/1, and to a lesser extent for the sample with thickness ratio 2/1, the one with the smaller thickness. We could say in any cases that the presence of a BTO film in LCMO/BTO bilayer enhances T_{M-I}, probably due to strains in LCMO are lower than those induced when LCMO film grows directly on the substrate. When t_{BTO} increases the strains in the structure relaxes, allowing release tension in the magnetic LCMO layer, improving conduction properties. Analyzing the width at half maximum of the curves, the sample with the lowest t_{BTO} (17.8 nm) shows a sharper transition than that for the thickest sample (98.5 nm), probably because of local structural changes at the interfaces causing competition between phases, originating an increase in the MI-width. This result is consistent with the BTO/LCMO/STO configuration reported elsewhere [6], where the T_{MI} temperature transition was 152 K (t_{BTO}=22.6 nm), 175 K (t_{BTO}=53.9 nm) and 177 K (t_{BTO}=98.5 nm) higher than that for samples with the same thickness. We have also observed a particularity in the resistance versus temperature curve at H=0 for bilayer with t_{BTO}=98.4 nm. The resistance exhibit a weak discontinuity displayed in insulator region at ~264 K, Figures 1(d), and in metal region at ~203 K, Figure 1(e). This jumps can possibly be attributed to strain effects of BTO layer due to two first-order structural transitions with temperature from tetragonal to monoclinic (T-M~275K) and monoclinic to rhombohedral (M-R~180K) [7]. This behavior has been found for temperatures ~190 K and 280 K in $La_{0.66}Sr_{0.33}MnO_3$ and $SrRuO_3$ thin films grown on BTO substrates, and it has been associated to strains attributed to structural transition of BTO substrate [7, 8] and recently on LCMO on BTO substrates [9]. Isothermal magnetoresistance curves are shown in Figure 1 (f) for the bilayer with t_{BTO} = 17.8 nm, (g) for that with t_{BTO} = 44.4 nm, and (h) for that with t_{BTO} =98.4 nm. We show in each case isothermal MR(H) curves for two temperatures, one in the insulator state close and above to the T_{MI} (~170 K, 220K and 200 K, respectively) and the other one in the metal state (135 K, 170 K and 160 K, respectively). All samples show hysteretic behavior at both measured temperatures, with colossal MR ratio at temperatures near T_C, MR decreases with increased applied magnetic field, as expected in optimally doped manganites. MR shows 80, 75, and 64%, for 17.8, 44.4, and 98.4 nm, respectively; this percentage decrease with increase of t_{BTO}, indicating that BTO film influence LCMO transport properties. On the other hand, bilayers display MR of 52, 15, and 40%, for the 17.8, 44.4, and 98.4 nm, respectively, at 160 or 170 K. The MR of all bilayers is lower than that for an LCMO/STO film (not shown), probably due to spin canting at the BTO/LCMO interface, agreeing with analysis reported by Murugavel *et al.*, [10]. Magnetotransport properties are extremely sensitive to changes in double-exchange interactions, which are, in turn, strongly affected by the variations in the angles between MnO_6 octahedra induced by strain at LCMO/ BTO interface or by changes in the crystal field (TiO_6 Octahedra).

3.2 Magnetic properties

Figure 2 (a) shows zero filed cooling (ZFC) thermal demagnetization curves, from 10 to 300 K, for LCMO (50nm)/BTO (t_{BTO}) bilayers, with t_{BTO} = 17.8 nm (open square), 44.4 nm (open circle) and 98.4 nm (open triangle). The calculated LCMO thicknesses in each bilayer are 50.9, 53.9 and 49.9 nm, respectively, close to the nominal value (50 nm). A magnetic field of 0.5 kOe was applied parallel to the plane of the film to measure magnetization during heating after ZFC. By fitting experimental data to a Gaussian distribution of long-range conducting FM regions, solid line, which has been used in LCMO films [11], we fit curves. Figure 2 (b) shows Thermal demagnetization curves display a typical behavior of an LCMO/STO film. Saturation magnetization increase non monotonic with increase in t_{LCMO} and t_{BTO}, probably due to strains on interface but for values of below found for LCMO/STO (600 emu/cm^3) and above of BTO/LCMO/STO (400 emu/cm^3) [6], coercive field decrease with t_{BTO} increase, then this configuration improve magnetic properties in contrast with the other configuration [6]. Figure 2 (c) shows a comparison between T_{MI} obtained by resistance measurements and T_c obtaniened by Gaussian fit on magnetization measurement, we observed a consistent behavior in this values and increase on the values for this configuration in comparison with LCMO/STO.

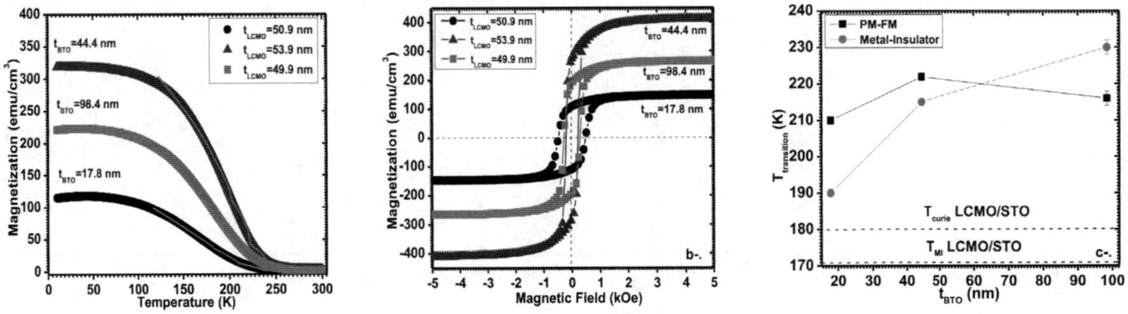

Figure 2 Magnetic properties of LCMO(50nm)/BTO(t_{BTO}) bilayers. a-. Temperature dependence on magnetization at ZFC for samples with t_{BTO} = 17.8 nm (square), 44.4 nm (circle), and 98.4 nm (triangle), b-. Isothermal magnetic hysteresis loops at 10 K for bilayers, c-. M-I transition temperature, and Curie temperature values for bilayers and LCMO/STO film. Lines are guides for the eye.

These results confirm the influence of BTO in the magnetic properties of the manganite (mismatch BTO/LCMO), that is, magnetic properties of the system are affected by the existence of the ferroelectric layer. It is expected that magnetization of the same volume of the same material must be the same at the same temperature. This result indicates that the presence of the BTO film on the LCMO film affects the magnetic moment of the ferromagnetic layer at the interface and this interaction depends on the thickness of the BTO layer. Probably, a magnetic contribution exists of the BTO induced by interfacial interactions at the interface, but also induced by crystalline strains originated during growth.

3.3 Conclusions

Bilayers show MR effect with metal insulator transition (T_{MI}), which shifts toward higher temperatures when increasing the BTO layer thickness. The saturation magnetization and coercive field show dependence on BTO thickness, which may be due to relaxation effects of the BTO film for thicker layers. This behavior shows the effect of strain on the transport and magnetic properties due to BTO lattice parameter, decreased with the thickness, reducing strain in the LCMO/BTO interface. The LCMO layer retains its magnetic properties, increase T_C and saturation magnetization decrease in presence of BTO layer. The behavior found in these bilayers allows us to predict the use of this system as a hybrid multiferroic structure.

4. Acknowledgement

This work was financially supported by "Instituto de Nanociencia de Aragón", Zaragoza, Spain, where the films were grown and partially characterized, Center of excellence for Novel Materials (CENM), Research Projects: No. 1106-48-925531 and CI 7917-CC 10510 contract 0002-2013 COLCIENCIAS-UNIVALLE.

5. References:

[1] Martin L, Chu Y-H, and Ramesh R 2010 *Mat. Sci. Eng R*, **68**, 89.
[2] Moreira dos Santos A, Cheetham A K, Atou T, Syono Y, Yamaguchi Y, Ohoyama K, Chiba H and Rao C N R 2002 *Phys. Rev. B* **66** 064425
[3] Grizalez M, Delgado E, Gómez M E and Prieto P 2007 *phys. stat. sol.* (c) **11** 4203
[4] Dagotto E, Hott T and Moreo A 2001 *Phys. Rep.*, **344**, 1.
[5] Petraru A, Schubert J, Schmid M, and Buchal C 2002 *Appl. Phys. Lett.*, **81** 1375.
[6] Ordoñez J E, Gómez M E, Lopera W and Prieto P 2013 *IEEE Trans. Mag.* **49**, NO. **8** **1.**
[7] Lee M K, Nath N K, Eom C B, Smoak M C and Tsui F 2000 *Appl. Phys. Lett.* **77**, No. 22, 27
[8] Dale D, Fleet A, Brock J D, Suzuki Y 2003 *Appl. Phys. Lett.*, **82**, No. 21, 3725.
[9] Alberca A, Munuera C, Tornos J, Mompean F J, Biskup N, Ruiz A, Nemes N M, Andres A. de, León C, Santamaría J and García-Hernández M 2012 *Phys. Rev.* B **86**, 144416
[10] Murugavel P and Prellier W 2006 *J. Appl. Phys.*, **100**, 023520.
[11] Campillo G, Berger A, Osorio J, Pearson J, Bader S, Baca E, and Prieto P 2001 *J. Magn. Magn. Mater.* **237** 61.

Journal of Physics: Conference Series **480** (2014) 012032 doi:10.1088/1742-6596/480/1/012032

Mössbauer study of the effect of gamma irradiation on the removal of pyrite from Colombian coals

J A Mejía[1], F Reyes Caballero[2], C A Palacio[1], E de Grave[3], H Olaya Dávila[2], S A Martínez Ovalle[2]
[1] Grupo PERSEO, Universidad Antonio Nariño-sede Tunja, Tunja, Colombia.
[2] Grupo de Física Nuclear Aplicada y Simulación,Universidad Pedagógica y Tecnológica de Colombia, Tunja, Colombia.
[3] Department of physics and astronomy, Ghent University, Ghent, Belgium
E-mail: juliethmejia@uan.edu.co

Abstract. The removal of sulfur from the coals is necessary before using it. It is due to the environmental and technological problems that it causes. In this work, the results of the study by Mössbauer spectroscopy of the gamma-irradiation effect on the pyrite in three Colombian coals are analyzed. They were exposed to different gamma-irradiation doses using a ^{60}Co source.

1. Introduction

The forms of sulfur in coals are elemental, organic and inorganic. Elemental sulfur is present only sporadically in trace amounts. The organic sulfur is chemically bound to the carbon matrix and it is complex to remove it by physical methods. Inorganically, it can be found as i) sulfide and ii) sulfate sulfur. In the first case it can be combined with iron in different ways; being pyrite (FeS_2) the most common. In the second case, it can be found as iron sulfate (e.g., Jarosite ($XFe_3(SO_4)_2(OH)_6$, ferrous sulfate $FeSO_4$, Rozenite ($FeSO_4{\bullet}4H_2O$), etc.) or calcium sulfate.

When studying coals, ^{57}Fe Mössbauer spectroscopy is widely accepted because it is a non-destructive technique with high sensitivity; useful for studying materials containing a mixture of iron compounds. However, the fitting of the spectra is not straightforward: in coal samples it is usual to have several Fe minerals that can show overlapping in the Mössbauer spectra. Also, the hyperfine parameter values can be affected by crystallinity, particle size and substitution. Therefore, these aspects should be taken into account when the Mossbauer spectra are fitted. A way to deal with these limitations is to obtain Mössbauer spectra at different temperatures.

The removal of iron-sulfur minerals from coals is necessary before their combustion because these minerals cause sulfur oxides SOx (i.e., SO_2, SO_3) associated with environmental and health problems. In addition, the relative amount of pyrite usually influences the formation of agglomeration, abrasion, corrosion, tackiness.

In preliminary works, methods used to remove Fe-S minerals have been studied. These methods consist of a treatment with hot nitric acid; and two physical methods, using flotation and hydrocyclone separation [1]. Some studies about a novel method for desulfurization of Indian and Egyptian coals using gamma irradiation have been published [2-4].

In this work, the purpose is to investigate the effect of gamma irradiation on the pyrite present in three Colombian coals by Mössbauer spectroscopy.

2. Experimental Procedure

Representative coal samples obtained from the mines Guachinte (Valle state), Las casitas and La Cañada (El Zalitre Zone, Boyacá state) were dried, milled (particle size 250 μm) and hermetically stored. Characterization studies of the samples from Guachinte and Las casitas, such as their chemical analysis and Mössbauer results were published elsewhere [1,5]. Information about ash values, total sulfur content, volatile matter and calorific value of the sample from La Cañada has been reported in this work according to the ASTM rule [6].

Table 1. Mössbauer results of the samples measured at 90 K. The quadrupole shifts ($2\varepsilon_Q$), quadrupole splittings (ΔE_Q) and isomer shifts (δ) are given in mm/s, the hyperfine fields (H_M) are in kOe and the areas (S) are given in % (* *Fixed parameter)*

Sample	Phase	H_M	$2\varepsilon_Q/\Delta E_Q$	δ	S
GUA284	Goethite	482.5	-0.25	0.47	32
	Pyrite		0.62	0.38	25
	Jarosite		1.2	0.46	28
	Kaolinite		2.73	1.41	15
GUA1800	Goethite	481.3	-0.23	0.49	31
	Pyrite		0.62	0.38	24
	Jarosite		1.2	0.46	30
	Kaolinite		2.72	1.41	15
CAS284	Pyrite		0.63	0.30*	100
CAS1800	Pyrite		0.63*	0.30*	4
	Iron Sulfate		1.78	0.87	96
CAN284	Pyrite		0.60*	0.36	75
	Jarosite		0.93	0.63	18
	Iron Sulfate		2.37	1.54	7
CAN1800	Pyrite		0.60*	0.28	1
	Jarosite		0.97	0.79	45
	Iron Sulfate		2.09	0.9	54

Homogeneous samples with approx. 20 g of mass of three coals were simultaneously irradiated with gamma photons from a ^{60}Co source. The used source is part of a Theratron 780, Model C146, which had an activity of 7143 Ci to November 2008. At the time of irradiation of the samples, May 20th, 2013, the equipment presented a reference rate of 118.48 cGy min^{-1} at 80 cm distance from the source. This equipment is located at the clinic for cancer of the city of Tunja (Boyacá).
The samples were placed at a distance of 40 cm from the source (minimum distance allowed by the equipment). Irradiation dose of 10 min was carried out to complete 380 min corresponding to 1800 Gy, the maximum dose given to the samples. Subsamples of the three coals, approx. 1000 mg, were extracted after being irradiated with 284 Gy; and a second set of approx. 1000 mg was taken out after receiving 1800 Gy. The samples were named GUAX (Guachinte mine), CASX (Las casitas mine) and CANX (La Cañada mine), being X the radiation dose.

For the Mössbauer analysis, the standard absorbers were prepared containing 100 mg of sample. Mössbauer measurements in transmission mode were performed at 90 K by using a conventional spectrometer with constant acceleration and using a source of 57Co/Rh. The velocity scale was regularly calibrated acquiring a spectrum of standard α-Fe2O3. The values given for the isomeric shift are referenced with respect to the α-Fe at room temperature. The measurements were made on speed ranges between -10 and 10 mm/s. DIST3E program was used for the analysis of the Mössbauer spectra, which is based on hyperfine parameter distributions. More details about this program can be found in [7]. The mineral phase identification was made by comparing the Mossbauer parameter values that were obtained with the values reported in the literature for coals and also for pure compounds [8-9].

3. Results and Discussion

All the Mössbauer spectra obtained from the Guachinte mine samples were fitted with three doublets and a sextet, corresponding to the phases Pyrite, Jarosite, Kaolinite and Goethite, respectively (see Figure 1 left). Regarding the spectrum of the natural sample [5], it shows that there is a reduction of Pyrite in favor of the Jarosite, Kaolinite sulfates and the Goethite oxyhydroxide for all the radiation doses. The presence of Goethite is due to the high oxidizing conditions. The hyperfine parameters of the samples GUA284 and GUA1800 are listed in Table 1.

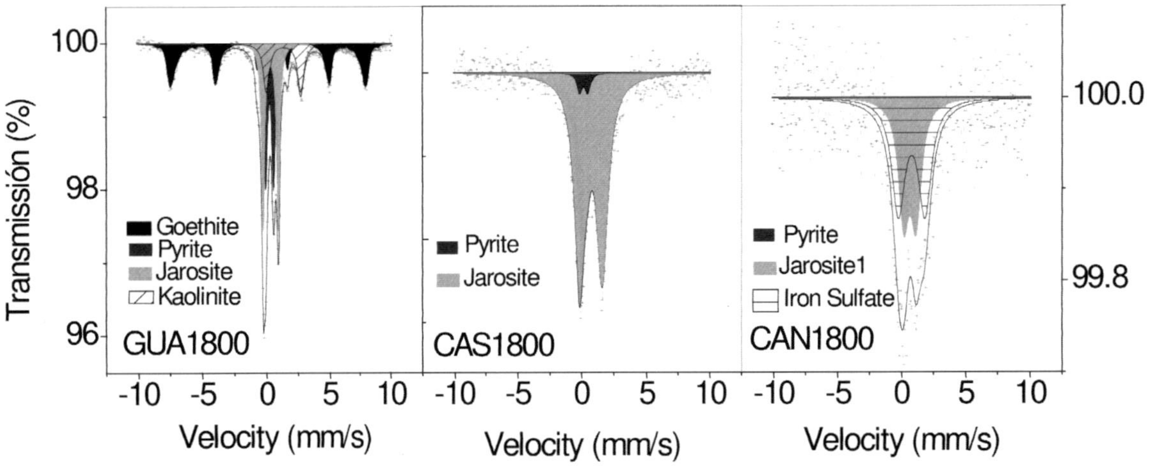

Figure 1. Mössbauer spectra obtained after gamma irradiation of 1800 Gy from coal samples.

The Mössbauer spectrum having the lowest (not shown here) irradiation dose (284 Gy) for the Las casitas sample, exhibit a unique phase associated to the mineral Pyrite (see Table 1). It is in agreement with the spectrum for the natural coal [1], meaning that applying the lowest irradiation dose; there is no reduction in the content of Pyrite for these samples.
The spectrum obtained for the highest gamma irradiation for the CAS sample shows a significant reduction of pyrite, accompanied by a formation of Jarosite as shown in Figure 1 (center). The prominent doublet corresponding to Jarosite has high values of the isomer shift and quadrupole splitting (Table 1). This can be explained by the presence in the structure of high spin Fe^{3+} [4].

Figure 1 (right) shows the typical spectrum obtained by Mössbauer spectroscopy for all the irradiation doses from the La Cañada mine samples. The spectra were fitted with three doublets, corresponding to the Pyrite, Jarosite and Iron Sulfate. The hyperfine parameters listed in Table 1 show a reduced pyrite accompanied by an increase of the Jarosite and Iron sulfate. Table 1 presents the values of the Mössbauer parameters corresponding to the samples from the La Canada mine.

Table 2 lists the results of the chemical analysis of the coals from the La Cañada mine according to the standard procedure ASTM D388. As shown in the table, it corresponds to the coal with lower ash content.

Table 2. Chemical analysis of the coal from the La Cañada mine.

Analysis	ASTM standard	Air-dried
Moisture [wt. %]		0.45
Volatile matter [wt. %]	D7582	52.02
Ash [wt. %]	D7582	3.64
Fixed Carbon [wt. %]	D3172	44.12
Calorific Power [cal/g]	D5865	6845.03
Total sulfur [wt. %]	D4239	0.71

Taking into account the differences between the Mössbauer results for each coal and comparing the chemical properties of them (see results in this paper, references [1,5] and Table 2), the effect of the irradiation is strongly correlated with the chemical properties of the coals such as mineral matter, ash, and sulfur content.

In the energy range of 100 keV -10 MeV the radiation absorption is primarily due to Compton effect. In this case, approximately 99% of gamma photons emitted from the Co-60 are of 1.17 MeV, which interact with the sulfur. Sulfur electronic configuration allows energy transfer with the photons achieving the desired effect.

The effect of the Gamma irradiation on the sulfur-iron minerals can be explained with the gamma radiolysis of the water present in the coals (see Eq. 1).

$$H_2O \overset{\gamma\ Rays}{\Rightarrow} H^*, e^-_{water}, OH^*, H_2O^+, H_2O_2, H_2^* \qquad (1)$$

Due to the presence of oxygen other products also can be formed. These radiolitic products generate an oxidizing atmosphere. It is expected that if the oxidation increases, the content of sulfides decrease in favor of ferrous and ferric sulfates and possibly a formation of oxyhydroxides could appear.

4. Conclusions

Coal samples of three different Colombian mines, with high and low values of pyritic sulfur content, have been irradiated with a high gamma radiation dose with the aim of achieving the desulfurization of them. The Mössbauer results show that with the given dose, the desired effect is achieved. However, it is not fully removed. These results suggest that the high costs and processes that are necessary to obtain radioactive sources and the locative physical changes for handling them would make impractical the use of this method for thermoelectric companies that use coal.

5. References

[1] Reyes F and Martinez Ovalle S A 2013 Mössbauer study of the inorganic sulfur removal from coals *Hyperfine interact.* 10.1007/s10751-013-0874-x.
[2] Ram L C, Tripathi P S M, Jha S K, Sharma K P, Singh G, and Mishra S 1997 *Fuel Process Technol.* **53** 1
[3] Tripathi P S M, Mishra K K,. Roy R R P, and Tewari D N 2001 *Fuel Process Technol.* **70** 77
[4] Ahmed M A, Mousa M A A,. Amami S M, Etmimi K M, and Ellid M S 2011 *Energ. Source Part A* **34** 227
[5] Reyes, F, Alcázar G P, Barraza J M, Bohórquez A, Tabares J A, and Speziali N L 2003 *Hyperfine interact.* **48(1-4)** 39
[6] ASTM., 1991. Annual Book of ASTM Standards.
[7] Vandenberghe R E, De Grave E and De Bakker P M A 1994 *Hyperfine Interact.* **83** 29
[8] Huffman G P, and Huggins F E 1978 *Fuel* **57**
[9] Gracia M, Marco J F, and Gancedo J R 1999 *Hyperfine Interact.* **122**(1-2) 97

Band structure of a 2D photonic crystal based on ferrofluids of $Co_{0.8}Zn_{0.2}Fe_2O_4$ nanoparticles under perpendicular applied magnetic fields

J. López, Luz E. González, M. F. Quiñonez, N. Porras-Montenegro, G. Zambrano, M. E. Gómez,

Departamento de Física, Universidad del Valle, A.A. 25360, Cali, Colombia

E-mail: nelsonporras.montenegro@gmail.com

Abstract. Using a ferrofluid of cobalt-zinc ferrite nanoparticles ($Co_{0.8}Zn_{0.2}Fe_2O_4$) coated with oleic acid and suspended in ethanol, we have fabricated a 2D photonic crystal (PC) by the application of an external magnetic field perpendicular to the plane of the ferrofluid. The 2D PC is made by rods of nanoparticles organized in a hexagonal structure. By means of the plane-wave expansion method, we study its photonic band structure (PBS) which depends on the effective permittivity and on the area ratio of the liquid phase. Additionaly, taking into account the Maxwell-Garnett theory we calculated the effective permittivity of the rods. We have found that the effective refractive index of the ferrofluid increases with its magnetization. Using these results we calculate the band structure of the photonic crystal at different applied magnetic fields, finding that the increase of the applied magnetic field shifts the band structure to lower frequencies with the appearance of more band gaps.

1. Introduction

Magnetic nanofluids or ferrofluids are part of a new class of magnetic materials. They exhibit magnetic and fluid properties simultaneously, becoming materials of technological importance [1]. Magnetic fluids have attracted a great deal of attention of scientist and engineers, no only because their structural patterns under external magnetic fields where periodic columns are formed in the ferrofluid by a applied perpendicular mangnetic field [2], but also by its optical properties, such as birrefringence, optical negative refraction, diffraction, light transmission and many potentially electro-optical characteristics of the magnetic fluids, for example, the optical switch, the tunable grating, and photonic crystals [3, 4, 5].

2. Experimental method

The size and physical properties of the nanoparticles depend on preparation parameters such as, reaction temperature, pH of the suspension, initial molar concentration, and others. In this case, the $Co_{1-x}Zn_xFe_2O_4$ magnetic nanoparticles were prepared using the chemical co-precipitation method, which is probably the most-common and the most-used synthesis of ferrofluids or magnetic fluids based on magnetic nanoparticles. The preparation of surfacted and ionic aqueous ferrofluids based on spinel ferrite nanoparticles has been reported by Massart [6]. The preparation of our ferrofluid samples are described in the previous work of J. Lopez et al. [7]. In their work they describe a complete process to the synthesis of $Co_{1-x}Zn_xFe_2O_4$ with nominal values of x = 0.25, 0.50 and 0.75.

3. Theoretical Framework

For the transversal magnetic mode (TM), the wave equation of the magnetic field $\mathbf{E_k}$ may be obtained from Maxwell's equations and expressed as

$$\frac{1}{\varepsilon(\mathbf{r})}\nabla \times (\nabla \times \mathbf{E_k}) = \frac{\omega^2}{c^2}\mathbf{E_k}, \tag{1}$$

where k represents the wave propagation vector of the mode and $\epsilon(r)$ is the position dependent dielectric constant of the periodic structure. From Bloch theorem and theorems of Fourier analysis, the electromagnetic fields in a periodic medium can be written as

$$\mathbf{E_k}(\mathbf{r}) = \sum_{\mathbf{G}} \mathbf{E_k}(\mathbf{G})e^{i(\mathbf{k}+\mathbf{G})\mathbf{r}}, \tag{2}$$

where $\mathbf{G}$ represents a lattice vector in reciprocal space, describing the periodic structure, similarly a periodic dielectric function can be expanded as

$$\frac{1}{\varepsilon(\mathbf{r})} = \sum_{\mathbf{G}} \chi(\mathbf{G})e^{i\mathbf{G}\cdot\mathbf{r}}, \tag{3}$$

where $\chi(\mathbf{G})$ are the Fourier expansion coefficients [8].

By substituing (2) and (3) into (1) a matrix eigenvalue problem is obtained, where for a fixed wave vector $\mathbf{k}$, the frequency ω of the allowed modes in the periodic strucutre are found as eigenvalues.

To characterize electromagnetic properties of composite media, we use the Maxwell-Garnett approximation (MGA). The following equation is the exact solution for the equivalent permittivity of the core-shell composite nanoparticle [9]

$$\frac{\varepsilon_r - \varepsilon_a}{\varepsilon_r + 2\varepsilon_a} = p\frac{\varepsilon_F - \varepsilon_a}{\varepsilon_1 + 2\varepsilon_a}, \tag{4}$$

where $p = \dfrac{r^3}{(r+d)^3}$ is the volume ratio of the nanoshell to the whole coated nanoparticle, r is the radius of the core ($Co_{0.8}Zn_{0.2}Fe_2O_4$) with permittivity ε_F, and d is the thickness of the shell (oleic acid) with permittivity ε_a.

4. Results and discussion

In order to consider the effects of the applied magnetic field on the PBS, we have used the data given in the work by Ramana [10], where it is stableshed the dependence of the dielectric constant of the nanoparticles with the Zn concentration. In our numerical calculations, we have taken ε_F to be 18 for $Co_{0.8}Zn_{0.2}Fe_2O_4$ ferrite, $\varepsilon_a = 2.34$, and $r = 4.8nm$ and $d = 5.8nm$ [7].

Since the structural pattern is very important for the calculation of the PBS and varies significantly with the perpendicular magnetic field, in Fig.1 we present the structural evolution of the magnetic ferrofluid film with a magnetic concentration of 0.16 emu/g, injected into a glass cell with an area of $18mm^2$ and a depth ranging from 10 microns, and under three values of the applied magnetic field: 0, 73 and 217 Oe. From Fig.1 it is observed that with the increasing of field strength, more magnetic columns are formed and the area occupied by the liquid phase is reduced.

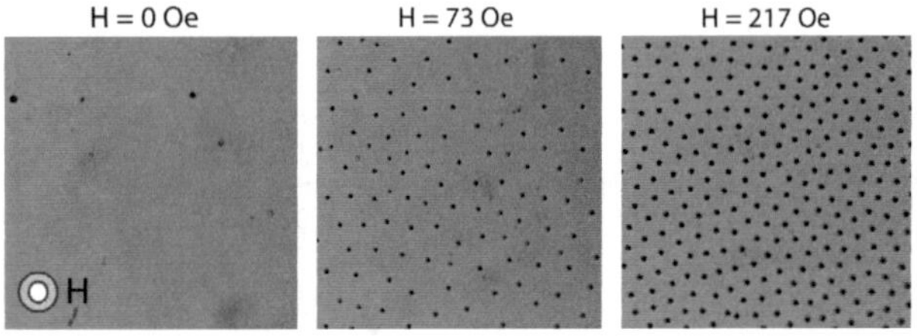

Figure 1. (Color online) Optical images of $Co_{0.8}Zn_{0.2}Fe_2O_4$ magnetic ferrofluid films under perpendicular magnetic field taken at room temperature.

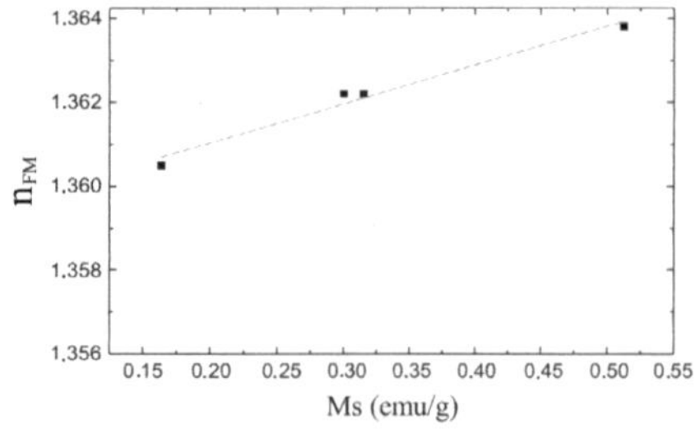

Figure 2. Linear relationship between the refractive index and the magnetic concentration of ferrofluid film under zero field.

With the values given above for ε_F, ε_a, and using Eq.5 we find $\varepsilon_r = 6.77$ for each column. For the PC structured in a hexagonal pattern, the ratio d/a of the diameter d to the period a of the magnetic columns is related to the area ratio through

$$\frac{d}{a} = \sqrt{\frac{2\sqrt{3}}{\pi}\frac{A_{col}}{A}}, \tag{5}$$

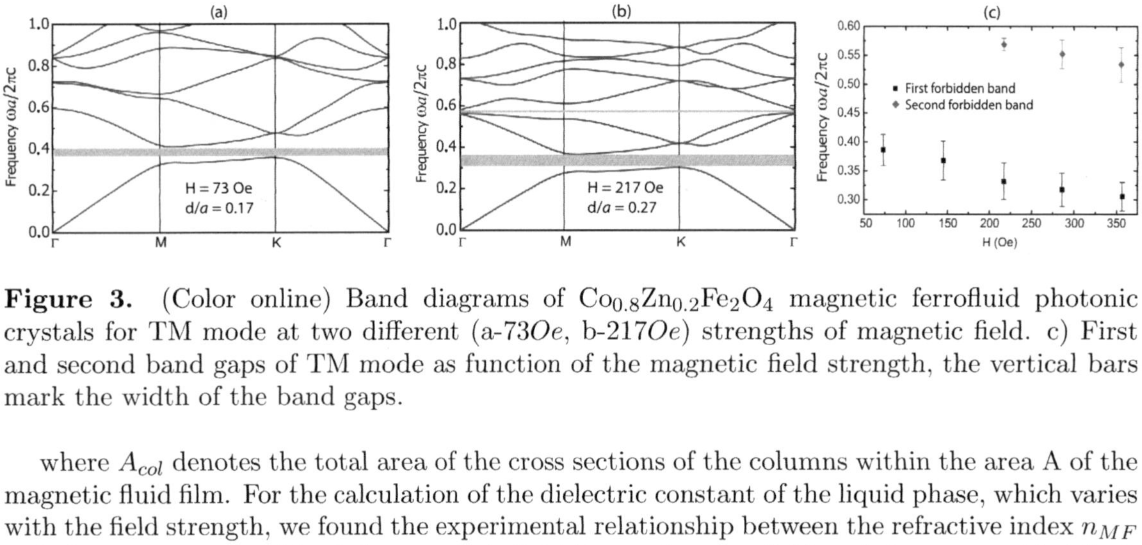

Figure 3. (Color online) Band diagrams of $Co_{0.8}Zn_{0.2}Fe_2O_4$ magnetic ferrofluid photonic crystals for TM mode at two different (a-73Oe, b-217Oe) strengths of magnetic field. c) First and second band gaps of TM mode as function of the magnetic field strength, the vertical bars mark the width of the band gaps.

where A_{col} denotes the total area of the cross sections of the columns within the area A of the magnetic fluid film. For the calculation of the dielectric constant of the liquid phase, which varies with the field strength, we found the experimental relationship between the refractive index n_{MF} and the concentration M_s of the magnetic fluid film under zero field, as it is presented in Fig.2, and given by

$$\varepsilon_{liq}(H) = n_{FM}^2(H=0) = (0.01M_s + 1.36)^2 \tag{6}$$

To obtain the effective permittity of the liquid phase, the refractive index of the magnetic ferrofluid with M_s= 0.16, 0.30, 0.32 and 0.51 emu/g, was measure under zero field. In the calculations the magnetic permeability of the ferrofluid is taken to be 1, due to that $Co_{1-x}Zn_xFe_2O_4$ magnetic ferrofluid presents a tendency to super-paramagnetic behaviour at room temperature

Using the plane wave method, the PBS for the TM mode of PCs based on ferrofluids with hexagonal structure were calculated. The PBS at two different strengths of magnetic field are plotted in Fig.3. The frequency has been normalized to $2\pi c/a$ where c is the light speed in vacuum. We found that for each of the directions $\Gamma M - MK - K\Gamma$ inside the PC there exist partial band gaps [8] as it is shown in Fig.3, where the radiation for a given wavelength cannot propagate inside the PC, namely, there are partial band gaps in the PC under some specific angles. Also, for this mode the number of the forbidden band gaps increases with the magnetic field strength from 1 to 2 band gaps, where the thickness of the first one does not vary in the range between 73 Oe and 217 Oe and it is shifted to lower frequencies. In addition, it is observed that the PBS is shifted to lower frequency regions. This behavior is relate to the increase of the filling factor with the magnetic field strength, where more rods appear in the PC structure, causing an area reduction of the liquid phase, which should cause an increment of the dielectric constant of the material that minimize the energy, in according to the electromagnetic variational theorem. From the Fig. 3c we can see the existence of band gaps for different magnetic field strengths in the TM mode. This behavior is explained by the high refractive index contrast in the PC.

5. Conclusions

Ferrofluids based $Co_{0.8}Zn_{0.2}Fe_2O_4$ ferrite nanoparticles coated by oleic acid and suspended in ethanol were prepared by co-precipitation method. Ferrofluid placed in an external magnetic field shows a 2D PC made by rods of nanoparticles organized in a hexagonal structure. Using the plane-wave method, we theoretically studied the PBS wich depends on the effective permittivity of the liquid phase, the area ratio of the ferrofluid, and the effective permittivity of the rods. We have found that the PBS is shifted to lower frequencies with the augment of the magnetic field. On the other hand, we found that the band gaps increases with the magnetic field strength.

6. Acknowledgements

This work was supported by fund project of internal convocation Universidad del Valle 2010 IC7829, IC7871, and CENM (Center of Excellence for Novel Materials). J. López thanks COLCIENCIAS for the doctoral fellowship "Francisco Jose de Caldas". The authors would like to thank professor Ivan I. Smalyukh and his Research Group "Liquid Crystal Materials Research Center" and Department of Physics, University of Colorado at Boulder.

References

[1] Rosensweig R. E., *Ferrohydrodynammics* (New York: Cambridge University Press, 1985).

[2] Shie-Yueh Yang, Herng-Er Horng, Chin-Yih Hong and Hong-Chang Yang, "Structures, optical properties and potentially electro-optical applications of magnetic fluid films," Tamkang J. of Science and Engineering **5**, (2002) 85-93.

[3] E. Yablonovitch, *Inhibited spontaneuos emission in solid-state physics and electronics.* Phys. Rev. Lett. **58**, (1987) 2059-2062 .

[4] Shengli Pu, Xuekkun Bai, Lunwei Wang, S. Pu, X. Bai, and L. Wang, "Temperature dependence of photonic crystals based on thermoresponsive magnetic fluids," J. Magn. Magn. Mater. **323**, (2011) 2866-2871.

[5] C. Z. Fan, G. Wang, J. P. Huang, "Magnetocontrollable photonic crystals based on colloidal ferrofluids," J. Appl. Phys. **103** (2008) 094-107.

[6] R. Massart, Preparation of aqueous magnetic liquids in alkaline and acidic media, IEEE Transactions on Magnetics **17** (1981) 1247-1248.

[7] J. López, L. F. González - Bahamon, J. Prado, J. C. Caicedo, G. Zambrano, M.E. Gómez, J. Esteve, P. Prieto. Study of magnetic and structural properties of ferrofluids based on cobaltzinc ferrite nanoparticles. J.Mag. Mag. Mater. **324** (2012) 394-402.

[8] Sukhoivanov I A, Guryev I V, *Photonic crystals: Physics and practical modeling* (Springer series in optical sciences, 2009).

[9] A. Sihvola, J. A. Kong, *Effective permittivity of dielectric mixtures.* IEEE Trans. on Geosci. Remote Sensing **26**, (1988) 420-429.

[10] S. Ramana Murty, *Dielectric behaviour of cobalt-zinc ferrites.* J. Mater. Sci. Lett. **3**, (1984) 1049-1051.

Journal of Physics: Conference Series **480** (2014) 012034 doi:10.1088/1742-6596/480/1/012034

Magnetic feature, compositional and structural analysis of the $La_2SrFe_2CoO_9$ complex perovskite

Francy M. Casallas[1,2], E. Vera-Lópeza[2], D. A. Landínez Téllez,[1] D.E. Saavedra Mesa[1], J. Roa-Rojas[1]
1. Grupo de Física de Nuevos Materiales, Departamento de Física, Universidad Nacional de Colombia, AA 5997, Bogotá DC, Colombia
2. INCITEMA, Universidad Pedagógica y Tecnológica de Colombia, Tunja, Colombia

E-mail: jroar@unal.edu.co

Abstract. In this work, we report the synthesis and characterization of the new $La_2SrFe_2CoO_9$ triple perovskite material. Samples were produced by the solid state reaction method from high purity precursor powders of La_2O_3, $SrCO_3$, Co_2O_3 and Fe_3O_4. XRD pattern was analyzed by Rietveld refinement through the GSAS code, revealing that observed diffraction peaks are indexed , to the complex perovskite system, and showed that $La_2SrFe_2CoO_9$ crystallizes in an orthorhombic structure, space group *Immm (#71)*. The lattice parameters were a=5.4314(3) Å, b=5.4583(3) Å and c=7.7018(2) Å. SEM micrographs evidence a strongly diffused granular morphology with mean grain size of $1\ \mu m$ and EDX spectra show that the chemical composition of samples are in good agreement with the nominal values of the stoichiometric formula. Measurements of magnetization as a function of temperature permitted to determine the ferromagnetic characteristic of material with an effective magnetic moment of $6.18\ \mu_B$, which is in agreement with the value calculated from the Hund's rules.

1. Introduction

Ceramic perovskite-like materials have been extensively investigated in last decades, because small structural distortions, vacancies and compositional modifications can induce a wide variety of physical and chemical properties [1]. When ideal perovskite formula ABO_3 is modified to introduce different kind of cations on into the octahedral site of the primitive perovskite unit cell, the cationic ordering leads to a complex triple perovskite, which is identified by the $A_2A'B_2B'O_9$ formula [2]. This circumstance permits to infer the possibility to produce new materials by the introduction of an alkaline or rare earth ion in the A, A' sites and transition metal ions in the B, B'. Depending on magnetic or electric characteristic of B and B', it is relatively easy to create new perovskite systems with half-metallic properties [3-4], magnetoelectric response [5] or magnetic ordering [6], which offer promissory perspectives in the recent spintronics technology [7].

However, our focus is on the production of new ferromagnetic materials whose response takes place at high temperatures, i.e., near room temperature, in order to enable their application in devices for storing information such as hard drives, portable memories, spin valves, magnetic readers and writers,

[1] jroar@unal.edu.co.

for example. In this work, we report the synthesis, structural characterization and magnetic studies of novel ferromagnetic triple perovskite $La_2SrFe_2CoO_9$, which was idealizes as the introduction of a lanthanide ion in the A site, an alkaline earth in the A' and a mixture of two magnetic transition metals in the B and B' sites of the $A_2A'B_2B'O_9$ formula, to construct a triple complex perovskite.

2. Experimental

Samples were synthesized by the solid state reaction recipe. Precursor powders of La_2O_3, $SrCO_3$, Co_2O_3 and Fe_3O_4 (Aldrich 99.9% purity) were mixed in stoichiometric proportions according to the chemical formula $La_2SrFe_2CoO_9$. Mixture was ground to form a pellet and annealed at *1250 ºC* for *12* hours. Then, samples were regrinded, repelletized and annealed at *1250 ºC* for 24 hours and *1350 ºC* for 24 hours. X-ray powder diffraction (XRD) pattern was collected using a *PW1710* diffractometer with $\lambda_{CuK\alpha}=1.5406$ *Å*. Rietveld refinement of diffraction pattern was made by the GSAS code [8]. Scanning electron microscopy (SEM) images were obtained by using a FEI QUANTA 200 microscope, which possess a system for Energy Dispersive X-ray (EDX) analysis. Zero Field Cooling (ZFC) and Field Cooling (FC) measurements of magnetization as a function of temperature were carried out by using a MPMS Quantum Design SQUID.

3. Results and Discussion

The analysis of XRD pattern showed in figure 1 reveals the presence of characteristic peaks of complex perovskite systems. In figure 1, crosses represent the experimental data and line corresponds to simulated pattern by means of GSAS code. Base line is the difference between theoretical and experimental results.

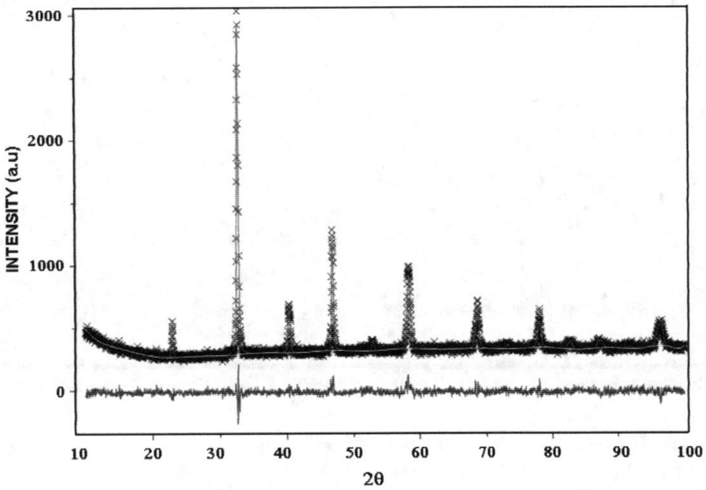

Figure 1. XRD pattern for $La_2SrFe_2CoO_9$ triple perovskite. Symbols represent the experimental diffraction data and base line corresponds to the difference between experimental and simulated patterns (continuous line).

Rietveld refinement permitted to establish that this material crystallizes in an orthorhombic triple perovskite with space group *Immm (#71)* with lattice constants: a=5.4314(3) Å, b=5.4583(3) Å and c=7.7018(2) Å. These results are 99.8% in agreement with the theoretical values obtained from the Structure Prediction Diagnostic Software *SPuDS* [9], which predicts that lattice constants a=5.458(8) Å, b=5.427(1) Å and c=7.688(9) Å for the $La_2SrFe_2CoO_9$ material. Parameters of refinement are: R_{F^2}=2.67 %; x^2=1.313; R_{WP}=3.82 % and R_P=3.20 %. Numeric results of the Rietveld analysis are listed in table 1.

Table 1. Atomic positions and site occupancy calculated for the $La_2SrFe_2CoO_9$ complex perovskite.

Ion	*x*	*y*	*z*	*Occupancy*
Sr	0.5000	0.0000	0,2476	0.3379
La	0.5000	0.0000	0.2476	0.6804
Co(1)	0.0000	0.0000	0.0000	0.6025
Fe(1)	0.0000	0.0000	0.0000	0.4371
Co(2)	0.5000	0.5000	0.0000	0.7491
Fe(2)	0.5000	0.5000	0.0000	0.2015
O(1)	0.2295	0.2809	0.0000	1.7620
O(2)	0.0000	0.0000	0.2509	1.3551

Surface morphology of $La_2SrFe_2CoO_9$ sample was studied by SEM images as shown in figure 2. Performed analysis reveals the occurrence of granular topology with sub-micrometric grain sizes. As observed in the micrograph, there are large topographic regions where the grains are strongly diffused between them and others nanosized regimes formed by few grains, which are surrounded by much porosity in the surface of the sample. However, it is important to notice that the sample evidences a single type of grain which is related with the obtaining of a single crystallographic phase.

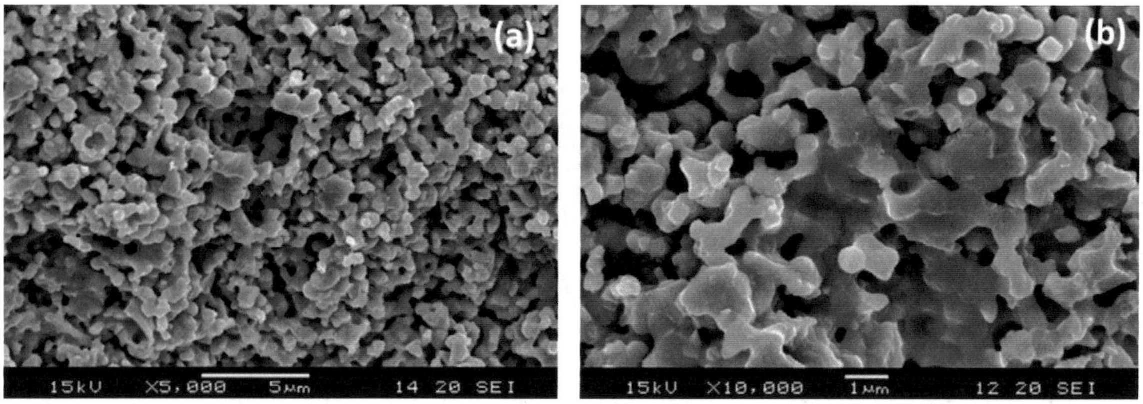

Figure 2. SEM micrograph of $La_2SrFe_2CoO_9$ obtained from the SEI detector (secondary electrons). Picture (a) shows s surface region with X5000 magnification and (b) X10000.

The EDX spectrum is shown in figure 3. No other chemical elements than expected were observed. Experimental composition of material was obtained from EDX semi-quantitative analysis, considering the spectrum areas for each element, and compared with the theoretical composition, calculated from the $La_2SrFe_2CoO_9$ stoichiometric formula.

The experimental composition values corroborate with 97% regard theoretical values, a crystalline single phase which corresponds to the stoichiometry of $La_2SrFe_2CoO_9$. The presence of two small Au peaks is observed in the spectrum. This is due to coating done on the samples as part of the preparation for SEM imaging experiment. From structural, morphologic and compositional characterizations, no other crystallographic phases or impurities were identified in the samples.

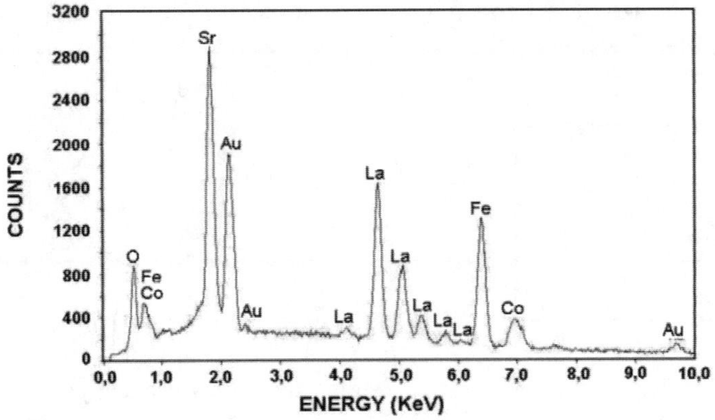

Figure 3. EDX spectrum obtained from the grains of $La_2SrFe_2CoO_9$.

Magnetic properties of $La_2SrFe_2CoO_9$ have been investigated by measuring the DC susceptibility in the temperature range from *2* to *300 K* under an applied magnetic field of *50 Oe*. Figure 4 shows the dependence of susceptibility as a function of temperature for $La_2SrFe_2CoO_9$ when it is measured by the ZFC and FC recipes.

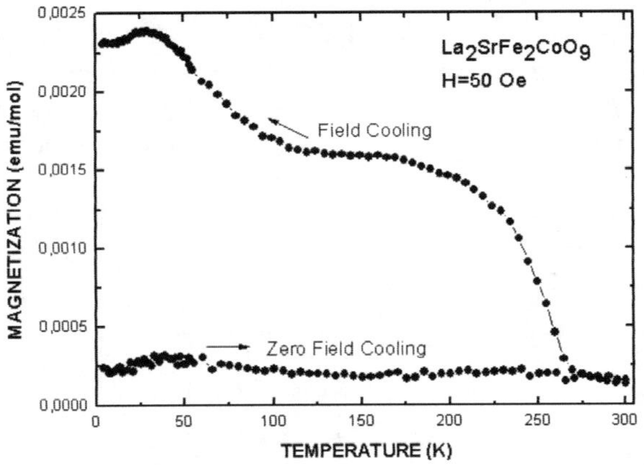

Figure 4. Magnetic behavior of $La_2SrFe_2CoO_9$ obtained from measurements of magnetization as a function of temperature by the ZFC and FC recipes.

It is observed in picture that the ZFC and FC curves present an irreversibility point for a temperature close to *270 K*. However, this behavior occurs in a temperature regime above the critical temperature of the paramagnetic-ferromagnetic transition, which is known as Curie temperature. This effect may be due to a cluster-glass percolative scenario [10]. When temperature is decreased from *300 K*, the ferromagnetic character is obtained in two steps: first a strong transition is observed close to *260 K*; then, at low temperatures (*50 K*) another ordering effect of the spin moments is detected. In order to verify the ferromagnetic behavior of the $La_2SrFe_2CoO_9$ triple perovskite, figure 5 shows the magnetization dependence with the applied magnetic field at *T=150 K*. In the picture it is observed that the magnetization shows a hysteretic behavior with a saturation magnetization value of *250 emu/cm³*. From this result and the volume of the crystallographic cell calculated from the Rietveld refinement, we obtain an effective magnetic moment of 6.18 μ_B. Theoretical calculations by the Hund's rule, with $P_{eff} = g\sqrt{J(J+1)}$, predict that magnetic moments of the isolated ions Co^{2+} and Fe^{4+} must be $\mu_{Co}{}^{2+}=3.0\mu_B$ and $\mu_{Fe}{}^{4+}=4.5\mu_B$, respectively [11-12]. The effective magnetic moment of $La_2SrFe_2CoO_9$ is obtained to be $\mu_{eff}=7.04\mu_B$, where we have used $\mu_{eff} = \sqrt{\mu^2_{Co^{2+}} + \mu^2_{Fe^{4+}}}$, by considering two Fe^{4+} cations in the crystallographic unit cell. Our experimental result corresponds to 88% of the theoretical value. This is because at *150 K* not all the magnetic moments are oriented in the direction of the applied magnetic field. Based on the results of the figure 5, we predict that at $T<50\ K$ the difference between theoretical and experimental values is less than 2%.

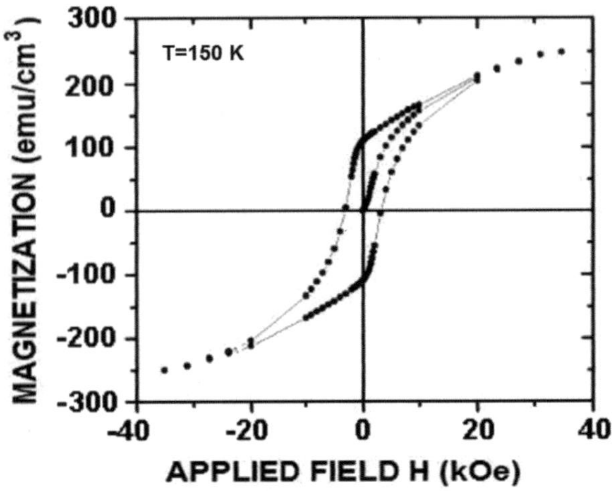

Figure 5. Hysteretic behavior of magnetization as a function of applied field for $La_2SrFe_2CoO_9$.

4. Conclusions

The synthesis and structural characterization of the new $La_2SrFe_2CoO_9$ perovskite-like material was performed. The Rietveld analyses reveal that this material crystallizes in an orthorhombic complex perovskite which corresponds to the *Immm (#71)* space group. Morphologic surface study by SEM images shows a single type of grains which is interpreted as an evidence of the formation of a single crystallographic phase. EDX analysis permitted to determine that the composition of material corresponds 97% with the expected from the stoichiometric formula. Measurements of magnetization as a function of temperature show the occurrence of a ferromagnetic ordering transition for a Curie temperature close to *260 K*. From the saturation magnetization in the hysteretic curve of magnetization as a function of applied field we obtain the magnetic moment of the unit cell to be $6.18\ \mu_B$.

5. References

[1] R M Hazen 1988 *Scientific American* **258** 54.
[2] Ortiz Diaz O, Roa-Rojas J, Landínez Téllez D A, Albino Aguiar J 2004 *Mod. Phys. Lett. B* **18** 1035.
[3] Philips J B, Majewski P, Alff L, Gross R, Graf T, Branst M S, Simon J, Walther T, Mader W, Topwal D, Sarma D D 2003 *Phys. Rev. B* **68** 144431.
[4] Bonilla C M, Landínez Téllez D A, Arbey Rodríguez J, Vera López E, Roa-Rojas J, *Physica B* **398** 208.
[5] Palkar V R, Malik S K 2005 *Solid Stat. Commun.* **134** 783.
[6] Lofland S E, Scabarozi T, Moritomo Y, Xu S H 2003 *J. Magn. Magn. Mater.* **260** 181.
[7] Di Trolio A, Larciprete R, Testa A M, Fiorani D, Imperatori P, Turchini S, Zema N 2006 *J. Appl. Phys.* **100** 13907.
[8] Larson A C, Von Dreele R B 2000 *General Structure Analysis System (GSAS)*, Los Alamos National Laboratory Report LAUR, pp. 86-748.
[9] Lufaso M W, Woodward P M 2001 *Acta Cryst.* **B 57** 725.
[10] Marcano N, Gomez Sal J C, Espeso J I, Fernandez Barquin L, Paulsen C 2007 *Phys. Rev. B* **76** 224419.
[11] Viola M C, Martínez Lope M J, Alonso J A, Martínez J L, De Paoli J M, Pagola S, Pedregosa J C, Fernández-Díaz M T, Carbonio R E 2003 *Chem. Mater.* **15** 1655.
[12] Ahmed M A, El-Dek S I 2006 *Mater. Sci. Eng.* **B 128** 30.

On the magnetic behavior of polycrystalline $RBaCo_2O_{5+\delta}$ synthesized by solid state and wet chemical routes

L Gómez[1*], V Galeano[1], E Vallejo[2], A Gómez[3], C Paucar[1] and O Morán[1].
[1] Universidad Nacional de Colombia, Campus Medellín, Facultad de Ciencias, Departamento de Física, Laboratorio de Cerámicos y Vítreos, Medellín, Colombia.

[2]Centro de Investigaciones en Materiales, Universidad de Nariño, Facultad de Ingeniería, Ciudad Universitaria Torobajo, Pasto, Colombia.

[3] Universidad Nacional de Colombia, Campus Medellín, Facultad de Minas, Laboratorio de caracterización de materiales, Medellín, Colombia.

E-mail: Legomezpl@unal.edu.co

Abstract. In this paper, polycrystalline $RBaCo_2O_{5+\delta}$ (R = La, Y, Eu) samples were obtained through solid state reaction and polymeric precursor method. The crystalline quality of the samples was carefully evaluated by means of X-ray diffraction. Depending on the R cation, the main phases were characterized in a tetragonal *P4/mmm*, orthorhombic *Pmmm* or trigonal R-3c structure. The oxygen content, being a fundamental parameter in the studied system, was determined by iodometric tritration. While no stoichiometric oxygen content was determined, the value of δ ended up being close to 0.5 for all the samples studied. VSM magnetization measurements were taken according to the zero field cooling protocol. The *M*(*T*) curves clearly shows the characteristic magnetic behavior of the cobaltate with stoichiometry 112. The temperature at which the paramagnetic-ferromagnetic-antiferromagnetic transition occurs varies with the cation R.

1. Introduction

The $RBaCo_2O_{5+\delta}$ cobaltates are transition metal oxides with perosvkite-like structure, which exhibit a variety of fascinating physical properties due to the strong interplay between spin, charge, orbital and lattice degrees of freedom[1]. Such cobaltates can content oxygen vacancies, which gives rise to variable ratios of Co^{2+}, Co^{3+} and Co^{4+} species. These species possess three possible spin states, namely low-spin (LS) state, intermediate-spin (IS) state and the high-spin (HS) state, which have a strong influence on the electrical and magnetic properties of the system [2]. Such spin states originate from a close competition between the atomic parameter Hund´s exchange energy and the crystal field energy. A small variation in temperature, pressure or magnetic field can produce a drastic change in the physical properties of these complex oxides [3]. The structural degree of freedom of this family of compounds provides a suitable playground to explore the interrelation between electronic, magnetic and structural properties. The crystal structure of these oxides can be described as an ordered sequence of $[CoO_2]$-$[BaO]$-$[CoO_2]$-$[LnO_\delta]$ layers along the c-axis, where oxygen vacancies exist in the lanthanide layer. The ideal crystallographic description consists of layers of CoO_6 octahedra along the (a,c)-planes. These layers are interconnected by two-leg ladders along the *a*-direction of CoO_5 pyramids. In between these ladders, the six-sided tunnels are occupied by Ln^{3+} cations [4]. The structure described above is very flexible i.e. sensitive to variations of the oxygen content and size of the R^{3+} cation so that long range order superstructures or even local distortions may be obtained, leading to dramatic variations of the magnetic properties from one sample to the other. Cobaltates exhibit successive paramagnetic-ferromagnetic-antiferromagnetic (PM-FM-AFM) transitions within a temperature range between 200 K and 300 K. Moreover, these compounds exhibit an insulator-metal transition in the temperature range between 300 K and 400 K [4]. The understanding of magnetic and electrical phases of $RBaCo_2O_{5.5}$ is a matter of strong debate. For instance, Taskin et al. [5] studied the Gd compound. They interpreted the successive PM-FM-AFM transitions based on a magnetic structure built up of S=1 Co^{3+} cations separated by non-magnetic S=0 Co^{3+} cations. The presence of LS Co^{3+} ion (S=0) implies that the Co ion should be in an octahedral site. However, an orthorhombic or a tetragonal distortion would destabilize the S=0 state due to the gain in crystal field and lowering of the electron-electron repulsion. On the other hand, Plakhty et al. [6] have studied the Tb-based compound and proposed a ferromagnetic structure at 260 K and two antiferromagnetic structures at 230 and 100 K, respectively.

Although the physical method known as solid state reaction has been extensively used in the synthesis of transitions metal oxides, the ordering of the cations in the form of alternating layers is rather difficult to achieve using this route. In order to favors the ordering of the R^{3+} and Ba^{2+} cations, the synthesis temperature must be as low as possible, and consequently a soft-chemistry synthesis route should be used instead since it allows a high reactivity at relatively low temperature [7]. Chemicals methods such as the polymeric precursor method or Pechini [8] route, are suitable for obtaining high-quality ceramics oxides at low cost and in a highly controlled way. The synthesis of $RBaCo_2O_{5+\delta}$ compounds with layered structures containing trivalent R^{3+} and Ba^{2+} cations distributions is complicated due to the differences in their ion sizes that can promote statistical distribution in the lattice [7]. In the present work, results of the structural and magnetic characterizations of $RBaCo_2O_{5+\delta}$ (R= Eu, La, Y) cobaltates are reported. Special attention is devoted to the discussion of the synthesis of these compounds via soft chemistry methods as reports on these alternative sinthesis methods are rather scarce.

2. Experiment.

A first set of polycrystalline $RBaCo_2O_{5+\delta}$ (R= Eu, La, Y) samples was obtained by standard solid state reaction. Stoichiometric quantities of Y_2O_3, La_2O_3, $BaCO_3$, Eu_2O_3 and Co_2O_3 reactants were mixed together and calcined at 1170 K for 12 h to achieve decarbonisation. The resulting powder was slowly heated in air (~ 5° C/min) up to a temperature of 1270 K and maintained there for 24 h. A second set of powders was prepared based on the classic Pechini method using $Y(NO_3)_3.6H_2O$, $Co(NO_3)_2.6H_2O$, $Ba(NO_3)_2$, La_2O_3 and Eu_2O_3 as starting precursors. First, precursors of lanthanum or europium were dissolved in nitric acid. Then the other precursors were added and combined with citric acid (CA) or ethylenediaminetetraacetic acid (EDTA). The final solution was heated to 350 K under strong stirring until a homogeneous, translucent (violet colour) solution was obtained. Then, ethylene glycol (EG) was added, and the polymerization process was carried out at 383-393 K under constant stirring until a resin was formed. This resin was heated at 390 K for 5 hours in order to eliminate the H_2O traces. Next, the resin was heated in a stepwise process to 520 K, 620 K and 770 K for 5 h, respectively. The solid obtained was ground and annealed in air at 1270 K for 15 hours and then slowly cooled to room temperature. X-ray powder diffraction patterns of the compounds were registered with a Panalytical X'Pert Pro MPD diffractometer using CuKα radiation (λ=0.15418 nm). Crystallographic parameters were obtained through Rietveld refinement using the FULLPROF package [9]. The oxygen content of the samples was determined by standard iodometric tritration. In turn, the magnetic measurements were performed using a vibrating sample magnetometer (VSM) coupled to a PPMS system (Quantum Design). The ZFC $M(T)$ data were recorded in a 100 Oe magnetic field over the temperature range 5-400 K.

Journal of Physics: Conference Series **480** (2014) 012035 doi:10.1088/1742-6596/480/1/012035

3. Results and discussion

Figure 1 show the XRD pattern of two $YBaCo_2O_{5+\delta}$ samples obtained by solid state reaction and Pechini method, respectively. The results show that the two methods are able to synthesize high quality $YBaCo_2O_{5+\delta}$ cobaltate. The Rietveld refinement of the XRD pattern confirm the formation of a single-phase material with tetragonal symmetry and *P4/mmm* space group. The lattice parameters ended up being a=3.85 Å and c=7.45 Å at room temperature which are very close to those reported by Akahoshi *et al* [10]. For the $LaBaCo_2O_{5+\delta}$ system, the analysis of the XRD results and the definition of a type of structure or superstructure are much more complex. The first structural study of an ordered oxygen deficient perovskite of type $LnBaCo_2O_{5+\delta}$ was presented for the series Ln = Pr to Ho [11]. Although, the series has been investigated in details for almost all lanthanide elements of the periodic table, interestingly there was no such report on the first member of this series, i.e. for $LaBaCo_2O_{5.5}$. Recently, $LaBaCo_2O_{5.5}$ was reported and characterized by neutron diffraction, electron microscopy and magnetic studies, showing that at room temperature the structure is 112- layered orthorhombic *ap* × *2ap* × *2ap* supercell with *Pmmm* space group [12]. In turn, XRD measurements on $EuBaCo_2O_{5+\delta}$ samples (not shown) show that the compound is characterized in an orthorhombic structure with lattice parameters (obtained by Rietveld analysis) a=3.92 Å, b=3.88 Å and c=7.53 Å (*Pmmm* space group). This result is very similar to that reported for M.M. Seikh *et al.* [12]. Indeed, according to E. L. Rautana, *et al.* [7], the system La-Ba-Co-O can exists in three phases, namely disordered $La_{0.5}Ba_{0.5}CoO_{3-\delta}$, ordered $LaBaCo_2O_{5+\delta}$ and nanoscale ordered $LaBaCo_2O_6$. The disordered phase $La_{0.5}Ba_{0.5}CoO_3$, as well as the ordered phase $LaBaCo_2O_6$ could be defined considering the structural data from the literature with the cubic *Pm-3m* or rhombohedral *R-3c* and orthorhombic *Pmmm* [13,14] space groups, respectively. The XRD pattern of the nanoscale-ordered $LaBaCo_2O_{5+\delta}$ is very similar to that observed for the disordered phase and it is indexed as a simple perovskite using the same space group and the same structural parameters. According to XRD measurements and Rietveld refinements, the $LaBaCo_2O_{5+\delta}$ samples obtained by Pechini method with citric acid (La-PC) and EDTA (La-PE) have rhombohedral structure and space group *R-3c*, with the lattice parameters a=5.44 Å and c=13.10 Å. In turn, measurements on samples obtained by solid state reaction (La-SS) show a cubic structure (space group *Pm-3m*). This result excludes the possibility of obtaining a cation ordering between LaO/BaO layers for samples synthesized by this route because the La/Ba ordering involves a slight deformation of the perovskite sublattice with a dilatation of the a_p parameter within the LaO/BaO layers and a compression along the LaO/BaO layers stacking direction. [7]. The main difficulty here is the possibility of obtaining random distribution of the cations Ba^{2+} and La^{3+} because the ionic radii are very close (considering an octahedral environment r_{Ba}=1.49 Å, r_{La}=1.17 Å and Δr = 0.32 Å) [15]. However, a better definition of these structures requires the use of more powerful characterization techniques such as neutron diffraction or transmission electron microscopy.

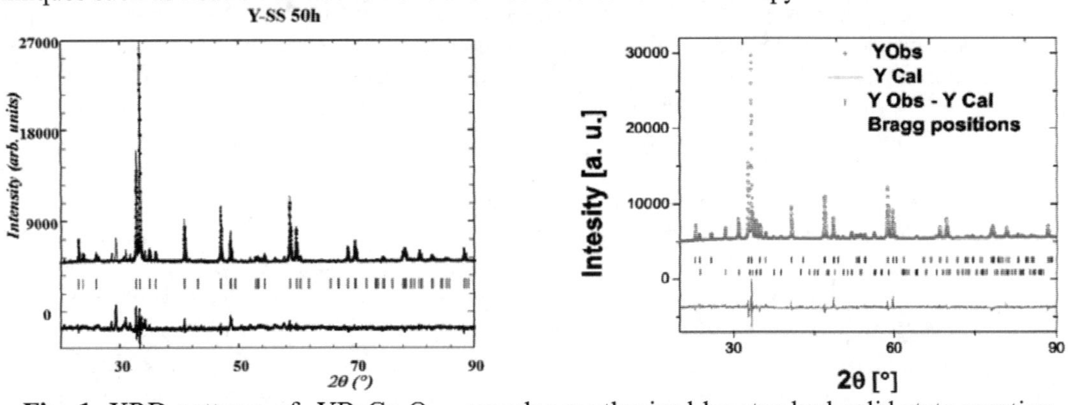

Fig. 1. XRD patterns of $YBaCo_2O_{5+\delta}$ samples synthesized by standard solid state reaction (a) and Pechini's method (b). The Rietveld refinement profiles of the samples are also included and marked by symbols. The solid lines correspond to the experimental data.

The lattice parameters of the polycrystalline $RBaCo_2O_{5+\delta}$ (R= Eu, La, Y) samples synthesized by the two mentioned routes are listed in Table 1. The results allow one to appreciate the effect of the size of the R cations on the structure of the cobaltates. The increasing in the size of the R^{3+}cation leads to increased lattice parameters and loss of symmetry. This is consistent with the fact that replacement of the cation in the R position by a larger cation provokes an increase in the dimensions of the unit cell. The ionic radii of Y^{3+}, Eu^{3+}, and La^{3+} cations are 101.9, 106.6 and 116, 0 Å, respectively.

Table 1. Lattice parameters of polycrystalline $RBaCo_2O_{5+\delta}$ (R= Eu, La, Y) samples obtained by Rietveld refinements. The oxygen content has also been included.

Sample	*space group*	*a(Å)*	*c(Å)*	χ^2	*Rwp*	**Oxygen (5 + δ)**
Y-PC	***P4/mm***	*3,84434(3)*	*7,43974(1)*	*3,08*	*15,8*	*5,57*
Y-PE	***P4/mm***	*3,85593(6)*	*7,46362(7)*	*2,46*	*13,55*	*5,52*
Y-SS	***P4/mm***	*3,86549(5)*	*7,48284(7)*	*2,37*	*12,03*	*5,55*
Eu-SS	***P mmm***	*3,91103(6)*	*7,55374(1)*	*1,76*	*13,88*	*5,54*
Eu-PE	***Pmmm***	*3,92085(8)*	*7,53703(0)*	*2,54*	*12,73*	*5,52*
Eu-PC	***Pmmm***	*3,92101(3)*	*7,53558(6)*	*2,19*	*12,69*	*5,51*
La-PE	***R-3c***	*5,44671(0)*	*13,10135(7)*	*2,05*	*5,16*	-
La-PC	***R-3c***	*5,44460(7)*	*13,09548(0)*	*2,27*	*8,55*	-

The oxygen content of the different samples is indicated in Table 1. The iodometric tritration analysis shows that the Y and Eu samples have oxygen contents very close to 5.5. This means that practically all the cobalt ions are Co^{3+}. However, for La samples the oxygen determination by iodometric tritration did not yield consistent results. By repeating the iodometric tritration analyzes, the reproducibility of the results ended up being ±0.02, demonstrating that chemical methods (except for La) can be used confidently to determine the oxygen content in the studied cobaltates [16].

Figure 2 shows the dependence of the magnetization on the temperature for $RBaCo_2O_{5+\delta}$ (R= Eu, La, Y) polycrystalline samples. The Eu-SS sample shows a para-ferro-antiferromagnetic transition in a narrow temperature window (~40 K) around ~250 K which is typical for this compound. The observed transitions are associated with magnetic ordering of the Co^{3+} ions. The additional transition observed at ~77 K has been attributed to the presence of ferromagnetism reentrant due to ordering of the ions Eu^{3+} [17]. Certainly, Eu as well as Gd possess intrinsic magnetic properties which should be separated from those displayed for the cobaltates. In turn, the Y-PC sample features the para-ferro-antiferromagnetic transition on a wider temperature window starting at ~300 K. A similar trend has been observed for Y-cobaltates with $\delta<0.5$ [1]. Hence, the wide para-ferro-antiferromagnetic transition observed for the Y-sample can be associated with oxygen variations. Finally, the La-PE sample exhibits a para-ferro-antiferromagnetic transition in a narrow temperature window around ~200 K (probably associated with a high oxygen content). Interestingly, this sample shows a diamagnetic component at low temperatures which could be assigned to the presence of impurities.

The results achieved are consistent with those reported in the literature for the studied system. Thus, it is concluded that for an oxygen content δ=0.5, the cobalt ions are purely trivalent, with half of cobalt cations in the octahedral sites and the remaining half in the pyramidal sites (supposing absence of charge ordering). However, δ may vary between 0 and 1, which leads to the mixed valence configurations Co^{2+}/Co^{3+} and Co^{3+}/Co^{4+}, respectively. Hence, the magnetic response of these cobaltates seems to be dominated mainly by three different ferromagnetic states, depending on the cobalt valence. Thus, for δ=0.5 (Co^{3+}) the ferromagnetic state could be called FM_1, for $\delta<0.5$ (Co^{2+}/Co^{3+}) FM_2, and for $\delta>0.5$ (Co^{3+}/Co^{4+}) FM_3. It has been demonstrated that the FM state of $YBaCo_2O_{5+\delta}$ has a Curie temperature, T_C, of ~ 300 K and a transition width of less than 40 K when the oxygen content $\delta \approx 0.50$ [12]. Nevertheless, when the oxygen deviates from the optimal doping, i.e., when the system is electron-doped ($\delta < 0.5$) or hole-doped ($\delta > 0.5$), the FM phase becomes more stable and the FM-AF transition shifts to lower temperatures [14].

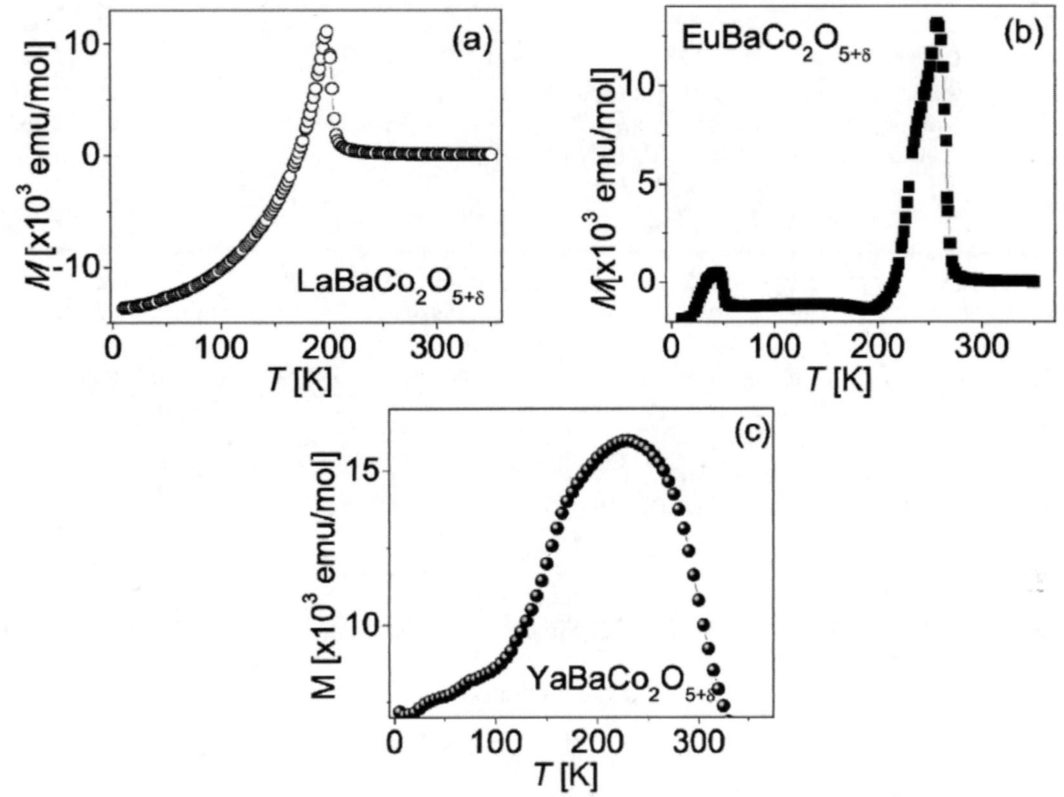

Fig. 2. *M*(*T*) dependence for La-PE (a), Eu-SS (b), and Y-PC (c) samples recorded in a magnetic field of 100 Oe

4. Conclusions

Polycrystalline $RBaCo_2O_{5+\delta}$ (R=Eu, La, Y) cobaltate were synthesized both by solid state reaction and soft chemistry method. The polymeric precursor route, using citric acid or EDTA as a complexing agent was chosen as chemical method to the synthesis. The XRD analysis showed that the compounds obtained by the two routes were single-phase with no appreciable affectation by foreign phases or impurities. The oxygen content as determined by iodometric tritration resulted to be close to the nominal value of 0.5 although deviations were observed for some samples. The non-stoichiometric oxygen content seemed to be related to the size of the cation R^{3+}. The lower oxygen content was obtained for the La-cobaltates. The size of the La ion is larger than the other ones (Y and Eu). The *M* (*T*) curves showed para-ferro-antiferromagnetic transitions with Curie temperatures varying according the cation R. In addition, it seemed that the presence of Co^{3+} ions in the lattice affected considerably the ferromagnetic response of the studied cobaltates.

5. Acknowledgment

This work was supported by Universidad Nacional de Colombia, sede Medellín. O. M. also acknowledges the financial support of the German Academic Exchange Service (DAAD) through the program “Wiederein-ladung”. The magnetic measurements were performed in the Laboratorio de Estado Sólido at the Universidad de Antioquia. The authors acknowledge this collaboration.

References

[1] D. Akahoshi and Y. Ueda, J. Phys. Soc. Jpn **68** (3).
[2] N. D. Todorov, M. V. Abrashev, V. G. Ivanov, G. V. Avdeev, and S. C. Russev, Journal of Physics: Conference Series **253** (1), 012071 (2010).
[3] D. D. Khalyavin, Physical Review B **72** (13), 134408 (2005).
[4] A. Maignan, C. Martin, D. Pelloquin, N. Nguyen, and B. Raveau, Journal of Solid State Chemistry **142** (2), 247 (1999).
[5] A. A. Taskin, A. N. Lavrov, and Y. Ando, Physical Review Letters **90** (22), 227201 (2003).
[6] V. P. Plakhty, Y. P. Chernenkov, S. N. Barilo, A. Podlesnyak, E. Pomjakushina, E. V. Moskvin, and S. V. Gavrilov, Physical Review B **71** (21), 214407 (2005).
[7] E.-L. Rautama, P. Boullay, A. K. Kundu, V. Caignaert, V. Pralong, M. Karppinen, and B. Raveau, Chemistry of Materials **20** (8), 2742 (2008).
[8] M. Pechini, US Patent **3** (97) (1967).
[9] J. Rodriguez-Carvajal, presented at the satellite meeting on powder diffraction of the XV congress of the IUCr, 1990 (unpublished).
[10] D. Akahoshi and Y. Ueda, J. Phys. Soc. Jpn **68** (3) (1999).
[11] Maignan A, Martin C, Pelloquin D, Nguyen N and Raveau B *J. Solid State Chem.***142**, 247 (1999).
[12] M. M. Seikh, V. Caignaert, V. Pralong, C. Simon, and B. Raveau, Journal of Physics: Condensed Matter **20** (1), 015212 (2008).
[13] M. M. Seikh, C. Simon, V. Caignaert, V. Pralong, M. B. Lepetit, S. Boudin, and B. Raveau, Chemistry of Materials **20** (1), 231 (2007).
[14] E. L. Rautama, V. Caignaert, P. Boullay, A. K. Kundu, V. Pralong, M. Karppinen, C. Ritter, and B. Raveau, Chemistry of Materials **21** (1), 102 (2008).
[15] K. Zhang, L. Ge, R. Ran, Z. Shao, and S. Liu, Acta Materialia **56** (17), 4876 (2008).
[16] M. Karppinen, M. Matvejeff, K. Salomaki, and H. Yamauchi, Journal of Materials Chemistry **12** (6), 1761 (2002).
[17] B. Raveau, C. Simon, V. Pralong, V. Caignaert, and F. X. Lefèvre, Solid State Communications **139** (6), 301 (2006).

Journal of Physics: Conference Series **480** (2014) 012036 doi:10.1088/1742-6596/480/1/012036

Irreversibility Curve on $Y_{1-x}Lu_xBa_2Cu_3O_{7-\delta}$ (x=0.4, 0.5 and 0.6) superconducting

J. F. Cepeda Grimaldos[1], M. V. Chiquillo[2], I. Supelano G.[1], D. Martínez B[2,3]., A. Sarmiento Santos[1] and C. A. Parra Vargas[2]

[1]Universidad Pedagógica y Tecnológica de Colombia, Grupo de Superficies Electroquímica y Corrosión, Tunja, Colombia.
[2]Universidad Pedagógica y Tecnológica de Colombia, Grupo Física de Materiales, Tunja, Colombia.
[3] Departamento de Ciência de Materiais, Universidade Federal de Pernambuco, 50740-560, Recife PE, Brazil

E-mail: asarmiento.santos@uptc.edu.co

Abstract. The irreversibility line in the H–T plane divides the irreversible and reversible behaviour of the magnetization which is of importance for the characterization of high T_c superconductors. In this work, we report the production of $Y_{1-X}Lu_XBa_2Cu_3O_{7-\delta}$ (X=0.4, 0.5 and 0.6) superconducting system using the usual solid state reaction method. The irreversibility line H-T plane for the $Y_{1-X}Lu_XBa_2Cu_3O_{7-\delta}$ polycrystalline sample was investigated. The curves of magnetization ZFC (cero field cooled)- FC (field cooled) were measured in magnetic fields between 100 Oe and 4000 Oe, and allowed to obtain the values for irreversibility and critical temperatures.

1. Introduction

The mixed state of high temperature superconductors (HTSC) reveals an unusual number of characteristic [1,3], within which we can denote the irreversibility temperature. In type II superconductors the magnetic irreversibility persists up to a well defined temperature limit that depends on the applied field as well as on certain thermodynamic properties of the superconductors [2]. This limit can most precisely be determined from ZFC and FC magnetization data as a function of temperature. The magnetic irreversibility in type II superconductors, originates starting from the movement of vortexes, those which consequently overcome the imprisonment forces. In the region magnetically reversible, the vortexes have enough energy and they can move freely, while to temperatures under T_{irr}, the mobility is restricted and the vortexes remain trapped in the pinning center.

The emergence of irreversibility in a system of superconducting grains is linked to the formation of structures, which are formed by frustrated coupled grains [4]. An arrangement of grains acts as a pinning center freezes locally topological degrees of freedom section of a vortex whose length is much smaller than the width of sample. Above T_{irr} the system is magnetically reversible due to thermal agitation. When temperature decreases below T_{irr}, enough so that many of the grains become coupled, strong irreversibility effects become visible in the magnetization curve.

Plotting the irreversibility data for a range of data in the (H,T), we can demarcate the irreversibility line $T_{irr}(H)$ of each sample. This line separates a region of high temperature magnetically reversible from another of low temperature magnetically irreversible, dividing the phase diagram H(T) in two different areas; one at high fields and temperatures where the magnetic properties are reversible and another region for low field and temperatures, where the magnetic properties are irreversible. Two main theoretical lines are used for the interpretation of the irreversibility line, one of which is that the vortexes are thermally actived. In this case, the reversible behaviour occurs when the flux-creep effects [5, 6] becoming dominant. The other line of reasoning suggests that associated whit T_{irr}, a phase transition occurs. In this line, we have the theories of lattice fusion of vortices (flux melting) [5, 6], glass superconductor (superconducting glass) [7] and glass of vortexes (vortex-glass) [8].

2. Experimental

Samples of $Y_{1-X}Lu_XBa_2Cu_3O_{7-\delta}$ material, were synthesized by the standard solid state reaction recipe, from Aldrich precursor powders of Y_2O_3 (99.9%), Lu_2O_3 (99.99%), $BaCO_3$ (99.997%) and CuO (99.9%). Powders were mixed, grinded, palletized and calcined at temperature of 750 °C for 12 h. Then, material was regrind, pressed as circular disks and sintered at 900 °C for 36 h, with two intermediate pulverizations. Crystalline structure was studied by X-ray diffraction, with a nickel-filtered Cu-Kα radiation (λ=1.5406 Å) of a SIEMENS D5000 equipment.

3. Results and discussion

In figure 1 the XRD patterns of $Y_{1-X}Lu_XBa_2Cu_3O_{7-\delta}$ are represented, being taken at room temperature. It is clear from figure 1 that all samples have the single phase with tetragonal RE:123 unit cell. Rietveld refinement was performed to determine the lattice parameters a= 3.8090 Å, b= 3.8770 Å y c= 11.6570 Å, whit symmetry of space group P4/mmm.

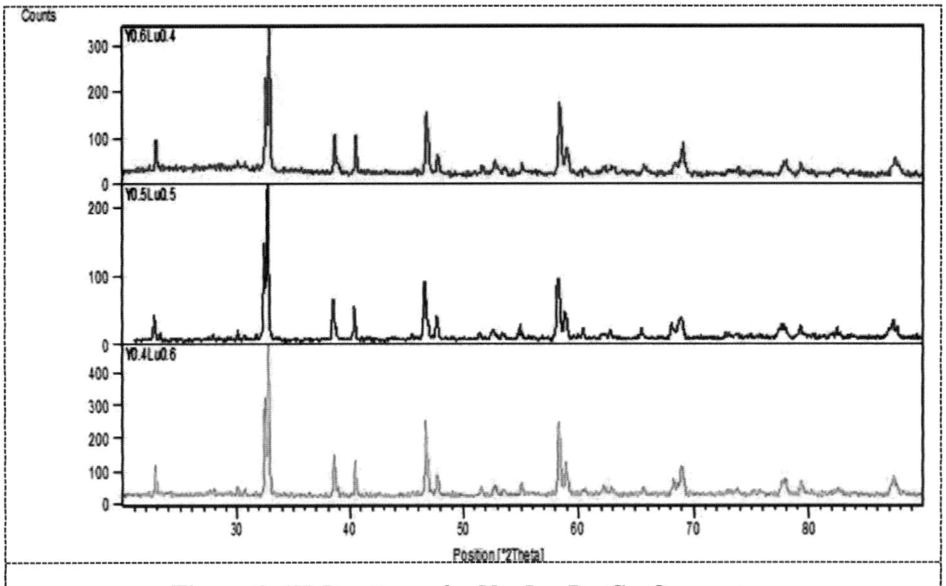

Figure 1. XDR patterns for $Y_{1-X}Lu_XBa_2Cu_3O_{7-\delta}$ system

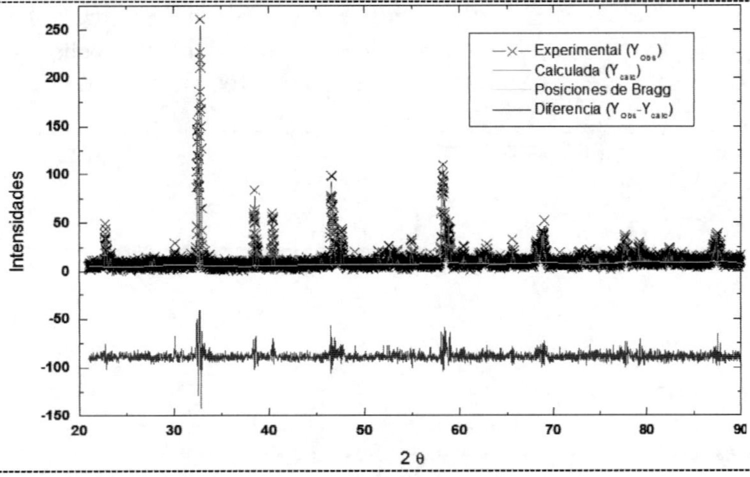

Figure 2. Rietveld refinement of sample $Y_{0.5}Lu_{0.5}Ba_2Cu_3O_{7-\delta}$ through the program General Structure Analysis System (GSAS)

By detecting secondary electrons and backscattered electrons are allowed to determine the morphological characteristics of the three samples, which were observed and analysed, at different magnifications: 2000X, 2500X y 5000X. Figure 3 shows the micrographs featuring homogeneous granularity including two zones (white and gray). The superconducting behaviour of the system $Y_{1-X}Lu_XBa_2Cu_3O_{7-\delta}$, was verified from measurements of magnetization, which is determined with critical temperatures ranging from 91.2 K ($Y_{0.6}Lu_{0.4}Ba_2Cu_3O_{7-\delta}$), 89.8 K ($Y_{0.4}Lu_{0.6}Ba_2Cu_3O_{7-\delta}$), and 90.2 K ($Y_{0.5}Lu_{0.5}Ba_2Cu_3O_{7-\delta}$). The T_c for these two samples was predicted by Skakle and West [9] and Anderson et al. [10], reporting temperatures lower than those obtained in this work. On the other hand, the critical temperatures obtained for the system are in close agreement with those predicted by other authors for CLBCO type system [11].

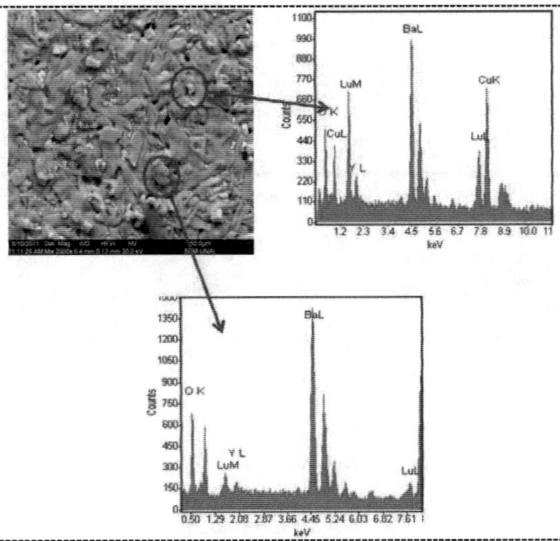

Figure 3. White and gray zone micrograph of the sample $Y_{0.5}Lu_{0.5}Ba_2Cu_3O_{7-\delta}$

The DC magnetization results for all $Y_{1-X}Lu_XBa_2Cu_3O_{7-\delta}$ system are shown in figure 4. The transition temperature T_C for systems has defined as the onset of the diamagnetic transition. The transition temperature and irreversibility were obtained through from Zero Field Cooled (ZFC) and Field

Cooled (FC) magnetization data as a function of temperature. In figure 5 the curve ZFC-FC is shown for the $Y_{0.6}Lu_{0.4}Ba_2Cu_3O_{7-\delta}$ sample. In figure 6 we show the behaviour of temperature irreversibility for each of the samples with fields ranging from 100 Oe to 4000 Oe.

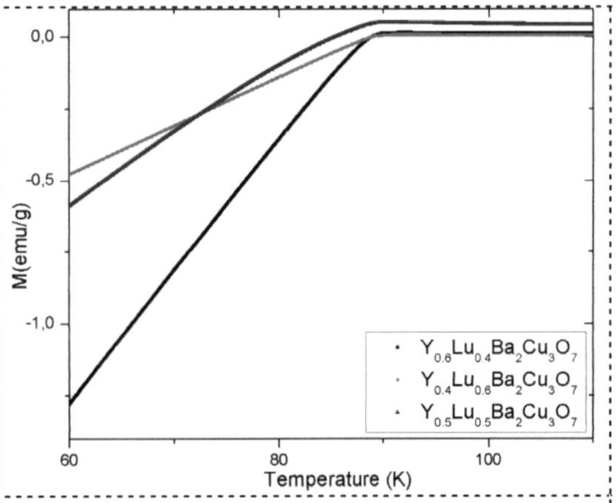

Figure 4. Magnetization result for all samples of the $Y_{1-X}Lu_XBa_2Cu_3O_{7-\delta}$ system.

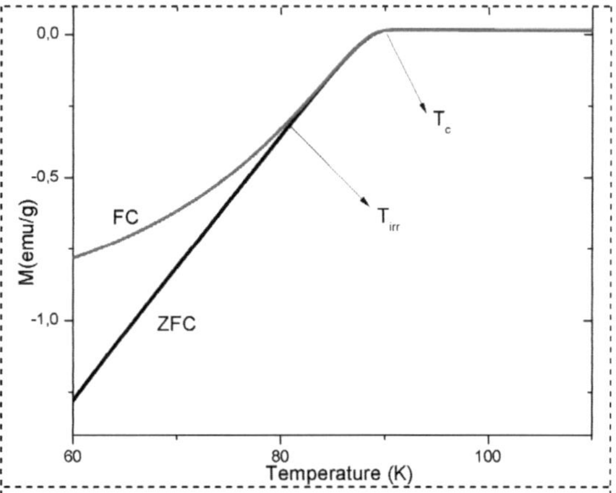

Figure 5. ZFC and FC magnetization by $Y_{0.6}Lu_{0.4}Ba_2Cu_3O_{7-\delta}$ sample.

Table 1. Structural results obtained from the Rietveld refinement method and critical temperature.

Sample	χ^2	a (Å)	b (Å)	c (Å)	T_C (K)
$Y_{0.6}Lu_{0.4}Ba_2Cu_3O_{7-\delta}$	1.183	3.813(9)	3.881(8)	11.678(7)	91.2
$Y_{0.4}Lu_{0.6}Ba_2Cu_3O_{7-\delta}$	1.348	3.814(3)	3.880(2)	11.671(7)	89.9
$Y_{0.5}Lu_{0.5}Ba_2Cu_3O_{7-\delta}$	1.527	3.813(1)	3.879(1)	11.684(1)	90.2

The results obtained by refinement (χ^2, structural parameters), as well as that T_C for the field but lower than they are in Table 1.

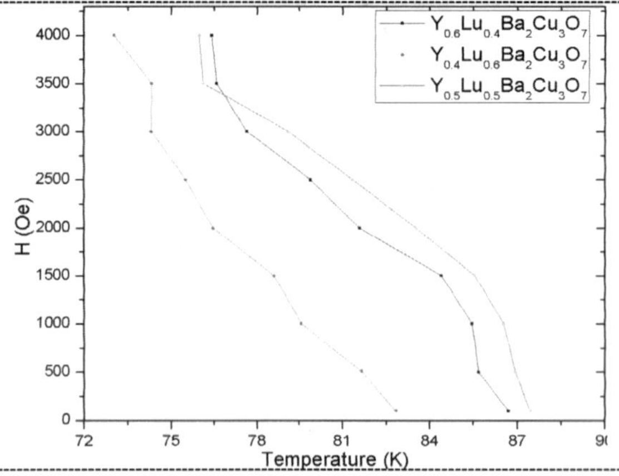

Figure 6. The magnetic irreversibility data of a $Y_{0.5}Lu_{0.5}Ba_2Cu_3O_{7-\delta}$ system

Journal of Physics: Conference Series **480** (2014) 012036 doi:10.1088/1742-6596/480/1/012036

4. Conclusion

The superconducting behaviour of the $Y_{1-X}Lu_XBa_2Cu_3O_{7-\delta}$ system was verified from measurements of magnetization, obtaining critical temperatures higher than those reports by Skakle and West [9] and Anderson et al. [10]. The obtained results allow concluding that in the $Y_{1-X}Lu_XBa_2Cu_3O_{7-\delta}$ system a characteristic bend of the Almeida–Thouless (AT) tendency is dominant for low fields and a bend Gabay–Toulouse (GT) behaviour for high magnetic fields. This feature of the irreversibility line has been reported as a characteristic of granular superconductors and it corroborates the topological effects of vortexes mentioned by several authors [14–16].

References

[1] Vieira V N, Pureur P and Schaf J 2001 Phys. C **353** 241.
[2] Roa-Rojas J, Menegotto Costa R, Pureur P and Prieto P 2000 Phys. Rev. B **61** 12457.
[3] Pavuna Davor, Cyrot Michel, Introduction to Superconductivity and High-Tc Materials, World Scientific Publishing Co. Pte. Ltd., 1992.
[4] Yagi T, Domon M, Okajima Y and Yamaya K 1991 Phys. C **173** 453.
[5] Anderson P W, Kim Y B Review of Modern Physics, (New York) 36 (1) (1964) 39.
[6] Barden J and Stephen M J 1965 Phys. Rev. 1197.
[7] Fischer M P A 1989 Phys. Rev. Lett. **62** 1415.
[8] Larson A C and Von Dreele R B General Structure Analysis System, Los Alamos National Laboratory Report LAUR, 2000, 86.
[9] Skakle J M S and West A R 1994 Phys. C **220** 187.
[10] Anderson P S, Kirk C A, Skakle J M S and west A R 2003 J. Solid State Chem. **170** 1.
[11] Peng J L, Klavins P et al. 1989 Phys. Rev. B **39** 9074.
[12] de Almeida J R L and Thouless D J 1978 J. Phys. A **11** 983.
[13] Gabay M and Toulouse G 1981 Phys. Rev. Lett. **47** 201.
[14] Barros F M, Vieira V N et al. 2004 Phys. C **632** 408.
[15] Parra Vargas C A, Landinez Téllez D A and Roa-Rojas J 2007 Phys. B **398** 301.
[16] Parra Vargas C A, Pimentel Jr J L, Pureur P, Landinez Téllez D A, and Roa-Rojas J 2009 Phys. B **404** 2766.

Journal of Physics: Conference Series **480** (2014) 012037 doi:10.1088/1742-6596/480/1/012037

The solution of the nonlinear Schrödinger equation using Lattice- Boltzmann

F Fonseca

Universidad Nacional de Colombia, Departamento de Física, Grupo de Ciencia de Materiales y Superficies, Ciudad Universitaria, Bogotá Av. Cra. 30 No 45-03, Colombia

E-mail: frfonsecaf@unal.edu.co

Abstract. We present the solution of the nonlinear Schrödinger equation using the lattice Boltzmann method. We show results for two dimensions using a **d2q9** lattice velocity scheme. To implement the expansion B.G.K. (Bhatnagar-Gross-Krook), we assume the distribution function as a complex valued function, whose real and complex components satisfy the Boltzmann equation. The strategy followed to obtain the motion equation is to define adequately the second moment of the distribution as a symmetric tensor. We obtain stable structures for given values of the nonlinear coupling constant.

1. Introduction

Nonlinear terms in Physics laws provide an amazing range of new dynamic effects. This is the case of the nonlinear Schrödinger equation (NLSEq), which plays a key role in describing the dynamics of Bose-Einstein condensation [1,2] In the past Lattice-Boltzmann has been successfully applied to the solution of the Schrödinger equation [3], and the Dirac equation its relativistic version of the problem [4]. Moreover, the NLSEq has brought a massive investigation into soliton behavior [5,6].

This paper proposes, using the lattice Boltzmann method a solution to the NLSEq. In section 2, we begin with a review of the lattice-Boltzmann technique and the moment relations of the equilibrium distribution functions applied to derivation of the NLSEq. In section 3, we obtain the NLSE equation, using a hypothesis in the Π^0 tensor. In section 4, we obtain the equilibrium distribution functions that we use on a **d2q9** lattice velocity scheme for the computational scheme. In section 5, we present results and at last, in section 6, we give conclusions.

2. The lattice-Boltzmann model

It is considered a bi-dimensional model where the velocities of particles are discretized on the grid into ***d*** direction. The lattice-Boltzmann equation is:

$$f_{i,j}(\vec{x}+\vec{e}_x\delta t, t+\delta t) - f_{i,j}(\vec{x},t) = -\frac{1}{\tau}(f_{i,j}(\vec{x},t) - f^{eq}{}_{i,j}(\vec{x},t)) \tag{1}$$

Where f_i is the probability density function of finding the group particle ***i***, of the ***j*** species, in the spatial point $\vec{x}$ and time ***t*** and δt is the time step. The left hand side of equation (1) is the B.G.K. approximation [7], where is the non-dimensional relaxation time that measures the approaching rated to the statistical equilibrium. Expanding the left-hand side of equation (1) up to second order, in a Taylor series, we have:

$$f_{i,j}(\vec{x}+\vec{e}_\alpha\delta t, t+\delta t) - f_{i,j}(\vec{x},t) = \delta t\left(\frac{\partial}{\partial t} + e_x\frac{\partial}{\partial x_1} + e_y\frac{\partial}{\partial y_1}\right)f_{i,j} + \frac{\delta t^2}{2}\left(\frac{\partial}{\partial t} + e_x\frac{\partial}{\partial x_1} + e_y\frac{\partial}{\partial y_1}\right)^2 f_{i,j} \quad (2)$$

Assuming the spatial and temporal derivatives as:

$$\frac{\partial}{\partial x} = \mathcal{E}\frac{\partial}{\partial x_1};\ \frac{\partial}{\partial y} = \mathcal{E}\frac{\partial}{\partial y_1};\ \frac{\partial}{\partial t} = \mathcal{E}\frac{\partial}{\partial t_1} + \mathcal{E}^2\frac{\partial}{\partial t_2} \quad (3)$$

The parameter $\mathcal{E} = \Delta x/L$, where Δx is the spatial size grid and L the spatial system size. Expanding the distribution function $f_{i,j}$ in a perturbative series:

$$f_{i,j} = f^0{}_{i,j} + \mathcal{E}f^1{}_{i,j} + \mathcal{E}^2 f^2{}_{i,j} \quad (4)$$

We obtain at first order in $\mathcal{E}$:

$$-\frac{1}{\tau}\left(\mathcal{E}f^1{}_{i,j}\right) = \delta t(\mathcal{E}\frac{\partial}{\partial t_1} + \mathcal{E}e_x\frac{\partial}{\partial x_1} + \mathcal{E}e_y\frac{\partial}{\partial y_1})f^0{}_{i,j} \quad (5)$$

And at second order in ε:

$$-\frac{1}{\tau}\mathcal{E}^2 f^2{}_{i,j} = \mathcal{E}^2\delta t(\frac{\partial}{\partial t_2}f^0{}_{i,j} + \frac{\delta t}{2}\left(\frac{\partial}{\partial t_1} + e_x\frac{\partial}{\partial x_1} + \mathcal{E}e_y\frac{\partial}{\partial y_1}\right)^2 f^0{}_{i,j} + \left(\frac{\partial}{\partial t_1} + e_x\frac{\partial}{\partial x_1} + e_y\frac{\partial}{\partial y_1}\right)f^1{}_{i,j}) \quad (6)$$

Also it is assumed

$$f^0{}_{i,j} = f^{eq}{}_{i,j} \quad (7)$$

Inserting equation (5) in equation (6), we obtain:

$$-\frac{1}{\tau}\left(f^2{}_{i,j}\right) = \delta t\left(\frac{\partial}{\partial t_1} + e_x\frac{\partial}{\partial x_1} + e_y\frac{\partial}{\partial y_1}\right)f^1{}_{i,j}\left(1-\frac{1}{2\tau}\right) + \delta t\left(\frac{\partial f^0{}_{i,j}}{\partial t_2}\right) \quad (8)$$

The moments of the distribution function are defined as:

$$\rho_j = \sum_i f^0{}_{i,j};\ \vec{u}_j = \sum_i \vec{e}_i f^0{}_{i,j};\ \Pi^0{}_j = \sum_i \vec{e}_i\cdot\vec{e}_i f^0{}_{i,j} \quad (9)$$

Also we assume the distribution functions $f_{i,j}$ satisfy the probability conservation condition with the equilibrium distribution $f^{eq}{}_i$ such that:

$$\sum_{i=0}^N f^{eq}{}_i = \sum_{i=0}^N f^0{}_i \quad (10)$$

3. The Schrödinger Equation

Doing algebra in equations (5) and (8) with the help of equations (10), we get:

$$\frac{\partial\rho_j}{\partial t} + \nabla\cdot\vec{u}_j = 0 \quad (11)$$

And

$$\frac{\partial\vec{u}_j}{\partial t} + \nabla\cdot\Pi^0{}_j = 0 \quad (12)$$

We assume $\Pi^0{}_j$ as a symmetric tensor given by:

$$\Pi^0{}_{1,\mu\nu} = \delta_{\mu\nu}\frac{\partial\rho_2}{\partial t} + (1-\delta_{\mu\nu})\lambda(\rho_1^2+\rho_2^2)\rho_2 \tag{13}$$

$$\Pi^0{}_{2,\mu\nu} = \delta_{\mu\nu}\frac{\partial\rho_1}{\partial t} + (1-\delta_{\mu\nu})\lambda(\rho_1^2+\rho_2^2)\rho_1 \tag{14}$$

Defining the complex function

$$\rho = \rho_1 + i\rho_2 \tag{15}$$

Where i is the complex number, and assuming zero the off diagonal components of $\Pi^0{}_j$, and doing some algebra, we have the NLSEq:

$$i\frac{\partial\rho}{\partial t} + \nabla^2\rho + \lambda|\rho|^2\rho = 0 \tag{16}$$

4. The equilibrium distribution function

We use the **d2q9** velocity scheme. For the directions $\vec{e}_i$ and weights w_i on each cell, we have:

$$w_i = \left\{\frac{4}{9} \to i = 0;\ \frac{1}{9} \to i = 1,2,3,4;\ \frac{1}{36} \to i = 5,6,7,8\right\} \tag{17}$$

$$\textstyle\sum_i w_i\, e_{i,\alpha} = 0;\ \sum_i w_i\, e_{i,\alpha}e_{i,\beta} = \frac{1}{3}\delta_{\alpha,\beta};\quad \sum_i w_i\, e_{i,\alpha}e_{i,\beta}e_{i,\gamma} = 0 \tag{18}$$

We use the equilibrium function as:

$$f^{eq}{}_{i,j} = \begin{Bmatrix} w_i(A + B\vec{e}_i\cdot\vec{u}) & i > 0 \\ w_0 C & i = 0 \end{Bmatrix} \tag{19}$$

Using equations (10), (18) and (19) we can determine A, B and C. Then, the equilibrium distribution functions that satisfies the NLSEq is:

$$f^{eq}{}_{i,1} = \begin{cases} 3w_i\left(\vec{e}_i\cdot\vec{u} + \frac{\partial\rho_2}{\partial t} - \lambda(\rho_1^2+\rho_2^2)\rho_2\right) & i > 0 \\ w_0\left(\frac{9}{4}\rho_1 - \frac{5}{4}\left(\frac{\partial\rho_2}{\partial t} - \lambda(\rho_1^2+\rho_2^2)\rho_2\right)\right) & i = 0 \end{cases} \tag{20}$$

$$f^{eq}{}_{i,2} = \begin{cases} 3w_i\left(\vec{e}_i\cdot\vec{u} + \frac{\partial\rho_1}{\partial t} - \lambda(\rho_1^2+\rho_2^2)\rho_1\right) & i > 0 \\ w_0\left(\frac{9}{4}\rho_2 - \frac{5}{4}\left(\frac{\partial\rho_1}{\partial t} - \lambda(\rho_1^2+\rho_2^2)\rho_1\right)\right) & i = 0 \end{cases} \tag{21}$$

5. Results

The derivative operator of $\rho(x,t)$ used in the distribution functions is discretized as:

$$\frac{\partial\rho}{\partial t} = \frac{\rho(x,y,t+\delta t) - \rho(x,y,t)}{\delta t} \tag{22}$$

The system is initialized with the function:

$$f_{i,j}(x, y, 0) = A_1 cos(D(x + 2)^2 + D(y + 2)^2) \tag{23}$$

The simulation starts at $t_0 = 0$, with the Equation (22) in all the points of the system, and for the two components. Figure (1) presents the simulation results using **d2q9** velocity scheme for grid size of **120x120** and **λ=0.7**. The two panels show the solution from different perspectives. In left panel, we have a top view colormap. A three-dimensional shaded surface is presented in the right panel.

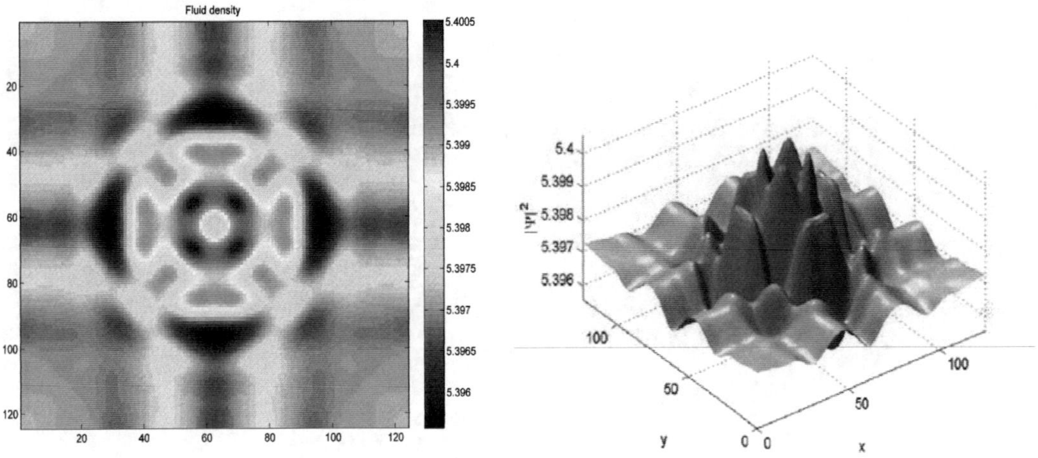

Figure 1. The numerical result of the **d2q9** lattice Boltzmann model. The two panels are the numerical results at times *t = 100*, lattice size *L=120, λ=0.7, Δx =Δy = 1/L.*

6. Conclusions

We have solved the nonlinear Schrödinger equation using a definition of the tensor $\mathbf{\Pi^0}$ defined in the Chapman-Enskog expansion. As a future work, the method can be extended to **3d,** employing **d3q15** and **d3q19** lattice schemes and using cubic-quintic nonlinear terms in the Schrödinger equation.

Acknowledgments

This work was supported by Universidad Nacional de Colombia (DIB-8003355).

7. References

[1] Anderson M. H., Ensher J.R., Matthews M.R., Wieman C.E., and Cornell E.A. 1995 *Science,* **269** 198
[2] Dunningham J., Burnett K., and Phillips W. D. 2005 *Phil Trans R Soc. A.* **363** 2165
[3] Succi S., 1996 *Physical Review E.,* **53(2)**, 1969
[4] Succi S., 2002 *Phil. Trans. R. Soc. Lond. A* **360** 429
[5] Brazovskii S. 2009 *Physica B,* **404** 482.
[6] Dmitriev S. V., Kivshar Y. S. and Shigenari T. 2002 *Physica B,* **316** 139
[7] Bathnagar P. L., Gross E. P. and Krook M. 1954 *Phys. Rev.* **94** 511

Journal of Physics: Conference Series **480** (2014) 012038 doi:10.1088/1742-6596/480/1/012038

The effect of the In concentration on the surface morphology of InGaAs-GaAs heterostructures grown by MBE on GaAs substrate

E Gómez-Barojas[1], R Silva-González[2], R M Serrano-Rojas[1], M A Vidal-Borbolla[3], M A Rodríguez-Moreno[1] and G Santamaría-Juárez[1].

[1]CIDS-IC, Benemérita Universidad Autónoma de Puebla. Apdo. Postal 196, Puebla, Pue. 72000. México
[2]Instituto de Física, Benemérita Universidad Autónoma de Puebla. Apdo. Postal J-48, Puebla, Pue. 72570 México
[3]Instituto de Investigación en Comunicación Óptica. Universidad Autónoma de San Luis Potosí, Av. Karakorum 1470, Col. Lomas Altas 4ª. Sección, San Luis Potosí, S. L. P., 78100, México

E-mail: egomez@ifuap.buap.mx

Abstract. A set of 3 heterostructures were formed by 10 periods of InGaAs-GaAs epitaxially grown on GaAs substrate by means of a molecular beam epitaxial system. Scanning electron microscopy (SEM) cross section images at high magnification show that the heterostructures present good periodicity. SEM micrographs of the surface morphology chemically etched show the coalescence effect of In due to an unequal etching rate of In and GaAs. Auger electron spectroscopy (AES) depth profiles show that the first GaAs layers in the 3 samples are off-stoichiometric and that the alloy layers present In square and triangular depth profiles.

1. Introduction

Since the decade of the 80s, considerable interest had been devoted to the growth of heterostructures based on lattice mismatch epitaxial layers in order to improve device performance and obtain new properties. In the case of strained (InAs)x(GaAs)y superlattices abrupt heterointerfaces are required because compositional fluctuations at heterointerfaces results in uncertainty in both composition and lattice constant [1]. These fluctuations are attributable to surface steps in each layer. The surface steps are composed of excess atoms, which remain on each layer because it is difficult to control the molecular beam intensities and the growth time precisely enough to form just one monolayer.
The aim of this work is to study exsitu the surface morphology, the periodicity and elemental composition of a set of 3 InGaAs-GaAs heterostructures grown on GaAs (100) substrate by a MBE system. The heterostructures are formed by 10 periods of InGaAs-GaAs epitaxially grown on GaAs substrate. The techniques used for this purpose are SEM and AES.

Journal of Physics: Conference Series **480** (2014) 012038 doi:10.1088/1742-6596/480/1/012038

2. Experimental procedure

2.1 Preparation of $In_xGa_{1-x}As$-GaAs heterostructures

Deposition was carried out in a RIBER, MBE32 system equipped with a high energy electron diffraction (RHEED) facility. The GaAs substrates were semi-insulating with (100) orientation. The GaAs substrates were degreased chemically. To remove native oxide from the surface substrate, after loaded in the growth chamber a heat cleaning was performed at 580 °C and the RHEED intensity oscillations of C and O were monitored until obtaining only the GaAs pattern. The substrate temperature T_s was measured by an infrared pyrometer which was calibrated by using the Gallium oxide desorption temperature 580 °C as a reference.

The heterostructures are formed by 10 periods of $In_xGa_{1-x}As$-GaAs epitaxially grown on GaAs substrates with nominal thickness of 500 and 1000 Å, respectively. Three different growth runs labeled A, B, C were done. The growth parameter for both GaAs and $In_xGa_{1-x}As$ layers of the three runs were: the molecular beam pressure (MBP) were: for As_4, 8.1×10^{-6} Torr at 205 °C; for Ga, 3.6×10^{-7} Torr at 914.5 °C and for the alloy layers the MBP were: for In of growth run A, 4×10^{-7} Torr and 807 °C, for run B, 2.1×10^{-7} and 768 °C and for the growth run C, 9.6×10^{-8} and 731 °C. SEM observations were performed in a JEOL 5400LV system and composition analysis in a JEOL (JAMP 7800) AES equipped with a secondary electron detector and an Ar ion sputtering system.

3. Results and discussion

3.1 Surface morphology by SEM

Figure 1 (a) shows a SEM cross section view at X50000 of sample A. The grooves pattern was revealed by dipping the sample into a solution NH_4OH:H_2O_2:H_2O (1:5:25) for 8 s. The etching rate is 400 Å/s for GaAs. Lighter fringes correspond to InGaAs, and the darker ones to GaAs layers. They show good periodicity, we could say good interface abruptness and the average thickness of the InGaAs and the GaAs layers determined from these SEM images, are in reasonable agreement with those obtained from the growth rate.

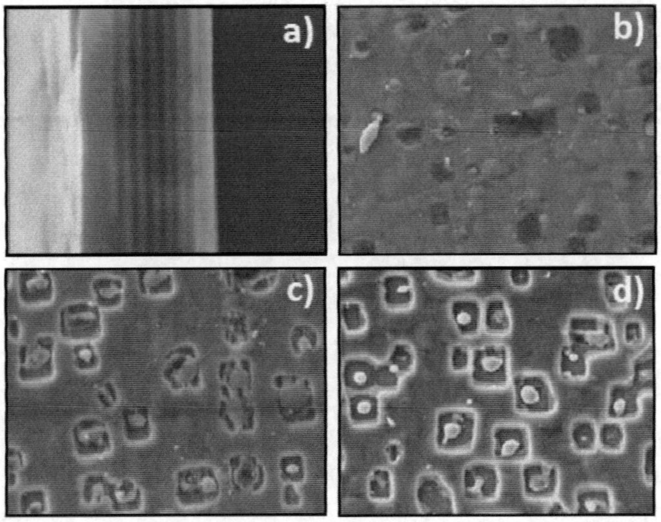

Figure 1. SEM micrographs of sample A: a) of a cross section, and of the surface morphology at 50000X: b) before etching, c) after 5 sec etching and d) after 8 sec etching.

Journal of Physics: Conference Series **480** (2014) 012038 doi:10.1088/1742-6596/480/1/012038

Also figure 1 shows SEM micrographs of the surface morphology of the uppermost layer of sample A b) before etching; this morphology looks like it contains bubbles some of them already exploted (darker areas); c) after 5 s etching; this morphology presents square dislocation pits with a spherical particle in the pits crater [2] and d) after 8 s etching; at this time the morphology has been better revealed.

The as "grown" surface morphology of sample B does not present any particular texture. However, after 10 s of chemical etching, the surface morphology looks like some parts of the film were not completely removed. Also, the surface morphology of the "as grown" heterostructure C does not present any texture but after 5 s of chemical etching very small etch pits are developed with no particle in the pits crater, SEM images not shown here.

3.2 AES analysis

Direct Auger spectra in the spot mode were recorded from the surface morphology of sample A. figure 2a shows the spectrum recorded from a particle in a crater. The average relative concentration was calculated using the formula: $C_i = (I_i/R_i)/\Sigma(I_j/R_j)$, where C is the concentration, I is the measured intensity and R is a sensitivity factor, the subscript i represents the element in the sample [3]. Thus, a particle in the pits crater contains 21.4 at % In, 23.1 Ga, 48.6 at % of As and 6.9 at % of O. In a region between pits the composition is 4.1 at % In, 46.7 Ga, 43.1 at % As, and 6.2 at % of O. We infer that the In coalesces because the etchant employed is not effective for In. Furthermore, we checked the stoichiometry of a brand new GaAs wafer by AES. This sample underwent the same chemical etching process for 8 s. and immediately analyzed by AES. The surface composition was 10.3 at % O, 36.6 at % Ga and 53.1 at % As which means that during the etching process the H_2O_2 used as an oxidizing agent causes Gallium oxidation mainly [4].

A personal computer program was done to calculate and plot the atomic percentage of the constituent elements of the 3 heterostructures from the Auger electron intensity signal vs Ar ion sputtering time spectra recorded by the Auger system [5]. These spectra are shown in figure 2 b, c, d, and were recorded from the "as grown" A, B and C samples. The Ar ion beam sputtering was operated at 3 keV and 25 mA of current. These depth profiles were recorded with similar measurement parameters for their later comparison.

The corresponding relative atomic percentage of elements was used to calculate the elemental mole fractions. The composition of the first layer of sample A is $Ga_{0.66}As_{0.34}$ then, the following layers almost consist of a series of alloy layers with a triangular variation in In since the GaAs layers between the alloy layers are very thin. Thus, the composition at the ridge of the first triangle is $In_{0.06}Ga_{0.59}As_{0.35}$. The Auger depth profile of sample B shows a square periodicity in the barrier-well structure. The composition of the first layer is $Ga_{0.63}As_{0.37}$ and that of the second layer is $In_{0.12}Ga_{0.52}As_{0.36}$. Finally, in the depth profile of sample C, the composition of the first layer is $Ga_{0.63}As_{0.37}$ and that of the second layer is $In_{0.25}Ga_{0.40}As_{0.35}$. From the Auger depth profiles data we infer that the first GaAs layers in the 3 samples are off-stoichiometric with similar values of mole fractions. The thickness of the first layer in B and C samples are equal since both were Ar ion etched in 10 min, nominally 1000 Å, while the first layer of sample A was etched in 25 min which would correspond to 2500 Å thickness. Analyzing the mole fractions of the first InGaAs layers in the three samples, we observe that the As mole fraction are very similar, this means that the under stoichiometric concentration for As is maintained in these alloy layers. Besides, since the In element is isoelectronic to Ga then the In atoms substitute the Ga atoms to form the alloy InGaAs. Now, the In mole fraction is the largest in C, smaller in B, and the smallest is the maximum mole fraction of sample A which as mention before show a triangular variation in concentration. Now, going back to the surface morphology of the chemically etched sample A (figure 1) we can make the following reasoning. Considering 400 Å/min as the chemical etching rate and 8 s of etching time, a GaAs layer of 3200 Å was removed chemically the corresponding surface morphology is shown in figure1d. Now, according to the Auger depth profile of sample A the thickness of the first GaAs layer is about 2500 Å.

A layer of this thickness would have been removed chemically in 6.2 s, in the next 1.8 s the etchant certainly found a series of thin InGaAs alloy layers with a triangular variation of In concentration. Since the In element is more resistant than the GaAs to this etchant then it coalesced forming the small spherical particles observed in figure 1d.

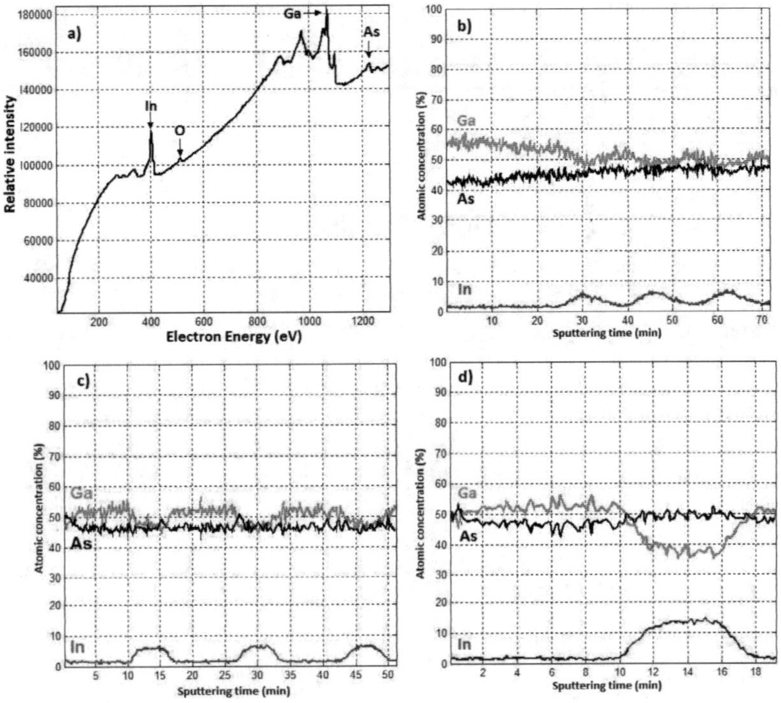

Figure 2 (a) Direct Auger spectrum of a particle in a crater of sample A. Auger depth profiles of the as grown $In_xGa_{1-x}As$-GaAs heterostructures grown on GaAs: (b) of sample A, (c) of B and (d) of sample C.

In summary, from SEM cross section micrographs at high magnification it has been possible to observe the constituting layers of the $In_xGa_{1-x}As$-GaAs multistructure grown on GaAs by an MBE system. Surface morphology SEM micrographs have shown the coalescence effect of In due to an unequal chemical etching rate of the etchant with respect to the one of GaAs. From Auger depth profiles we have inferred that the GaAs layers of the $In_xGa_{1-x}As$-GaAs heterostructure deposited on GaAs substrate are off stoichiometric.

4. Acknowledgments

Work supported by Benemérita Universidad de Puebla by means of a VIEP-Project and by Secretaría de Educación Pública: CA190, México.

References

[1] Y Matsui, H Hayashi, and K Yoshida 1986 *Appl. Phys.* **1** 219
[2] I Tarján and M Mátrai 1972 Quality Control by Chemical Etching p 74
[3] Handbook of Auger Electron Spectroscopy (JEOL) 1982
[4] D B J Lee 1969 *Appl. Phys.* **40** 4569
[5] D Briggs and M P Seah 1990 Practical Surface Analysis. Vol. 1 Auger and X-ray photoelectron spectroscopy. (John Wiley and Sons)

Production and Characterization of BDD/Si Electrodes for Environmental Analysis

N. M. Santos, N. G. Ferreira and M. R. Baldan

National Institute for Space Research, Sào José dos Campos, Sao Paulo, Brasil.

Email: nazir.santos@las.inpe.brr, neidenei@las.inpe.br, baldan@las.inpe.br

Abstract. Boron-Doped Diamond electrodes were produced on silicon substrate (BDD/Si) by Hot Filament Chemical Vapor Deposition reactor (HFCVD) using CH_4/H_2 gas mixtures. A series of experiments varying the levels of boron doping from 5000 to 25000 ppm were performed. Morphological, structural and electrochemical studies were conducted whose aim was the understanding about the film properties. The morphological and structural characterizations of these materials were made by Scanning Electron Microscopy and by Raman Scattering Spectroscopy techniques. The evaluation of the electrodes was performed using cyclic voltammetry. The electrochemical response showed the influence of boron content in the work potential window. The doped diamond films produced and characterized in this work are efficient as an electrode for environmental applications.

1. Introduction

BDD/Si electrodes have been extensively studied for their ability to oxidize a wide range of water contaminants. Contaminants can be oxidized by a combination of direct electron transfer and reaction with hydroxyl radicals (OH*) produced from water oxidation [1-3]. BDD/Si electrodes are considered as one of the most ideal electrodes for electro analysis and electrolysis [4, 5]. To date, conductive diamond has several superior characteristics compared with other carbon materials and metals, including very low background current density and wide potential window in aqueous solution [6, 7]. In this work, the experiments were carried out on a boron incorporated diamond film sample, which was produced with boric oxide (B_2O_3) dissolved in methanol (CH_3OH) in the gas phase in a hot-filament reactor during high pressure diamond deposition. The amount of incorporated boron is linear to the B/C ratio in the gas phase.

Previous papers mentioned that the boron concentration is homogeneous all over the coating film thickness [8]. The existence of boron dopants is expected to modify the atomic structure and the morphology of the diamond films. A different distribution of atomic structure in diamond films would further affect the physical properties of the diamond films.

2. Experimental procedure

The diamond electrodes doped with boron were produced by CVD (Chemical Vapor Deposition) technique in a reactor assisted by hot filament. The experimental setup is composed by the reactor comprises a stainless steel cylindrical shape with Pyrex windows and copper coil through which cooling water circulates. The internal layout of the reactor consist of five filament of tungsten with diameter of 125 μm supported with two brackets of molybdenum and a gas inlet tube located just

Journal of Physics: Conference Series **480** (2014) 012039 doi:10.1088/1742-6596/480/1/012039

above the region of the filaments. In the first step the substrate was scratched with 0.50 μm diamond-suspension for 60 min to enhance diamond nucleation.

Afterwards, the sample was ultrasonically cleaned in acetone and the substrate was placed 5 mm below the filaments and then the diamond deposition started. For the growth of BDD/Si electrodes a gas mixture of H_2 and CH_4 were used at a constant flow rate of 197 sccm and 3 sccm, respectively. The reactor temperature was of 800 °C and pressure was maintained at 40 Torr, and the deposition time was of 8 hours. For the doping of the electrodes an additional row of hydrogen was passed through a bubbler containing a boric oxide (B_2O_3) dissolved in methanol (CH_3OH), which was introduced into the gas inlet of the reactor. The temperature of the flask was maintained at 30 °C and the hydrogen flow to drag the inside of the reactor was controlled by a rotameter and kept at 35 sccm. When dissolving boric oxide (B_2O_3) in methanol (CH_3OH), trimethyl borate $(CH_3)_3BO_3$ is formed which is the doping substance during the film growth. The solutions were prepared at different concentrations of B_2O_3 dissolved in methanol ranging from 5000 to 25000 ppm of boron atoms with respect to the carbon atoms (B/C) of methanol. The electrode morphology and quality were analyzed by SEM using a JEOL JSM- 5310 model, and by micro-Raman scattering spectroscopy (Renishaw microscope system 2000) using the 514.5 nm line of an argon ion laser and the spectra covered the range from 300 to 3000 cm^{-1}.

3. Results and discussion

Images of the samples presented in figure 1 refer to diamond film un-doped and the electrodes of doped diamond produced with B/C ratio with 5000 and 25000 sccm. It has been found that the films grown in the entire length of the substrate homogeneously with the formation of microcrystalline grains randomly oriented. From these images it is possible to infer that the different doping levels caused little change in the surface morphology. This behavior confirms the results observed by other researchers in the literature to BDD/Si electrodes [9,10].

Diamond film un-doped   5000 ppm   25000 ppm

Figure 1. SEM photographs of the BDD/Si electrodes with B/C ratio of the 5000 to 25000 ppm. Magnification 5000 times.

Figure 2 shows the Raman spectra of the BDD/Si electrodes that exhibit a characteristic diamond peak at 1332 cm^{-1} and a band at 1200 cm^{-1} attributed to the disorder induced in the diamond lattice due to the incorporation of boron [11]. It is observed from this figure that the band at 1200 cm^{-1} becomes steeper and wider by increasing the doping concentration, and this effect is associated to a reduction of the diamond peak and related to the locally disordered structures induced by the heavily boron doping [10, 12]. A band around 500 cm^{-1} is also observed, which can be attributed to the vibration modes of pairs of boron, causing a distortion in the diamond network [13]. A complete absence of peaks around 1500 cm^{-1} can also be noticed related to the graphitic bonds, indicating the quality of the electrode [14].

Journal of Physics: Conference Series **480** (2014) 012039 doi:10.1088/1742-6596/480/1/012039

The cyclic voltammograms (CV) for the BDD/Si electrodes at different B/C ratio were performed in an aqueous solution of H_2SO_4 0.5 M, with a scan rate of 10 mVs^{-1}. We obtained a value of window work potential around 3.0 V for the various BDD/Si electrodes. This broad potential window value (E) allows the detection of redox reactions in addition to the potential range of working potential of conventional electrodes, which is around 2.0 V for platinum and graphite [15].

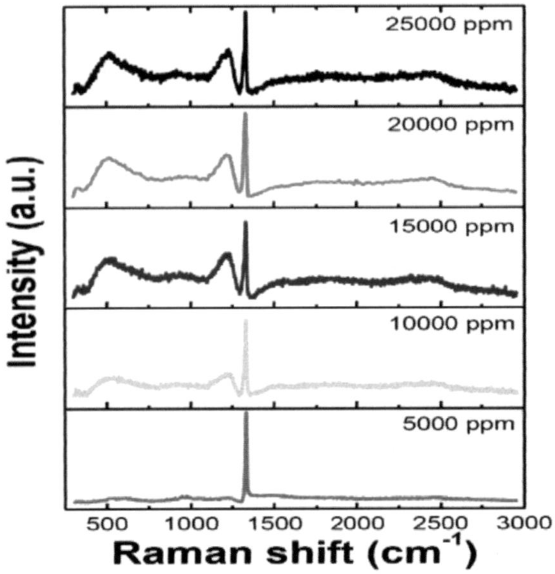

Figure 2. Raman spectra of BDD/Si electrodes with B/C ratio of 5000 to 25000 ppm.

A band around 500 cm^{-1} is also observed, which can be attributed to the vibration modes of pairs of boron, causing a distortion in the diamond network [13]. A complete absence of peaks around 1500 cm^{-1} can also be noticed related to the graphitic bonds, indicating the quality of the electrode [14]. The cyclic voltammograms (CV) for the BDD/Si electrodes at different B/C ratio were performed in an aqueous solution of H_2SO_4 0.5 M, with a scan rate of 10 mVs^{-1}. We obtained a value of window work potential around 3.0 V for the various BDD/Si electrodes.

This broad potential window value (E) allows the detection of redox reactions in addition to the potential range of working potential of conventional electrodes, which is around 2.0 V for platinum and graphite [15]. Figure 3 shows the voltammograms for the BDD/Si electrodes at different scanning speeds in the presence of $[K_4Fe(CN)_6]$ 1 mM in H_2SO_4 0.5 M electrolyte. In this figure, it is observed that the increase in the scan rate of potential causes an increase in the intensity of the peak currents for the cathodic and anodic BDD/Si electrodes. It is found that the anodic peak shift occurs to more positive potentials by increasing the anodic current, and the displacement direction for the cathodic peak potential becomes smaller with the increase in the cathodic current. The anodic and cathodic peaks occur due to the charge transfer reactions (oxidation and reduction) of one electron during the application cycle potential.

Figure 4 shows the variation of potential (ΔEp) as a function of the scan rate (V) solution of ferrocyanide 1 mM/H_2SO_4 0.5 M to BDD/Si electrodes. It is observed in Figure 4 that there is an increase in the amount of ΔEp analyzed for all electrodes when the scan speed is increased. The BDD/Si electrodes produced at 5000 to 15000 ppm doping of B/C ratio showed lower values of ΔEp (60-70 mV) when compared to BDD/Si electrodes at 20000, and 25000 ppm at low speed scanning (10-50 mV/s), which shows that, under these conditions, these electrodes are reversible. ΔEp values can be taken as a criterion of reversibility, even when they deviate from the theoretical value, since this deviation is not too large [16].

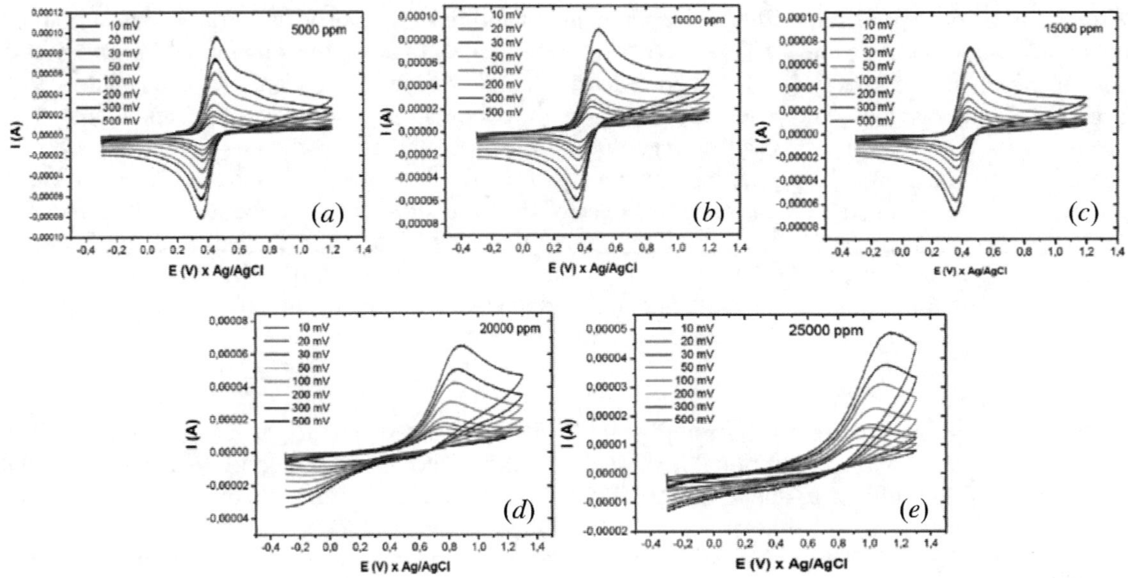

Figure 3. CV of electrodes at different scanning speeds and doping concentrations in K4[Fe(CN)$_6$]1 mM/H_2SO_4 0.5 M electrolyte.

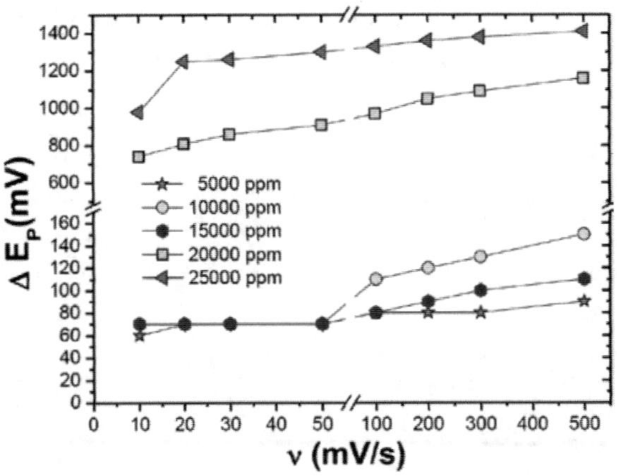

Figure 4. Variation of potential (ΔEp) as function of the scan range of the BDD/Si electrodes produced at different doping concentrations.

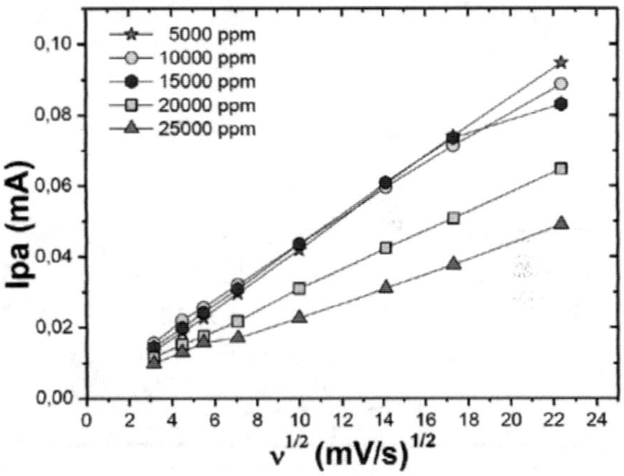

Figure 5. Anodic peak current (Ipa) as a function of the square root of the scan speed (V1/2) in ferrocyanide solution to the BDD/Si electrodes.

Therefore, it can be inferred that according to this criterion BDD/Si electrodes produced at 20000, and 25000 ppm are almost reversible, i.e., they have a slower kinetic charge transfer reactions on ferrocyanide. Taking this criterion only as a reference, it can be stated that the BDD/Si electrodes produced at 5000 to 15000 ppm are more reversible. The behavior of the anodic peak current (Ipa) as a function of the square root of scan speed ($V^{1/2}$) in ferrocyanide solution for the BDD/Si electrodes is shown in figure 5. In this figure, it can be seen that there was a linear increase in the anodic peak current as a function of the square root of the scan speed for all electrodes analyzed. The fact that the graph (Ipa x $V^{1/2}$) present a line for all electrodes examined, confirms that the ratio Ipa / $V^{1/2}$ independent from the scan rate, which is a criterion of reversibility. This result suggests that these electrodes may be reversible only at low scan speeds.

According to Yagi et al. (1999), the diamond electrodes in hydrogen terminations are more reversible with respect to the system $[Fe(CN)_6]^{3-}/[Fe(CN)_6]^{4-}$ than to the diamond film endings in oxygen, which can be understood when considering the nature of the connections established in each case [17]. In the diamond electrode formed by a high-carbon sp^2 hybridized as the functional group formed by carbon atoms bonded with oxygen tend to form bonds at the ends of the flat sites linked together by carbon double bond, i.e. that contain sp2 bonds, which results in diamond lattice oxygen terminations. The terminal oxygen atoms thus established are sites available for the establishment of new connections among the species present in the solution, which leads to the establishment of parallel reactions and/or for electron capture. The high electronegativity of the terminal oxygen atoms may be responsible for obstructing the exit of electrons associated with them, which in turn has a direct influence on the low-speed transfer of these electrons in the system.

4. Conclusions

In all Raman spectra is present the strong band at 1332 cm^{-1}, characteristic of the first order Raman scattering due to the optical vibrations of sp^3-bonded carbon atoms in diamond crystal structure. All characteristic peaks of boron-doped diamond films were found features in the Raman spectrum including the broad band around at 500 cm^{-1} and 1200 cm^{-1} with a reduction of the diamond peak. The BDD/Si electrodes produced in this work can be considered as great anodic materials for the electro degradation of because of the high value of the anodic threshold the extensive electrochemical window and low values of capacitive currents.

5. Acknowledgments

Special thanks to FAPESP (Proc. 2012/07326-6), CAPES and INCT for Climate Change funded by CNPq (573797/2008-0).

References

[1] Chaplin B P, Scharader G and Farrell J. 2010 Env. Sci. and Technol. 44 4263
[2] Kapalka A, Foti G, Comninellis C 2008 J. of the Electrochem. Soc. 155 E27
[3] Carter K E and Farrell J 2008 Env. Sci. and Technol 42, 6111
[4] Gandini D et al 2000 J. Appl. Electrochem. 30 1345.
[5] Suffredini H B, Machado S A S, Avaca L A, Braz J 2004 Chem. Soc. 15 16
[6] DeClements R and Mswain G 1994 J. Electrochem. Soc. 144 3382
[7] Kondo T, Honda K, Tryk D A, Fujishima A 2003 Electrochim. Acta 48 2739
[8] Spicka H et al 1996 Diam. and Rel. Mat. 5 383-387
[9] Zhang R J, Lee S T, Lam Y W 1996 Diam. and Rel. Mat. 5 1288 24
[10] Azevedo A F, Barros R C M, Serrano S H P, Ferreira N G 2006 Surf. Coat. Technol 200 5973
[11] Wei J J, Li Ch M, Gao X H, Hei L F, Lvun F X 2012 Appl. Surf. Sci. 258 6909- 6913
[12] Jia F, Bai Y, Qu F, Zhao J, Zhuang C, Jiang X 2010 Vac. 84 930-934
[13] May P W et al 2008 Diam. and Rel. Mat. 17 105-117
[14] Trava-Airoldi V J et al 1995 Diam. and Rel. Mat. 4 1255-1259
[15] Azevedo A F and Ferreira N G 2006 Quim. Nova, 29 129-136
[16] Pleskov Y V et al 2005b J. Appl. Electrochem. 35 857-864
[17] Yagi I et al 1999 J. of Electroanal. Chem. 443 173-178

Journal of Physics: Conference Series **480** (2014) 012040 doi:10.1088/1742-6596/480/1/012040

Giant-vortex state induced by negatives values of deGennes parameter in a mesoscopic square

J D González[1 2], J L Aguilar[1], J Barba-Ortega[2]

[1] Departamento de Física, Universidad del Magdalena, Santa Marta, Colombia

[2] Departamento de Física, Universidad Nacional de Colombia, Bogotá, Colombia

E-mail: `jedgonzalezac@unal.edu.co`

Abstract. Using different negatives values of deGennes parameter in the boundary condition of a square sample superconducting type II is possible to change its vorticity. Due to the proximity effects, the cooper pairs density snapshot show how a giant vortex occurs when the external magnetic field is increased. Solving numerically the coupled nonlinear time dependent Ginzburg-Landau equations we obtain the vorticity, magnetic induction, cooper pair density, magnetization and phase of the order parameter as a function of the external magnetic field for this sample. We found that a giant vortex state occurs when a sample is surrounded by another superconductor at higher critial temperature.

1. Introduction

The Ginzburg-Landau (GL) equations [1] are arguably the most convenient and frequently employed tool in studying of the vortex matter and thermodynamics properties of superconducting samples with general boundary condition. The size of the sample is usually considered to be large enough that the influence of finite dimensions on their properties is negligible; however, in many situations of interest, i.e. mesoscopic superconductivity, metallic nanoparticles [2, 3], microfluidics,[4, 5] the size effects are very important. Advances in engineering have been possible to enhance critical fields, critical temperature and critical currents to have a better control in the thermodynamic properties of superconducting samples. The possibility to control the dynamics and vortex density has made of this physics one of the favorite experimental and theoretical topic for studies of the solid state [6, 7, 8, 9, 10, 11]. In this paper, we solve the Ginzburg Landau equations for a mesoscopic square sample in presence of an external magnetical field applied perpendicular at its surface, we consider a superconducting/superconducting at higher critical temperature boundary conditions. We found that is possible to obtain a Giant-vortex state induced due to the presence of the another superconductor.

2. Theoretical Formalism

We considered a small superconducting square in the presence of a perpendicular uniform magnetic field H_e. The general form of the time dependent Ginzburg-Landau equations is

$$\frac{\partial \psi}{\partial t} = -(i\nabla + \vec{A})^2\psi + \psi(1 - |\psi|^2) \quad (1)$$

$$\frac{\partial \vec{A}}{\partial t} = \mathrm{Re}\left[\bar{\psi}(-i\nabla - \vec{A})\psi\right] - \kappa^2\nabla \times \nabla \times \vec{A} \quad (2)$$

In Eqs. (1) and (2) dimensionless units were introduced as follows: $|\psi|$ is the order parameter in units of $\psi_\infty(0) = \sqrt{-\alpha(0)/\beta}$, where $\alpha(0)$ and β are two phenomenological constants; lengths in units of the coherence length $\xi(0)$; time in units of $t_0 = \pi\hbar/8K_BT_c$; $\mathbf{A}$ in units of $H_(0)\xi(0)$, where $H_{c2}(0)$ is the second critical field, temperature in units of the critical temperature T_c, $\lambda(0)$ is the penetration depth [12]. We studied a mesoscopic superconducting square of size $3\xi(0) \times 3\xi(0)$ with a superconductor/superconducting at higher critical temperature external interface simulated by $b/\xi(0) = -1.7$, with $\kappa = 5.0$, $T = 0$, $a_x = a_y = 0.125$.

2.1. Numerical Method

The full discretization of the TDGL equations can be found in more detail in Ref. [13]. We used $U\psi$ method [14] for solve the TDGL equations in a discrete grid. Complex link variables $\mathcal{U}^x$ and $\mathcal{U}^y$ are introduced to preserve the gauge-invariant properties of the discretized equations. $\mathcal{U}^x$ and $\mathcal{U}^y$ are related to A by:.

$$\mathcal{U}^x(x,y,t) = \exp\left(-i\int_{x_0}^{x} A_x(\xi,y,t)\,d\xi\right) \Leftrightarrow \mathcal{U}^y(x,y,t) = \exp\left(-i\int_{y_0}^{y} A_y(x,\eta,t)\,d\eta\right) \quad (3)$$

The link variable method is used since a better numerical convergence is obtain in high magnetic fields [15]. The TDGL equations 1 and 2 can be written in the following form:

$$\frac{\partial \psi}{\partial t} = \bar{\mathcal{U}}_x\frac{\partial^2(\mathcal{U}_x\psi)}{\partial x^2} + \bar{\mathcal{U}}_y\frac{\partial^2(\mathcal{U}_y\psi)}{\partial y^2} + (1-T)\,\psi\left(1 - |\psi|^2\right) \Leftrightarrow \mathbf{J}_{\mathbf{s}\alpha} = (1-T)\,Im\left[\bar{\mathcal{U}}_\alpha\bar{\psi}\frac{\partial(\mathcal{U}_\alpha\psi)}{\partial\alpha}\right] \quad (4)$$

where $\alpha = (x, y)$, and Im indicates the imaginary part. We used this method to obtain our results. The outline of this simulation procedure is as follows: the sample is divided in a rectangular mesh consisting of $N_x \times N_y = 36$ cells, with mesh spacing $a_x \times a_y = 0.5$. To derive the discrete equations let us define by $x_i = (i-1)a_x$, $y_i = (i-1)a_y$, an arbitrary vertex point in the mesh and:

$$U^x_{i,j} = \bar{\mathcal{U}}^x_{i,j}\mathcal{U}^x_{i+1,j} = \exp\left(-i\int_{x_i}^{x_i+1} A_x(\xi, y_i)\,d\xi\right) \Leftrightarrow U^y_{i,j} = \exp\left(-i\int_{y_j}^{y_j+1} A_y(x_i,\eta)\,d\eta\right)$$

$$L_{i,j} = U^x_{i,j}U^y_{i+1,j}\bar{U}^x_{i,j+1}\bar{U}^y_{i,j} = \exp(-ia_xa_yH_e)$$

Then the discretized version of the TDGL equations maintaining second order accuracy in space are gyven by:

$$\frac{\partial \psi_{i,j}}{\partial t} = \frac{U^x_{i,j}\psi_{i,j} - 2\psi_{i,j} + \bar{U}^x_{i-1,j}\psi_{i-1,j}}{\eta a_x^2} + \frac{U^y_{i,j}\psi_{i,j+1} - 2\psi_{i,j} + \bar{U}^y_{i,j-1}\psi_{i,j-1}}{\eta_y^2} - \frac{1-T}{\eta}(\bar{\psi}_{i,j}\psi_{i,j} - 1)\psi_{i,j}$$

$$\frac{\partial U^x_{i,j}}{\partial t} = -i(1-T)U^x_{i,j}Im\left(\bar{\psi}_{i,j}U^x_{i,j}\psi_{i+1,j}\right) - \frac{\kappa^2}{a_y^2}U^x_{i,j}\left(\bar{L}_{i,j-1}L_{i,j} - 1\right)$$

$$\frac{\partial U^y_{i,j}}{\partial t} = -i(1-T)U^y_{i,j}Im\left(\bar{\psi}_{i,j}U^x_{i,j}\psi_{i,j+1}\right) - \frac{\kappa^2}{a_x^2}U^y_{i,j}\left(\bar{L}_{i,j}L_{i-1,j} - 1\right)$$

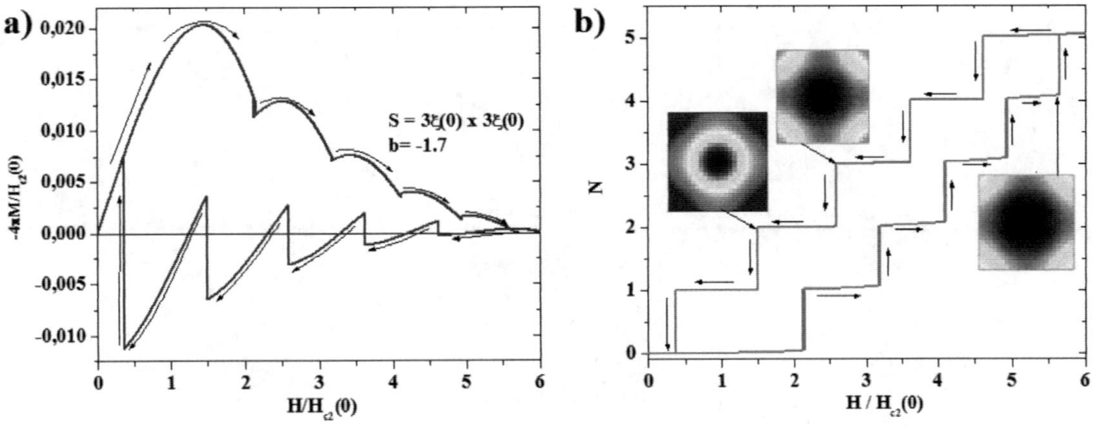

Figure 1. a). Magnetization and b). number of fluxoid (the cooper pairs density inset) as a function of external magnetic for a square $3\xi(0) \times 3\xi(0)$ with following value of deGennes parameter $b/\xi(0) = -1.7$.

3. Results and Discussion

Fig. 1 a). We present the magnetization as a function of the external magnetic field H_e for a mesoscopic superconducting square of size $3\xi(0)\times3\xi(0)$ with a superconductor/superconducting at higher critical temperature external interface. These curves present a typical profile of a magnetization curve of a mesoscopic superconductor. It exhibits a series of discontinuities, in which each jump signals the entrance of vortices into the sample.

Fig. 1 b). We have vortex transition from N to $N+1$ in part of the upward branch of the magnetic field. The magnetic field for vortices entries are $H_e/H_{c2}(0) = 2.608, 3.618, 4.478,$ 5,219 and 6.027 with $N = 1, 2, 3, 4, 5, 6$ for using $b/\xi(0) = -1.7$ and inset the cooper pairs density snapshot show how a giant vortex occurs when the external magnetic field is increased as we can see the order parameter which is mostly $\Psi = 0$ in the sample (blue region) and this condition allows the penetration of vortices and the number of vortices can be determined in a given region, by counting the number of fluxoid variation.

Fig. 2 we shown the contour plot of the the magentic induction and the phase of order parameter for three negatives values of deGennes parameter $b/\xi(0) = \infty, -2.5, -12.5$, for mesoscopic sample with $3\xi(0) \times 3\xi(0)$ dimension. Following the panels from the left to the right, in this order, it shown that is possible to obtain a giant-vortex configuration. That is why, although the vorticity grows from $N = 1$ to $N = 3$, as we can see in the phase of order parameter which changes from dark (red) to bright (blue) regions represent values of the modulus of the order parameter (as well as $\Delta\phi/2\pi$,changes from 0 to 1). The color code for values of the phase close to zero are given by dark (red) regions and close to 2π by light (blue) regions. Despite the vortices are indistinguishable in the contour plot of the magnetic field induction, leading to the formation of the giant vortex. This kind of vortex can be achieved if the deGennes parameter values and applied magnetic field harmonize for every giant-vortex state in this high confinement system.

4. Conclusions

We studied the effect of the different negative values of the deGennes parameter (b) in the thermodynamics properties for a small superconducting square of size $3\xi(0) \times 3\xi(0)$ in presence of a magnetic field is investigated theoretically, simulating this sample with boundary conditions

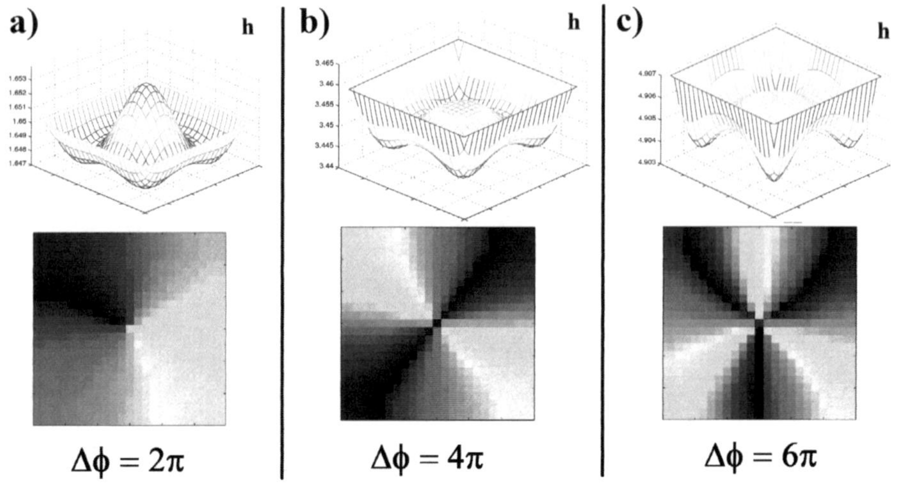

Figure 2. Snapshot of the magnetic field h and phase of the order parameter $|\Psi|^2$ as a function of negative values of deGennes parameter for mesoscopic sample with $3\xi(0) \times 3\xi(0)$ dimension.

in which its lateral surface has a superconducting/superconducting at higher critical temperature by solving the coupled nonlinear Ginzburg-Landau equations. Our results shown that the giant-vortex state and their principal caracteristics such as magnetic induction, cooper pair density, magnetization and the phase of order parameter depend strongly of the chosen boundary condition and the applied magnetic field.

5. Acknowledgments

JBO thanks Alejandro Barba for useful discussions. ES thanks the Brazil agency FAPESP for financial support. This work was partially financed by the Universidad del Magdalena (Fonciencias) and the Colombian Agency COLCIENCIAS through doctoral scholarships 567.

6. References

[1] Ginzburg V L and Landau L D 1950 *Zh. Eksp. i Teoret. Fiz.* **20**, 1064.
[2] Halperin W P 1986 *Rev. Mod. Phys.* **58**, 533.
[3] Sun Y G and Xia Y N 2002 *Science* **298**, 2176.
[4] Harrison J D, Fluri K, Seiler K, Fan Z H, Effenhauser C S, and Manz A 1993 *Science* **261**, 895.
[5] Baletto F and Ferrando R 2005 *Rev. Mod. Phys.* **77**, 371.
[6] Koshelev A E and Vinoku V M 2001 *Phys. Rev. B* **64**, 134518.
[7] Golubovic D S, Milosevic M V, Peeters F M, and Moshchalkov V V 2005 *Phys. Rev. B* **71**, 180502.
[8] Milosevic M V and Peeters F M 2004 *Phys. Rev. Lett.* **93**, 267006.
[9] Milosevic M V and Peeters F M 2003 *Phys. Rev. B* **B** 68, 024509.
[10] Berdiyorov G R, Baelus B J, Milosevic M V, and Peeters F M 2003 *Phys. Rev. B* **68**, 174521.
[11] Milosevic M V, Berdiyorov G R, and Peeters F M 2007 *Appl. Phys. Lett.* **91**, 212501.
[12] Zeldov E, Larkin A I, Geshkenbein V B, Konczykowski M, Majer D, Khaykovich B, Vinokur V M, and Shtrikman H 1994 *Phys. Rev. Lett.* **73**, 1428.
[13] Buscaglia G C, Bolech C and López A,*Connectivity and Superconductivity*, J. Berger, and J. Rubinstein (Eds.), Springer,(2000).
[14] Gropp W D, Kaper H G, Leaf G K, Levine D M, Palumbo M, and Vinokur V M 1996 *J. Comput. Phys.* **123**, 254.
[15] Kato R, Enomoto Y, Maekawa S 1993 *Phys. Rev. B* **47**, 8016.

Electronic Structure of the Double Perovskite $Ba_2Er(Nb,Sb)O_6$

A. V. Gil Rebaza[1], C.E. Deluque Toro[2], D.A. Landínez Téllez[3] and J. Roa-Rojas[3]

[1] *Depto. de Física, Facultad de Ciencias Exactas - UNLP, Instituto de Física La Plata IFLP-CONICET, cc 67 (1900) La Plata, Argentina*
[2] *Grupo de Nuevos Materiales, Departamento de Física, Universidad Popular del Cesar, Valledupar, Colombia*
[3] *Grupo de Física de Nuevos Materiales, Departamento de Física, Universidad Nacional de Colombia, AA 5997, Bogotá DC, Colombia*

Received 4 October 2013, in final form XX Novembrer 2013

Abstract. In this work, we present a detailed study of the structural and the electronic structure of the double perovskite $Ba_2Er(Nb,Sb)O_6$. All calculations were performed with the Full-Potential Linear Augmented Plane Wave method (FP-LAPW) based on the Density Functional Theory (DFT). From the minimization of energy as a function of volume using the Murnaghan's state equation has been obtained the equilibrium lattice parameter and the bulk modulus of these compounds. The study of the electronic structure was based in the analysis of the electronic density of states (DOS), and the density of charge, showing that these compounds have a total magnetic moment of 3.0 μ_B per formula unit due to Er atoms.

1. Introduction

Currently the complex perovskites with generic formula $A_2MM'O_6$ [1-3] have been subject of enhanced scientific studies because the wide range of applications that these materials present on an industrial scale, double perovskites belonging to the series $Ba_2LnB'O_6$ (Ln=lanthanide and Y^{3+} and B'=Nb^{5+}, Ta^{5+}, Sb^{5+}) are of interest due to their potential use as substrates for high-Tc superconductors [4,5] and their likely high chemical compatibility with the structurally analogous, oxygen deficient double perovskites, $Ba_2LnSnO_{6-\delta}$; a series of interest for use as solid state electrolytes due to their high ionic conductivity [6,7]. Substitution of Nb^{5+}, Ta^{5+} or Sb^{5+} could provide a method of controlling the oxygen stoichiometry of these compounds, potentially minimizing the problems caused by reduction of $Ba_2LnSnO_{6-\delta}$ that is thought to be a consequence of the large level of oxygen vacancies in these compounds [6]. Phase transitions in perovskites have long been of interest to solid-state chemists [8]. Previous studies [9, 10], using laboratory X-ray and medium resolution neutron diffraction, reported that the structures in the series Ba_2LnSbO_6 change from R-3 rhombohedral (tilt system $a^-a^-a^-$) to Fm-3m cubic ($a^0a^0a^0$) symmetry with decreasing size of the Ln^{3+} cation. Conversely the niobate series Ba_2LnNbO_6 exhibits the sequence: I2/m monoclinic ($a^-a^-c^0$), I4/m tetragonal ($a^0a^0c^-$) to Fm-3m cubic symmetry. That the symmetry increase as the ionic radius of the lanthanide decreases, this is consistent with the increase of the tolerance factor *t*. An increase in the tolerance factor indicates that the volume of the BO_6 octahedron is better matched to the size of the AO_{12} polyhedron reducing the need for the octahedral tilting to accommodate this A-site cation. Since octahedral tilting is

Journal of Physics: Conference Series **480** (2014) 012041 doi:10.1088/1742-6596/480/1/012041

responsible for the lowering of the symmetry from cubic, the symmetry tends to increase as the B-type cation gets smaller.

The aim of this work is to carry out a detailed *ab initio* theoretical study of the complex perovskites Ba_2ErSbO_6 and Ba_2ErNbO_6, which were experimentally reported as cubic structures with spacial group Fm3m [11-13]. Our calculations shown that these compounds have a total magnetic moment of 3.0 μ_B per formula unit due to Er atoms

2. Computational Details

The calculations were performed using Full Potential Linearized Augmented Plane Wave (FP-LAPW) method, within the Density Functional Theory formalism [14], Wien2k code [15]. The exchange-correlation potential was described using the Perdew–Burke–Ernzerhof [16] expression, based in the Local Spin Density Approximation (LSDA) and Generalized Gradient Approximation (GGA) [17]. In all calculations, we used the next muffin-tin radius $R_{MT}(Ba)=2.5$, $R_{MT}(Er)=2.3$, $R_{MT}(Nb, Sb)=2.05$ and $R_{MT}(O)=1.60$ bohr; and the parameter $R_{MT}\times K_{max}$ was kept equal to 7 and $G_{max} = 12$. The irreducible Brilloiun zone was described with 286 k-points, a mesh of 21×21×21. Furthermore, in all calculations we have considered spin polarization and the parameter of convergence was the variation of charge 10^{-5} a.u., the core and valence states are treated in the fully and scalar relativistic scheme, respectively.

In the present work, we have considered these compounds with cubic crystalline structure Fm-3m (225), see Figure 1.

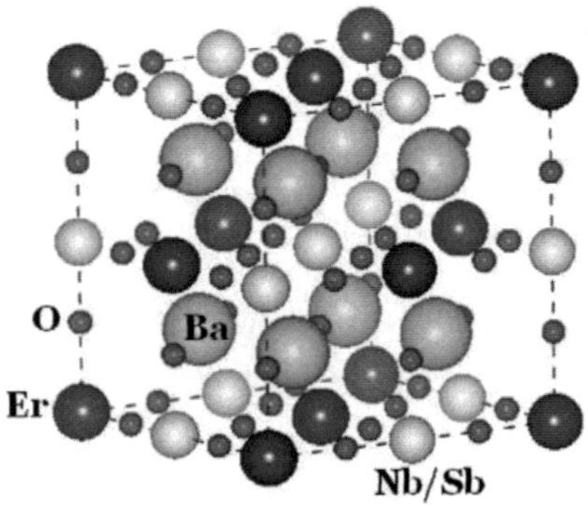

Figure 1: Cubic crystal structure Fm-3m of the compound $Ba_2(Nb,Sb)ErO_6$.

3. Results and discussion

In order to obtain the equilibrium structure, the total energy vs. lattice parameter curves were obtained for BaEr(Nb,Sb)O compounds using LSDA and GGA. These curves were fitted using the Birch-Murnaghan's state equation [18] to determine the equilibrium lattice parameters (*a*), bulk modulus (B) and their derivate (B'). In Table 1 we shows these values together experimental data obtained from the bibliography [11-13]. For both compounds the ground state belong to GGA calculations, where the *a* value for these compounds are in agree with experimental one.

Table 1: Theoretical value of a (Å), B (GPa) and B' of the compound $Ba_2Er(Nb,Sb)O_6$.

		a (Å)	B (GPa)	B'
Ba_2ErNbO_6	LDA	8.3176	161.3	5.1
	GGA	8.4956	131.2	5.6
	Exp.	8.4270[11]		
		8.4193[12]		
Ba_2ErSbO_6	LDA	8.3124	146.7	3.9
	GGA	8.5048	123.8	5.9
	Exp.	8.3960[12]		
		8.3904[13]		

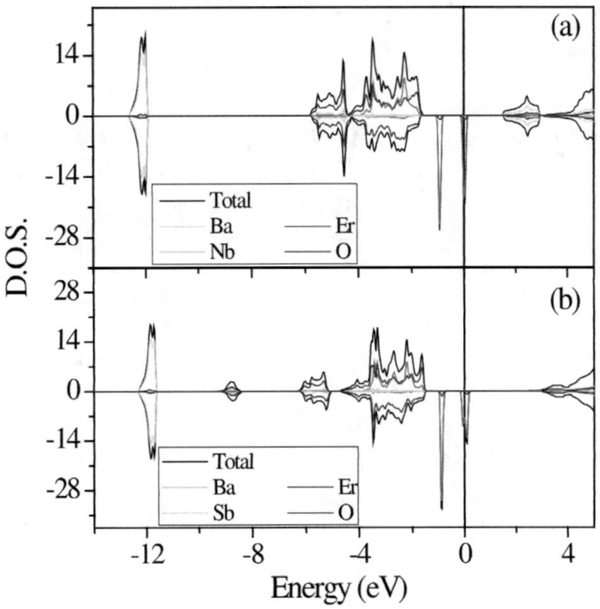

Figure 2: Total density of states and projected per atom of the compound (a) Ba_2ErNbO_6 and (b) Ba_2ErSbO_6.

In figure 2, we show the total density of states (D.O.S.) of these compounds where and by atom. In both compounds the DOS belong to Ba atoms are locate in depth energies values without any hybridization with other atomic orbitals. Between -6.0 and -4.0 eV there are hybridization between the *4d*-(Nb, Sb) and the *2p*-O orbitals whereas between -4.0 and -1.5 eV we found hybridization of the *2p*-O and *4f*-Er orbital, but only in the majority DOS. Only in the Ba_2ErSbO_6 compound is possible see a hybridization *5s*-Sb and *2s*-O orbitals. In both compounds, the DOS of the Er atom show a dislocation and different population between the majority and minority DOS, producing a magnetic moment in this atom close to 2.86 μ_B. In the figure 3, we show the difference of the density of charge up and down in the plane (100) for both compounds, where is possible to see a mayor concentration of this difference around to the Er atoms, showing the magnetic character of these atoms. This is the reason why these compounds have a magnetic ground state with a magnetization of 3.0 μ_B per formula unit, agreeing with experimental studies for the case of the Ba_2ErSbO_6 [13].

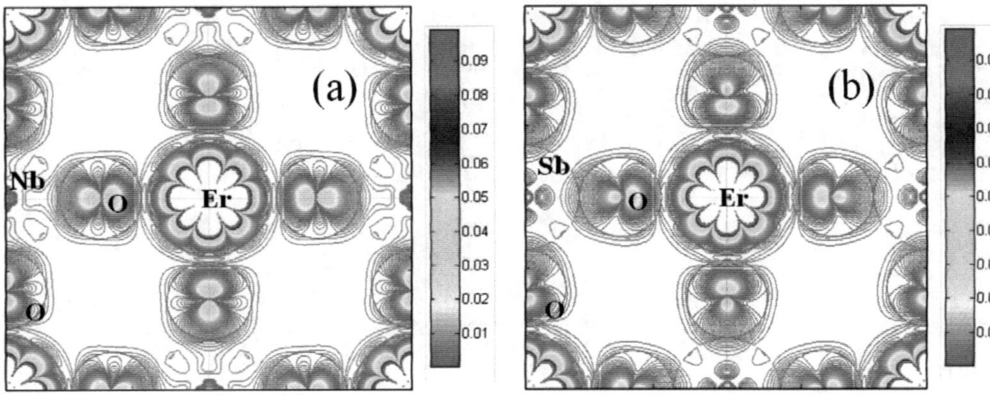

Figure 3: Difference between the charge density up and down in the plane (100) of the (a) Ba_2ErNbO_6 and (b) Ba_2ErSbO_6.

4. Conclusions

The structural and electronic properties of the $Ba_2Er(Nb,Sb)O_6$ perovskite compound in the cubic structures were calculated using the method of Linearized Augmented Plane Wave on the Generalized Gradient (GGA) and Local Density (LDA) approximations with similar results. By minimizing the energy as a function of volume, structural characteristics such as the total energy, the equilibrium volume, the volume module, and the lattice parameters that minimize energy of the system were determined, where the equilibrium lattice parameter is in agree with experimental data. Furthermore, from analysis of the DOS and the density of charge, was possible explain the magnetic ground state of these compound, with a magnetization of 3.0 μ_B per formula unit due mainly to the contribution of Er atom.

Acknowledgments

This work was supported in part by the Universidad Popular del Cesar.

References

[1] L.T. Corredor, J. Roa-Rojas, D.A. Landínez Téllez, R. Beltrán, P. Pureur, F. Mesquita, J. Albino Aguiar 2013 J. Appl. Phys. **113** 17E302
[2] D.P. Llamosa, D.A. Landínez Téllez, J. Roa-Rojas 2009 Physica B **404** 2726
[3] R.B. Macquart, Q.D. Zhou, B.J. Kennedy 2009 Journal of Solid State Chem. **7** 1691
[4] A.M. Glazer 1972 Acta Crystallogr. Sect. B **28** 3384
[5] A.M. Glazer 1975 Acta Crystallogr. Sect. A **31** 756
[6] P.M. Woodward 1997 Acta Crystallogr. Sect. B **53** 32
[7] C.J. Howard, B.J. Kennedy, P.M. Woodward 2003 Acta Crystallogr. B **59** 463
[8] M. Wakeshima, D. Harada, Y. Hinatsu 1999 J. Alloy and Compd. **287** 130
[9] M. Wakeshima, D. Harada, Y. Hinatsu, N. Masaki 1999 J. Solid State Chem. **47** 618
[10] M. Wakeshima, D. Harada, Y. Hinatsu 2000 J. Mater. Chem. **10** 419
[11] F.S. Galasso, Structure, Properties and Preparation of Perovskite-type Compounds, Pergamon Press, Oxford 1969
[12] P. J. Saines, B. J. Kennedy, M. M. Elcombe 2007 J Solid State Chem. **180** 401
[13] S. Calder, T. Fennell, W. Kochelmann, G. C. Lau, R. J. Cava, S. T. Bramwell 2010 J. Phys.: Condens. Matter **22** 116007
[14] P. Hohenberg, W. Kohn 1964 Phys. Rev. **136** 864
[15] P. Blaha, K. Schwarz, G.K.H. Madsen, D. Kvasnicka, J. Luitz, WIEN2k_11.1, Techn. Universitat Wien, Austria, 2001
[16] J.P. Perdew, Y. Wang 1992 Phys. Rev. B. **45** 13244
[17] J.P. Perdew, S. Burke, M. Ernzerhof 1996 Phys. Rev. Lett. **77** 3865
[18] F. D. Murnaghan 1944 Proc. Natl. Acad. Sci., USA. **30** 244

AUTHOR INDEX

AUTHOR INDEX